Academic Press Rapid Manuscript Reproduction

Proceedings of a Symposium on Quantitative Approximation
held in Bonn, West Germany, August 20-24, 1979

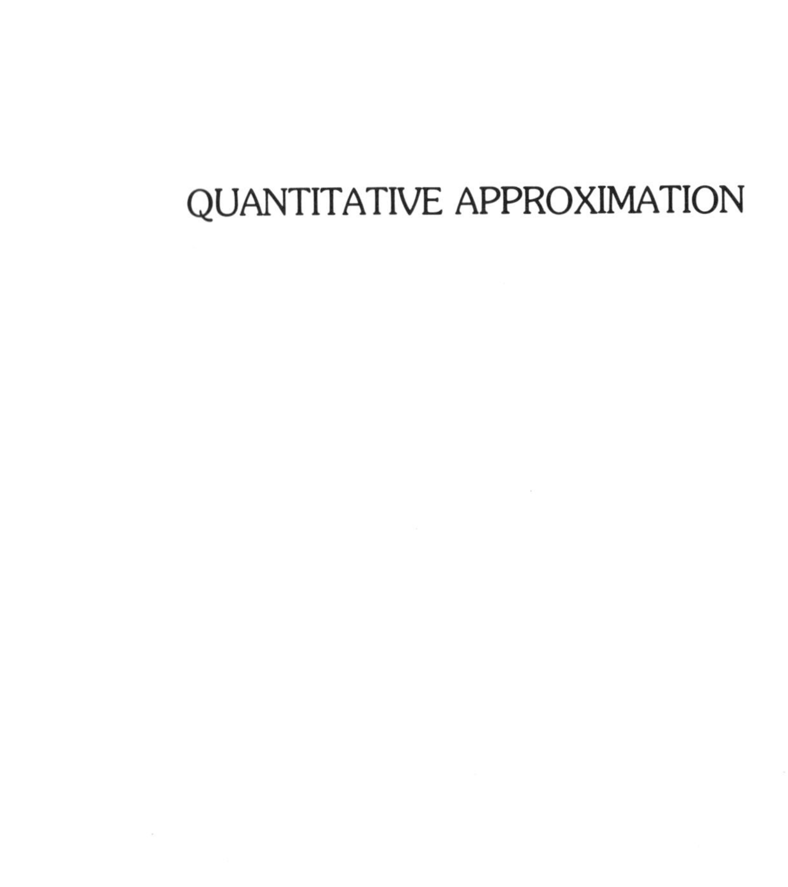

QUANTITATIVE APPROXIMATION

QUANTITATIVE APPROXIMATION

edited by

RONALD A. DEVORE

Department of Mathematics and Statistics
University of South Carolina
Columbia, South Carolina

KARL SCHERER

Institut für Angewandte Mathematik
der Universität Bonn
Bonn, F.R. Germany

ACADEMIC PRESS

A Subsidiary of Harcourt Brace Jovanovich, Publishers

New York London Sydney Toronto San Francisco 1980

ACADEMIC PRESS, INC.
111 Fifth Avenue, New York, New York 10003

United Kingdom Edition published by
ACADEMIC PRESS, INC. (LONDON) LTD.
24/28 Oval Road, London NW1 7DX

Library of Congress Cataloging in Publication Data
Main entry under title:

Quantitative approximation.

Proceedings of an international symposium held in Bonn, West Germany, Aug. 20-24, 1979.
1. Approximation theory—Congresses. 2. Spline theory—Congresses. 3. Numerical analysis—Congresses. I. DeVore, Ronald A. II. Scherer, Karl.
QA221.036 511'.4 80-17554
ISBN 0-12-213650-0

PRINTED IN THE UNITED STATES OF AMERICA

80 81 82 83 9 8 7 6 5 4 3 2 1

Contents

Contributors

C. Bennett, Department of Mathematics and Statistics, University of South Carolina, Columbia, South Carolina 29208

H. Berens, Mathematisches Institut, Universität Erlangen-Nürnberg, D-8520 Erlangen, West Germany

C. de Boor, Mathematics Research Center, 610 Walnut Street, Madison, Wisconsin 53706

Yu. A. Brudnyi, Yaroslavl State University, Department of Mathematics, 150 000 Yaroslavl, USSR

H. G. Burchard, Department of Mathematics, Oklahoma State University, Stillwater, Oklahoma 74074

A. S. Cavaretta, Jr., Department of Mathematics, Kent State University, Kent, Ohio 44242

Charles K. Chui, Department of Mathematics, Texas A&M University, College Station, Texas 77843

Wolfgang Dahmen, Institut für Angewandte Mathematik, Universität Bonn, D-5300, Bonn, West Germany

S. Demko, School of Mathematics, Georgia Institute of Technology, Atlanta, Georgia 30332

Ronald A. DeVore, Department of Mathematics & Statistics, University of South Carolina, Columbia, South Carolina 29208

G. Gasper, Department of Mathematics, Northwestern University, Evanston, Illinois 60201

M. v. Golitschek, Institut für Angewandte Mathematik und Statistik, Universität Würzburg, D-8700 Würzburg, Germany.

B. Güssman, Abt. FE 324, Messerschmitt-Bölkow-Blohm GmbH, D-8000 München 80, West Germany

K. Höllig, Institut für Angewandte Mathematik / SFB, D-5300 Bonn, West Germany

B. S. Kashin, Stecklov Institute, 43 Vavilov, 117966, GSP-1, Moscow, USSR

P. E. Koch, Institutt for informatikk, Universitetet i Oslo, Oslo 3, Norway

G. G. Lorentz, Department of Mathematics, The University of Texas, Austin, Texas 77183

T. Lyche, Institutt for informatikk, Universitetet i Oslo, Oslo 3, Norway

A. A. Melkman, Department of Mathematics, Ben-Gurion University, Beer-Sheva, Israel

Charles A. Micchelli, IBM Research Center, PO Box 218, Yorktown Heights, New York 10598

I. P. Mysovskikh, Mat.-Mech. Fak., Leningr. Gosudarstvenn. Univ., Bibliotecnaja Pl., 2, Staryi Petergof, Leningrad, 198904, USSR

G. Nürnberger, Institut für Angewandte Mathematik, Universität Erlangen-Nürnberg, D-8520 Erlangen, West Germany

A. Pinkus, Department of Mathematics, Technion-Israel Institute of Technology, Technion City, 32000, Haifa, Israel

V. A. Popov, Mathematical Institut of Bulgarian Acad. of Sci., Sofia, Bulgaria

S. D. Riemenschneider, Department of Mathematics, University of Alberta, Edmonton, Alberta, Canada T6G 2G1

U. Sattes, Institut für Angewandte Mathematik, Universität Erlangen-Nürnberg, D-8520 Erlangen, West Germany

Walter Schempp, Lehrstuhl für Mathematik I, Universität Siegen, Hölderlinstrasse 3, D-5900 Siegen 21, West Germany

Karl Scherer, Institut für Angewandte Mathematik, Universität Bonn, D-5300 Bonn, West Germany

H. J. Schmid, Mathematisches Institut, Universität Erlangen-Nürnberg, D-8520, Erlangen, West Germany

A. Sharma, Department of Mathematics, University of Alberta, Edmonton, Alberta, Canada T6G 2G1

R. Sharpley, Department of Mathematics & Statistics, University of South Carolina, Columbia, South Carolina 29208

Philip W. Smith, Department of Mathematics, Texas A&M Unversity, College Station, Texas 77843

M. Sommer, Institut für Angewandte Mathematik, Universität Erlangen-Nürnberg, D-8520 Erlangen, West Germany

W. Trebels, Fachbereich Mathematik, Technische Hochschule, D-6100 Darmstadt, West Germany

R. S. Varga, Dept. of Mathematics, Kent State University, Kent, Ohio 44242

R. A. Waldo, Phillips Petroleum Company, Bartlesville, Oklahoma 74004

Joseph D. Ward, Department of Mathematics, Texas A&M University, College Station, Texas 77843

U. Westphal, Institut für Mathematik, Technische Universität Hannover, D-3000 Hannover, West Germany

D. Wulbert, Department of Mathematics, University of California-San Diego, La Jolla, California 92093

Preface

An international symposium on "Quantitative Approximation" was held in Bonn, West Germany, August 20–24, 1979. Leading mathematicians from several nations participated in this symposium; this proceedings volume represents the majority of the invited lectures.

While the contributions to these proceedings represent a wide range of topics, the following emerged as the main themes of the symposium:

(i) n-widths especially mixed norm cases;
(ii) spline approximation with the most emphasis on multivariate and knot independent questions;
(iii) nonlinear approximation including rational approximation and optimal knot spline approximation.

These are the deeper areas where much still needs to be done, as is indicated by the number of important unsolved problems in these subjects posed at the conference.

Many people from SFB 72 and the Institut für Angewandte Mathematik in Bonn helped with the organization of the symposium and the handling of associated details. Several people in Columbia helped with the preparation of these proceedings. Also, H. Berens and H. J. Schmid in Erlangen helped with the translation and preparation of the Mysovskikh article. Our sincere thanks go out to all these people.

ON AN INEQUALITY FOR THE SHARP FUNCTION

C. Bennett and R. Sharpley

The purpose of this note is to refine a rearrangement inequality for $f^{\#}$ in terms of the maximal rearrangement f^{**} for locally integrable functions on $\mathbb{R}^n$. One consequence of this inequality is an improvement and clarification of the proof that interpolation between L^1 and BMO "coincides" with that between L^1 and L^∞.

The "oscillation" of a locally integrable function f on $\mathbb{R}^n$ is gauged by its sharp function (cf [1])

$$(1)\quad f^{\#}(x) = \sup_{Q \ni x} \left\{ \frac{1}{|Q|} \int_Q |f(y) - f_Q|\, dy \right\}, \qquad x \in \mathbb{R}^n .$$

Here f_Q denotes the average $|Q|^{-1} \int_Q f(y)\,dy$ of f over a cube Q with sides parallel to the coordinate axes, and the supremum in (1) is taken over all such cubes Q containing the point x. Observe that the sharp function ignores constants: if f and g differ by a constant, then $f^{\#}$ and $g^{\#}$ coincide.

The decreasing rearrangement of f will be denoted by f^*, and its average $t^{-1}\int_0^t f^*(s)\,ds$ by $f^{**}(t)$. The latter function is "equivalent" to the decreasing rearrangement of the Hardy-Littlewood maximal function Mf of f in the sense that the ratio of f^{**} and $(Mf)^*$ is contained between positive constants independent of f (cf [1, Theorem 1.3]).

It follows immediately from (1) that $f^{\#} \le 2Mf$ and hence, by the preceding remarks, that $f^{\#*} \le cf^{**}$. There is also a result in the opposite direction, that is, it is possible to estimate f^{**} in terms of $f^{\#*}$. In fact, it was shown in [1, Corollary 4.2] that the inequality

$$(2)\quad f^{**}(t) \le c \int_t^\infty f^{\#*}(s) \frac{ds}{s} + f^{**}(+\infty), \qquad 0 < t < \infty ,$$

holds.

ISBN 0-12-213650-0

The importance of this result was demonstrated in [1] where a "local" version of (2) easily produced the well-known John-Nirenberg lemma. Nevertheless, the presence of the term $f^{**}(+\infty)$ on the right of (2) is bothersome in several of the applications in [1], and this led us to seek a more convenient formulation. In view of a remark made earlier, the first term on the right of (2) is unchanged when a constant γ is subtracted from f. In this note it will be shown that the finiteness of this integral guarantees the existence of a unique constant γ such that $(f-\gamma)^{**}(+\infty) = 0$, and hence, if f is replaced by $f-\gamma$ in (2), the troublesome term at infinity does not arise. The main result is thus as follows.

THEOREM 1. *Let* f *be locally integrable on* $\mathbb{R}^n$ *and suppose*

$$(3) \quad \int_1^\infty f^{\#*}(s)\,\frac{ds}{s} < \infty .$$

Then there is a unique constant γ $\left(= \lim_{|Q|\to\infty} f_Q\right)$ *such that*

$$(4) \quad (f-\gamma)^{**}(t) \le c\int_t^\infty f^{\#*}(s)\,\frac{ds}{s}, \qquad 0 < t < \infty .$$

The proof requires a pair of lemmas.

LEMMA 2. *If* $Q_0 \subset Q_1 \subset \mathbb{R}^n$, *then*

$$(5) \quad |f_{Q_0} - f_{Q_1}| \le c\int_{\frac{1}{2}|Q_0|}^\infty f^{\#*}(s)\,\frac{ds}{s},$$

where c *depends only on the dimension* n.

Proof. From inequality (4.23) of [1] it follows that

$$(6) \quad [(f-f_Q)\chi_Q]^{**}(t) \le c\int_t^{|Q|} f^{\#*}(s)\,\frac{ds}{s}, \qquad 0 < t < \frac{|Q|}{2},$$

for any cube Q. But if $0 \le t \le |Q|$, then $[\lambda\chi_Q]^{**}(t) = \lambda$, so

$$\begin{aligned} |f_{Q_0} - f_{Q_1}| &= [(f_{Q_0}-f_{Q_1})\chi_{Q_0}]^{**}\left(\frac{|Q_0|}{2}\right) \\ &\le [(f-f_{Q_0})\chi_{Q_0}]^{**}\left(\frac{|Q_0|}{2}\right) + [(f-f_{Q_1})\chi_{Q_1}]^{**}\left(\frac{|Q_0|}{2}\right), \end{aligned}$$

since $(g+h)^{**} \le g^{**} + h^{**}$ and $Q_0 \subset Q_1$. Together with (6) (applied for $Q = Q_0$ and $Q = Q_1$) this gives the desired inequality (5).

LEMMA 3. <u>If condition</u> (3) <u>holds, then</u> $\lim_{|Q|\to\infty} f_Q$ <u>exists.</u>

<u>Proof.</u> If $Q(k)$ denotes the cube with side length 2^k and centered at the origin, then it follows directly from (3) and (5) that $\{f_{Q(k)}\}_{k=1}^{\infty}$ is a Cauchy sequence. Let γ be its limit. For each $\varepsilon > 0$ it is possible, by (3), to choose a corresponding $M > 0$ such that

$$(7)\qquad c\int_{M/2}^{\infty} f^{\#*}(s)\frac{ds}{s} < \frac{\varepsilon}{3},$$

where c is the constant in (5). Then k may be chosen so large that $|f_{Q(k)} - \gamma| < \varepsilon/3$ and $|Q(k)| > M$. In that case, if Q is any cube with $|Q| > M$ and if Q' is any cube containing both $Q(k)$ and Q, then it follows from (5), (7), and the choice of k that

$$|f_Q - \gamma| \le |f_Q - f_{Q'}| + |f_{Q'} - f_{Q(k)}| + |f_{Q(k)} - \gamma| < \varepsilon .$$

<u>Proof of Theorem 1.</u> Let $\gamma = \lim_{|Q|\to\infty} f_Q$, which exists by virtue of the preceding lemma. Fix $t > 0$ and let $\varepsilon > 0$ be arbitrary. Then there is a cube Q with measure exceeding $2t$ and satisfying $|f_Q - \gamma| < \varepsilon$. Consequently, by (6),

$$\begin{aligned}[(f-\gamma)\chi_Q]^{**}(t) &\le [(f-f_Q)\chi_Q]^{**}(t) + |f_Q - \gamma| \\ &\le c\int_t^{\infty} f^{\#*}(s)\frac{ds}{s} + \varepsilon .\end{aligned}$$

If now $Q \uparrow \mathbb{R}^n$, the monotone convergence theorem shows that $[(f-\gamma)\chi_Q]^{**}(t) \uparrow (f-\gamma)^{**}(t)$. Hence the preceding estimate, after ε is allowed to decrease to zero, produces the desired inequality (4). For the uniqueness, note that if (4) holds for each of two constants γ_1 and γ_2, then for any $t > 0$,

$$|\gamma_1 - \gamma_2| = |\gamma_1 - \gamma_2|^{**}(t) \le (f-\gamma_1)^{**}(t) + (f-\gamma_2)^{**}(t)$$

$$\le 2c \int_t^{\infty} f^{\#*}(s)\frac{ds}{s} ,$$

and this, by (3), tends to 0 as $t \to \infty$.

Theorem 1 has a bearing on the identification of the interpolation spaces between the Lebesgue space L^1 and the space BMO of functions of bounded mean oscillation. The latter space consists of all equivalence classes (denoted by F) modulo constants of functions f for which $f^{\#}$ is bounded. Since the sharp function is invariant under addition of constants the notation $F^{\#} = f^{\#}$ for any representative f of F is meaningful. Hence, with the norm

$$(8)\quad \|F\|_{BMO} = \|F^{\#}\|_{L^{\infty}} ,$$

BMO is a Banach space.

It was established in [1,§6] that the Peetre K-functional

$$K(f,t;L^1,BMO) = \inf_{f=g+h} (\|g\|_{L^1} + t\|H\|_{BMO})$$

is given by

$$(9)\quad K(f,t;L^1,BMO) \sim tf^{\#*}(t) , \qquad 0 < t < \infty ,$$

for any f in $L^1 + BMO$.

Notice, however, that (L^1,BMO) is not, strictly speaking, a compatible Banach couple in that one space consists of functions and the other of equivalence classes of such functions modulo constants. The difficulty is resolved by introducing the space $\mathcal{L}^1$ consisting of all equivalence classes F for which the norm $\|F\|_{\mathcal{L}^1} = \inf_{f\in F} \|f\|_{L^1}$ is finite. Of course there will be precisely one representative f in L^1 of each equivalence class F in $\mathcal{L}^1$. With this, it is clear from (9) that the analogous result

$$(10)\quad K(F,t;\mathcal{L}^1,BMO) \sim tF^{\#*}(t)$$

holds for the Banach couple (L^1, BMO). Hence, with the aid of Theorem 1, the (θ,q)-interpolation spaces may be identified as follows.

COROLLARY 4. *Suppose* $0 < \theta < 1$, $0 < q \le \infty$, *and let* $1/p = 1-\theta$. *Then, for any* F *in* $L^1 + \mathrm{BMO}$,

$$\|F\|_{(L^1,\mathrm{BMO})_{\theta,q}} \sim \|F^\#\|_{L^{pq}} = \|f^\#\|_{L^{pq}} \sim \|f-\gamma\|_{L^{pq}},$$

where f *is any representative of* F *and* $\gamma = \lim_{|Q|\to\infty} f_Q$. *Hence*

$$(L^1,\mathrm{BMO})_{\theta,q} = (L^1,L^\infty)_{\theta,q} = L^{pq},$$

in the sense that if F *belongs to* $(L^1,\mathrm{BMO})_{\theta,q}$, *then there is a unique representative* $g = f-\gamma$ *of* F *such that*

$$\|g\|_{L^{pq}} \le c\|F\|_{(L^1,\mathrm{BMO})_{\theta,q}},$$

and, conversely, that if f *belongs to* L^{pq}, *then*

$$\|F\|_{(L^1,\mathrm{BMO})_{\theta,q}} \le c\|f\|_{L^{pq}},$$

where F *is the coset containing* f.

REMARKS. (i) The identification of the interpolation spaces, as in Corollary 4, does not require the complete knowledge (9) of the K-functional (cf [1, Corollary 4.4]).

(ii) The Hilbert transform and its n-dimensional analogues (the Riesz transforms) are well defined on L^p $(1 \le p < \infty)$ as the principal value integrals

$$(11)\quad R_j f(x) = \lim_{\varepsilon\to 0^+} \int_{\mathbb{R}^n} f(y)K_\varepsilon(x-y)\,dy$$

where $K_\varepsilon(y) = K(y)\chi_{(\varepsilon,\infty)}(|y|)$ and $K(y) = c_n y_j/|y|^{n+1}$. On L^∞, however, these integrals will normally not exist and so must be modified [2] according to

(12) $$\tilde{R}_j f(x) = \lim_{\varepsilon \to 0^+} \int_{\mathbb{R}^n} f(y)[K_\varepsilon(x-y) - K_1(-y)]dy .$$

To use the facts that the "Hilbert transform" maps L^∞ into BMO and say, H^1 into L^1 (or, L^1 into weak L^1) together with interpolation in order to obtain intermediate results, the ambiguity between the definitions (11) and (12) must be resolved. The inequality one is able to obtain [1, Corollary 5.6] is

(13) $$(R_j f)^*(t) \leq c\left\{\frac{1}{t}\int_0^t f^*(s)ds + \int_t^\infty f^*(s)\frac{ds}{s}\right\}$$

where the left hand side makes sense whenever the right hand side of the inequality is finite. This inequality is derived by using the finiteness of $\int_t^\infty f^*(t)\frac{dt}{t}$ to establish that $\int_{R^n} |f(y)|\ |K_1(-y)|dy < \infty$ and so (11) and (12) differ by at most a constant γ which is controlled according to Theorem 1.

REFERENCES

1. Bennett, C. and R. Sharpley, Weak-type inequalities for H^p and BMO, Proc. Sympos. Pure Math. 35I (1979), 201-229.

2. Fefferman, C. and E. M. Stein, H^p spaces of several variables, Acta Math. 129 (1972), 137-193.

C. Bennett*
R. Sharpley*

Department of Mathematics, Computer Science, and Statistics
University of South Carolina
Columbia, South Carolina 29208

* Supported in part by Grant MCS-7703666 of the National Science Foundation.

ON THE BEST CO-APPROXIMATION IN A HILBERT SPACE

H. Berens and U. Westphal

The following theorem is proved: For the best co-approximation in a Hilbert space the existence and uniqueness sets are the closed flats.

1. PRELIMINARY REMARKS

Let E be a real normed vector space. For $x \in E$ and $r \in \mathbb{R}^+$, $b_r(x)$ denotes the open ball centered at x with radius r and $\bar{b}_r(x)$ denotes its closure.

Let K be a nonempty set in E. The best approximation of $x \in E$ by elements of K can be defined as

$$P_K(x) = C_K(x) \cap K,$$

where

$$C_K(x) = \bigcap_{k \in K} \bar{b}_{\|x-k\|}(x).$$

Clearly, $C_K(x)$ equals $\bar{b}_{d_K(x)}(x)$, $d_K : E \to \mathbb{R}$ being the distance function.

For the <u>best co-approximation</u> we have

$$R_K(x) = B_K(x) \cap K,$$

where

$$B_K(x) = \bigcap_{k \in K} \bar{b}_{\|x-k\|}(k).$$

Correspondingly, the set-valued mapping $R_K : E \to K$ is said to be the <u>metric co-projection</u> from E onto K.

The notion of b. co-appr. goes back to P. L. Papini and I. Singer [3]. Motivated by a paper of C. Franchetti and M. Furi on characteristic properties of the real Hilbert space,

ISBN 0-12-213650-0

published in 1972, the authors studied this notion for linear subspaces of a real or complex norm. vs.

In his study of contractive mappings on bounded, closed and convex subsets of a real Hilbert space into itself F. E. Browder introduced the notion of an approximation region to describe the fixed point set of such mappings, as early as 1967. In our setting the <u>approximation region between</u> $x \in E$ <u>and</u> K is defined by

$$A_K(x) = \{y \in E : 0 \leq \langle x - y, y - k \rangle_s \quad \forall k \in K\} .$$

Here, $\langle y,x \rangle_s$, $(x,y) \in E \times E$, is the semi-inner product on $E \times E$ defined as the right-hand derivative of $\|\cdot\|^2/2$ at x in the direction y. For a Hilbert space H, $A_K : H \to H$ has an obvious geometrical interpretation. For details see R. E. Bruck, Jr. [2].

It is easy to verify that for an arbitrary set K in E

$$A_K \subset B_K$$

and that for an arbitrary flat K

$$\forall x \in E \qquad R_K(x) = A_K(x) \cap K.$$

Hence, for the metric co-projection on flats the following characterization holds

$$E \ni x \to R_K(x) = \{k \in K : 0 \leq \langle x - k, k - k' \rangle_s \quad \forall k' \in K\},$$

while for the metric projection we have

$$E \ni x \to P_K(x) = \{k \in K : 0 \leq \langle k - k', x - k \rangle_s \quad \forall k' \in K\}.$$

So much to clarify the notion and to give a motivation for the study of the best co-approximation. It isn't however our intention to make the reader acquainted with its basic properties, we just want to present and prove the little result stated in the abstract.

2. PROOF OF THE THEOREM

In 1961 V. Klee conjectured that there are nonconvex existence and uniqueness sets for the b. appr. in infinite dimensional Hilbert spaces. Although several fine results on this subject have been proved within the last two decades, we are still far from being able to give a definite answer. For details see e.g. [1]. In contrast to this, for the b. co-appr. the existence and uniqueness sets in a Hilbert space are easily characterized.

Let K be such a set in the real Hilbert space H. Existence implies that K is closed and convex. While the first property is obvious, let us prove the second one. Assume, K not to be convex. Then there are points k and k' in K for which $x = (k+k')/2$ is not in K. But

$$B_K(x) = \bar{b}_{\|x-k\|}(k) \cap \bar{b}_{\|x-k'\|}(k') = \{x\}$$

$$(\|x-k\| = \|x-k'\| = \|k-k'\|/2).$$

(More generally, in a strictly convex norm. vs every existence set for the b. co-appr. is closed and convex.)

As a consequence, the metric projection P_K defines a contractive retraction of H onto K. Hence, $P_K \subset R_K$. Moreover, by uniqueness we have equality, i.e.,

$$P_K = R_K.$$

It is well-known, that for closed convex sets in H the mapping $Q_K = 2P_K - I$ defines a contraction on H into itself with K as its fix-point set. For an $x \in H$, the image of x under Q_K is the reflexion of x in the supporting hyperplane of K at $k = P_K(x)$. Clearly, $Q_K \subset B_K$.

Let $(x,k) \in P_K$, $x \in H \setminus K$, and let $y = Q_K(x)$. By uniqueness, $y \notin K$. We claim, k is the element of b. appr. of y in K. Indeed, for all $k' \in K$

$$\|P_K \circ Q_K(x) - k'\| \leq \|Q_K(x) - k'\| \leq \|x - k'\|,$$

i.e.,

$$P_K \circ Q_K(x) \in R_K(x) = P_K(x) = \{k\}.$$

Hence, for all $x \in H \setminus K$

$$K \subset k + (x - k)^{\perp},$$

and finally,

$$K = \bigcap_{\substack{x \in H\setminus K \\ (x,k)\in P_K}} \{k + (x - k)^{\perp}\}.$$

This completes the proof.

REFERENCES

1. Berens, H. und U. Westphal, Kodissipative metrische Projektionen in normierten Vektorräumen, in Linear Spaces and Approximation, ed. by P. L. Butzer and B. Sz.-Nagy, ISNM 40, Birkhäuser Verlag, Basel 1978, 119-130.

2. Bruck, Jr., R. E., Nonexpansive projections on subsets of Banach spaces, Pacific J. Math. 47 (1973), 341-355.

3. Papini, P. L. and I. Singer, Best co-approximation in normed linear spaces, Monatsh. Math., submitted for publication.

H. Berens
Mathematisches Institut
Universität Erlangen-Nürnberg
D-8520 Erlangen

U. Westphal
Institut für Mathematik
Technische Universität Hannover
D-3000 Hannover

WHAT IS THE MAIN DIAGONAL OF A BIINFINITE BAND MATRIX ?

Carl de Boor

It is shown how to single out a particular diagonal of a biinfinite band matrix A as its main diagonal, using the decomposition of the solution set of $A\underline{x} = \underline{0}$ into those which are bounded at ∞ and those which are bounded at $-\infty$. As an application, it is proved that the inverse of the coefficient matrix for the system satisfied by the B-spline coefficients of the cubic spline interpolant at knots is checkerboard and that, under certain assumptions, the local mesh ratio must be bounded.

INTRODUCTION

The study of approximation by splines on a biinfinite knot sequence leads to linear systems

$$A\underline{x} = \underline{b}$$

with a banded biinfinite coefficient matrix A. Questions as to the existence of A^{-1}, its boundedness or its possible checkerboard nature need to be answered, and these, in turn, raise the question of which diagonal of A may be the main diagonal.

For example, it is well known that the inverse of a finite totally positive matrix A is checkerboard, i.e.,

$$(-)^{i+j} A^{-1}(i,j) \geq 0 , \quad \text{all } i,j$$

and, in particular, the entries of the main diagonal of A^{-1} are positive. One would expect the same statement to be true when A is biinfinite, but it is not clear a priori which diagonals of A^{-1} will be positive and which negative.

Again, in one approach toward proving that the inverse of a biinfinite totally positive matrix A is checkerboard, one would try to show that the inverse is approximated in some pointwise sense by the inverses of finite sections of A, whose checkerboard pattern is then known. Now, as we will show

ISBN 0-12-213650-0

below, if there is any convergence at all, then these finite sections are necessarily principal with respect to one fixed diagonal of A . That diagonal then has earned the epithet "main".

Here is an outline. In Section 1, even-order spline interpolation at knots is discussed, since I was led to wonder about the main diagonals of biinfinite matrices because of an argument with C.A. Micchelli concerning mesh ratio restrictions for that scheme. Section 2 contains a discussion of biinfinite band matrices, in particular some propositions regarding existence and character of their inverses. In the last section, some of these results are applied to cubic spline interpolation at a biinfinite knot sequence, giving me an opportunity to correct a mistake in [4].

1. MESH RATIO RESTRICTIONS IN EVEN-ORDER SPLINE INTERPOLATION AT KNOTS

Let $m\$_{k,\underline{t}}$ be the normed linear space of bounded splines of order k with knot sequence $\underline{t} = (t_i)$ and the sup-norm. We take the knot sequence $\underline{t}$ to be biinfinite and strictly increasing. Also, let $\underline{\tau}$ be a strictly increasing biinfinite sequence. The problem is to determine for given $\underline{a} \in \ell_\infty$ an f in $m\$_{k,\underline{t}}$ with $f(\tau_i) = a_i$, all i. I call this interpolation problem <u>correct</u> (others have called it "poised") if it has exactly one solution for every $\underline{a} \in \ell_\infty$, i.e., if

$$R : mS_{k,\underline{t}} \longrightarrow \ell_\infty : f \longmapsto f\big|_{\underline{\tau}} := (f(\tau_i))_{-\infty}^{\infty}$$

is invertible. Note that R^{-1} is necessarily bounded if it exists. The case k=2 being trivial, I will only consider k>2.

This interpolation problem has received particular attention in the special case of even-order spline interpolation at knots, i.e.,

$$\underline{\tau} = \underline{t} \quad \text{and} \quad k \text{ even} .$$

See, e.g., [6] for a recent survey. I showed in my talk [4] at the last Bonn conference that this interpolation problem is correct in case the global mesh ratio

$$M_{\underline{t}} := \sup_{i,j} \Delta t_i / \Delta t_j$$

is finite. I also stated there without proof that, for a correct problem, the local mesh ratio

$$m_{\underline{t}} := \sup_{|i-j|=1} \Delta t_i/\Delta t_j$$

would have to be finite, since it is possible to bound this mesh ratio in terms of $\|R^{-1}\|$. It was this claim which Micchelli doubted when we discussed various possible sufficient conditions for the correctness of the interpolation problem last summer. Now, my claim was based on the corresponding result in case of a finite knot sequence, in [3]. Here is an adaptation of the argument there to the present biinfinite context.

Supposing the problem correct, write the interpolant $R^{-1}\underline{f}$ to the particular data sequence $\underline{f} := ((-)^i)$ in terms of the normalized B-splines of order k for $\underline{t}$,

$$R^{-1}\underline{f} = \sum_j \alpha_j N_{j,k}$$

Then, from [2],

$$D_k^{-1}\|\underline{\alpha}\|_\infty \leq \|R^{-1}\underline{f}\|_\infty \leq \|\underline{\alpha}\|_\infty$$

for some positive constant D_k independent of $\underline{t}$. Now, for any particular i ,

$$\frac{2(-)^{i+1}}{\Delta t_i} = [t_i, t_{i+1}]R^{-1}\underline{f} = (R^{-1}\underline{f})'(\xi) = \sum \frac{\alpha_j - \alpha_{j-1}}{t_{j+k-1} - t_j} N_{j,k-1}(\xi)$$

Since $(N_{j,k-1})$ forms a partition of unity and $N_{j,k-1}$ has its support in (t_j, t_{j+k-1}) while $\xi \in (t_i, t_{i+1})$, this implies that

$$\min_{i-k+2 \leq j \leq i} \frac{t_{j+k-1} - t_j}{\Delta t_i} \leq D_k \|R^{-1}\|$$

and it was from this inequality that I had drawn a bound for $m_{\underline{t}}$ in terms of $\|R^{-1}\|$. But, actually this inequality is not strong enough for such a conclusion in case, e.g., $t_{j+k-1} - t_j$ is a decreasing function of j . The desired conclusion can be drawn, though, if we are certain that the α_j alternate, and alternate in the right way.

PROPOSITION 1. <u>If the B-spline coefficients $\underline{\alpha}$ of the interpolant $R^{-1}\underline{f}$ to the data sequence $\underline{f} = ((-)^i)$ alternate in such a way that $(-)^j \alpha_j \geq 0$, all j, then</u> $m_{\underline{t}} \leq D_k \|R^{-1}\|$.

Proof. Under this additional assumption, we have $(-)^{i+1}(\alpha_j-\alpha_{j-1}) \le 0$ for $j = i$ and also for $j = i-k+2$ (since k is even), hence from the above

$$\frac{2}{\Delta t_i} \le \sum_{j=i-k+3}^{i-1} (-)^{i+1} \frac{\alpha_j-\alpha_{j-1}}{t_{j+k-1}-t_j} N_{j,k-1}(\xi)$$

$$\le 2\|\underline{\alpha}\|_\infty / \min\{t_{j+k-1}-t_j : i-k+3 \le j \le i-1\}$$

and therefore

$$\frac{\Delta t_{i-1} + \Delta t_{i+1}}{\Delta t_i} \le \min_{i-k+3\le j\le i-1} \frac{t_{j+k-1}-t_j}{\Delta t_i} \le D_k \| R^{-1} \|$$

which does imply the desired result.

This leaves open the question as to when we can expect the alternation assumption to hold. In a finite-dimensional situation, the alternation is immediate because of the total positivity of the coefficient matrix of the linear system

$$\sum_j \alpha_j N_{j,k}(t_i) = (-)^i , \text{ all } i$$

But the sense of the alternation depends on just what the range of i and j here is. Specifically, if $i = I+1,\ldots,I+n$, $j = J+1, \ldots, J+n$, then (assuming that $J < I < J+k$)

$$(-)^{I+j} \alpha_{J+j} > 0 , \text{ all } j .$$

In the biinfinite case, it is not even clear that $\underline{\alpha}$ alternates. In order to investigate this question further, we now turn to an analysis of biinfinite band matrices.

2. THE INVERSE OF A BIINFINITE BAND MATRIX

In order to give our definition of main diagonal, we need notation to describe sections of biinfinite matrices and their relationship to the linear map represented by that matrix.

Let I, J be integer intervals. Then

$$A_{I,J} := A|_{I\times J} = (A(i,j))_{i\in I, j\in J}$$

denotes the corresponding section of the biinfinite matrix A. We can think of $A_{I,J}$ simply as a $|I|\times|J|$ matrix. But, $A_{I,J}$ also describes the nontrivial part of the linear map

$$P_I A P_J$$

with

$$(P_I \underline{a})(i) \quad := \begin{cases} \underline{a}(i) & , \; i \in I \\ 0 & , \; i \notin I \end{cases}$$

More precisely, $A_{I,J}$ is the matrix representation (with respect to the canonical basis) of the linear map

$$P_I (A|_{\operatorname{ran} P_J})$$

and we will not distinguish between these two.

Here and below, we will use the alternative notation $\underline{a}(i)$ for the i-th entry of the sequence $\underline{a}$, which is consonant with the notation $A(i,j)$ for the (i,j)-entry of the matrix A.

DEFINITION. The bounded and boundedly invertible biinfinite matrix A (as a map on $\ell_\infty(\mathbb{Z})$, say) has its r-th diagonal as main diagonal := the matrices $(A_{I,I+r})^{-1}$ converge stably to A^{-1} as $I \longrightarrow \mathbb{Z}$, i.e., $A_{I,I+r}$ is invertible for all sufficiently large finite intervals I and $\overline{\lim} \, \|(A_{I,I+r})^{-1}\| < \infty$, so that

$$A^{-1}(i,j) \quad = \quad \lim_{I \to \mathbb{Z}} \; (A_{I,I+r})^{-1}(i,j) \; , \quad \text{all } i,j \; .$$

Here, we have used the abbreviation

$$I+r \quad := \quad \{i+r : i \in I\} \; .$$

For example, suppose that both A and A^{-1} are upper triangular in the usual meaning of the word, i.e.,

$$A(i,j) \quad = \quad A^{-1}(i,j) \quad = \quad 0 \quad \text{for} \quad i > j \; .$$

Then

$$\delta_{ij} \quad = \quad \sum_{k=i}^{j} A(i,k) A^{-1}(k,j)$$

showing that $(A^{-1})_{I,I} = (A_{I,I})^{-1}$. In this case then, diagonal 0 is the main diagonal of A (as usual!). But now let E be the map or matrix of the left shift,

$$(E\underline{a})(i) \quad := \quad \underline{a}(i+1) \; , \; \text{all } i \; ,$$

and let $r \in \mathbb{Z}$. Then $E^r A$ is also invertible, with inverse $A^{-1}E^{-r}$. But now

$$(E^r A)(i,j) = A(i+r,j) \; , \qquad (E^r A)^{-1}(i,j) = A^{-1}(i,j+r)$$

hence now

$$((E^rA)_{I-r,I})^{-1} = (A_{I,I})^{-1} = A^{-1}_{I,I} = (E^rA)^{-1}_{I,I-r}$$

In other words, E^rA has diagonal r as its main diagonal (while $(E^rA)^{-1}$ has diagonal $-r$ as its main diagonal).

For matrices which are not triangular with triangular inverse, it is much more difficult to ascertain whether or not they even have a main diagonal, let alone which diagonal it might be. We now discuss this question in the context of banded matrices.

DEFINITION. We say that the biinfinite matrix A is m-banded if

(1) $A(i,j) = 0$ for $j \notin [i,i+m]$

(2) $A(i,i)A(i,i+m) \neq 0$, all i.

Thus an m-banded matrix has at most $m+1$ nonzero bands. If A is m-banded, then, for any r, E^rA also has just $m+1$ possibly nonzero contiguous bands and so could, with justification, also be called m-banded. But we will use the term "m-banded" only as described in order to suppress an additional inessential parameter. The other assumption, viz. the nonvanishing of the first and last band, is a nontrivial one. It makes certain statements simpler and is satisfied in the case of spline interpolation at knots.

An m-banded matrix A gives rise to a linear map on $\mathbb{R}^{\mathbb{Z}}$ which we will identify with A. This map has an m-dimensional nullspace or kernel,

$$\mathfrak{N} := \mathfrak{N}_A := \{\underline{f} \in \mathbb{R}^{\mathbb{Z}} : A\underline{f} = 0\}$$

In particular,

(3) for every i, $\mathfrak{N} \longrightarrow \mathbb{R}^m : \underline{f} \longmapsto \underline{f}|_{[i+1,i+m]}$ is 1-1 and onto

because of (2). Conversely, if $\mathfrak{N}$ is an m-dimensional subspace of $\mathbb{R}^{\mathbb{Z}}$ for which (3) holds, then there is, up to left multiplication by a diagonal matrix, exactly one m-banded matrix having $\mathfrak{N}$ as its nullspace

We also introduce two subspaces of $\mathfrak{N}$,

$$\mathfrak{N}^+ := \{\underline{f} \in \mathfrak{N} : \overline{\lim_{i\to\infty}}\, \underline{f}(i) < \infty\}, \quad m^+ := \dim \mathfrak{N}^+$$

$$\mathfrak{N}^- := \{\underline{f} \in \mathfrak{N} : \overline{\lim_{i\to-\infty}}\, \underline{f}(i) < \infty\}, \quad m^- := \dim \mathfrak{N}^-$$

From now on, we assume that A is m-banded and bounded on ℓ_∞. Then the i-th column $A^{-1}(\cdot,i)$ of its inverse, if it exists, would solve the linear system

$$A\underline{x} = (\delta_{ij}) .$$

The following proposition is therefore a first step toward understanding the inverse of an m-banded matrix.

PROPOSITION 2. <u>For all</u> i, <u>there exists exactly one</u> $\underline{L}_i \in \ell_\infty$ <u>such that</u> $A\underline{L}_i = (\delta_{ij})$ <u>if and only if</u> $\mathfrak{N} = \mathfrak{N}^- \oplus \mathfrak{N}^+$.

<u>Proof</u>. Since $\mathfrak{N}^- \cap \mathfrak{N}^+$ is the kernel of $A|_{\ell_\infty}$, there is at most one solution (for any particular i) if and only if $\mathfrak{N}^- \cap \mathfrak{N}^+ = \{0\}$. Hence it is sufficient to prove that, given uniqueness, there is a solution for every i iff $m^- + m^+ = m$.

For this, note that $\underline{L}_i \in \ell_\infty$ satisfies $A\underline{L}_i = (\delta_{ij})$ iff

$$(4) \quad \underline{L}_i(j) = \begin{cases} \underline{L}_i^-(j), & j<i+m \\ \underline{L}_i^+(j), & j>i \end{cases}, \text{ with } \begin{array}{l} \underline{L}_i^- \in \mathfrak{N}^- \\ \underline{L}_i^+ \in \mathfrak{N}^+ \end{array}$$

and

$$(5) \quad \sum_{j=i}^{i+m} A(i,j)\underline{L}_i(j) = 1 .$$

Here, $\underline{L}_i^-$, $\underline{L}_i^+$ is the extension of $\underline{L}_i|_{[i,i+m-1]}$ and $\underline{L}_i|_{[i+1,i+m]}$, respectively, to an element of $\mathfrak{N}$.

Now to see that $m^- + m^+ = m$ implies existence of $\underline{L}_i$, note that (4) and (5) constitute a linear system

$$(6) \quad B_i(\underline{L}_i|_{[i,i+m]}) = \begin{bmatrix} 0 \\ 0 \\ \cdot \\ 0 \\ 1 \end{bmatrix}$$

in the $m+1$ unknowns $\underline{L}_i(i), \ldots, \underline{L}_i(i+m)$, with the first $m-m^-$ homogeneous conditions ensuring that the extension $\underline{L}_i^-$ of $\underline{L}_i|_{[i,i+m-1]}$ to an element of $\mathfrak{N}$ lies in $\mathfrak{N}^-$, and the next $m-m^+$ homogeneous conditions ensuring that the extension $\underline{L}_i^+$ of $\underline{L}_i|_{[i+1,i+m]}$ to an element of $\mathfrak{N}$ lies in $\mathfrak{N}^+$, and the last, the only inhomogeneous, condition being (5). But if now $m^- + m^+ = m$, then (6) has as many equations as unknowns and, as we already know that it has at most one solution, the existence of a solution follows.

Conversely, assuming the existence of a solution for every i, consider the maps

$$\phi^* : \mathbb{R}^{\bar m+1} \longrightarrow \mathfrak{N} : \underline{a} \longmapsto \sum_{j=0}^{\bar m} a_j \underline{L}^*_{i+j}$$

with $*$ standing for $+$ or $-$, and $\bar m := m^- + m^+$. Then

$$\dim\ker\phi^* = \bar m+1 - \dim\operatorname{ran}\phi^* \geq \bar m+1-m^* = \begin{cases} m^-+1 , & * = + \\ m^++1 , & * = - \end{cases}$$

Consequently, there exists $\underline{a} \in \ker\phi^+ \cap \ker\phi^- \setminus \{0\}$. For this $\underline{a}$,

$$\underline{M} := \sum_{j=0}^{\bar m} a_j \underline{L}_{i+j} \neq 0$$

since $(\underline{L}_j)$ is obviously linearly independent. On the other hand, since

$$\underline{L}_{i+j}(s) = \begin{cases} \underline{L}^-_{i+j}(s) , & \text{for } s < i+m \\ \underline{L}^+_{i+j}(s) , & \text{for } s > i+\bar m \end{cases} , \quad j=0,\dots,\bar m ,$$

we find

$$\underline{M} = \begin{cases} \Sigma_0^{\bar m}\, a_j \underline{L}^-_{i+j} & \text{on }]\infty, i+m[\\ \Sigma_0^{\bar m}\, a_j \underline{L}^+_{i+j} & \text{on }]i+\bar m, \infty[\end{cases}$$

and therefore, by choice of $\underline{a}$, $\underline{M}(s) = 0$ for $s < i+m$ and $s > i+\bar m$. This implies $\bar m \geq m$, and therefore, since by assumption $\mathfrak{N}^+ \cap \mathfrak{N}^- = \{0\}$, i.e., $m^+ + m^- \leq m$, the conclusion $m^+ + m^- = m$ follows.

Next, we characterize bounded invertibility of a bounded m-banded matrix A in terms of $\mathfrak{N}$.

PROPOSITION 3. A <u>is boundedly invertible if and only if (i)</u> $\mathfrak{N} = \mathfrak{N}^- \oplus \mathfrak{N}^+$, <u>(ii) the elements of</u> $\mathfrak{N}^-$ <u>and</u> $\mathfrak{N}^+$ <u>decay exponentially and, (iii) for each</u> i, <u>the matrix</u> B_i <u>in (6) can be chosen so that</u> $\sup_i \|B_i^{-1}\| < \infty$.

Concerning the exponential decay, I had proved at the last Bonn conference that, for a bounded and boundedly invertible m-banded matrix A on some ℓ_p with $p<\infty$, there exists const so that, for all $\underline{f} \in \mathfrak{N}^-$ and all i,

$$\|\underline{f}^{(r)}\|_p^p \geq \text{const}\ \Lambda^r\ \|\underline{f}^{(0)}\|_p^p , \quad r=1,2,3,\dots$$

with

$$\Lambda := \frac{\kappa^p + 1}{\kappa^p - 1} , \quad \kappa := \|A\|\,\|A^{-1}\|$$

and

$$\underline{f}^{(r)} := \underline{f}|_{[i+rn,i+(r+1)m-1]},$$

and an analogous statement for $\underline{f} \in \mathfrak{N}^+$. The exponential decay mentioned in condition (ii) of the proposition is meant in this sense.

The proof of the necessity of the exponential decay rests on Demko's [5] nice ideas. As a footnote, I would like to record here that, in response to my talk, S. Demko, at the end of his talk at the present conference, made the point that he had been materially helped by a referee's report authored as it happens, by T. Lucas.

Finally, the proof of the necessity of the last condition is a bit tricky.

On the other hand, the sufficiency of the three conditions is immediate since they insure that the $\underline{L}_i$'s, constructed in Proposition 2 on the strength of (i), are ℓ_∞-bounded uniformly in i and decay exponentially, hence the matrix $[\ldots,\underline{L}_i,\ldots]$ is a bounded map on ℓ_p for every p, etc.

The complete proof of Proposition 3 (and of Proposition 4 to follow) can be found in [7]. Finally, we state a necessary and sufficient condition for such a bounded m-banded matrix A to have a main diagonal.

PROPOSITION 4. <u>A has a diagonal</u> r <u>as its main diagonal if and only if</u> (i) A <u>is boundedly invertible</u>, (ii) r=m+, (iii) <u>there exists a positive</u> const <u>so that, for all large</u> n <u>and</u>

$$\text{for all } \underline{f} \in \mathfrak{N}^+, \quad \|\underline{f}|_{[-n,-n+m^+-1]}\| \geq \text{const}\, \|\underline{f}|_{[-n,-n+m]}\|$$

<u>and</u>

$$\text{for all } \underline{f} \in \mathfrak{N}^-, \quad \|\underline{f}|_{[n+m^++1,n+m]}\| \geq \text{const}\, \|\underline{f}|_{[n,n+m]}\| .$$

In effect, under these assumptions, we can construct the column $\underline{L}_i^{(I)}$ of $(A_{I,I+r})^{-1}$ for all large I as a modification of the corresponding column $\underline{L}_i$ of A^{-1},

$$\underline{L}_i^{(I)} = \underline{L}_i - \underline{L}^+ - \underline{L}^- \quad \text{on } I+r$$

with

$$\underline{L}^{+} + \underline{L}^{-} = \underline{L}_i \quad \text{on} \quad I\setminus(I+r) \cup (I+m)\setminus(I+r)$$

and $\underline{L}^{*} \in \mathfrak{N}^{*}$. This guarantees that $\|\underline{L}_{*i}^{(I)}\| \sim \|\underline{L}_i\|$ while, because of the exponential decay of $\underline{L}^{*}$, $\underline{L}_i^{(I)} \sim \underline{L}_i$ away from the boundary of $I+r$.

3. CUBIC SPLINE INTERPOLATION AT KNOTS

In this section, we establish the checkerboard pattern for A^{-1} in case

(7) $\quad A \; = \; (N_{j,4}(t_i))$

of cubic spline interpolation at knots.

Given that the interpolation problem is correct, we now know that there are just three possibilities: $m^{+} = 0,1,2$.

Case $m^{+}=0$. In this case, $\mathfrak{N} = \mathfrak{N}^{-}$, and from (4) we see that A^{-1} is upper triangular. Therefore, as discussed in Section 2, the first diagonal of A is main, i.e., diagonal 1 in our way (7) of writing A . It follows that

$$(-)^{i+j+1}A^{-1}(i,j) \geq 0 \;, \quad \text{all } i,j$$

and, in particular, the solution $\underline{\alpha}$ of $A\underline{\alpha} = ((-)^{i})$ satisfies $(-)^{i+1}\alpha_i > 0$, all i. Thus, $\underline{\alpha}$ alternates but in the wrong sense if we are after bounding the local mesh ratio in terms of $\|R^{-1}\|$ using the argument of Proposition 1. In fact, it is not difficult to construct a knot sequence $\underline{t}$ for which the interpolation problem is correct and for which $t_i \longrightarrow 0$ as $i \longrightarrow -\infty$ so strongly that $m_{\underline{t}} = \infty$. My statement at the last Bonn conference ([4;p.48]) that $m_{\underline{t}} \leq \text{const} \|R^{-1}\|$ must therefore be qualified to exclude the case $m^{+} = 0$ and the analogous case $m^{+} = 2$.

Case $m^{+}=1$. We find it convenient to associate with the element $\underline{L} \in \mathfrak{N}$ the nullspline

$$L \;:=\; \sum_j \underline{L}(j)\, N_{j,4}$$

for which it supplies the B-spline coefficients. We claim that

(8) $\quad \underline{L} \in \mathfrak{N}^{-}\setminus\{0\}$ implies $\; L'(t_i)L''(t_i) > 0$, all i .

Indeed, $\mathfrak{N}$ contains a sequence $(\underline{L}_{[i]})$ so that $L'_{[i]}(t_i)L''_{[i]}(t_i) > 0$, hence $L'_{[i]}(t_j)L''_{[i]}(t_j) > 0$ for $j \geq i$, by [1]. Since $\mathfrak{N}$ is finite-dimensional, a properly normalized subsequence then has a limit L for which $L'(t_i)L''(t_i) \geq 0$, all i. But then, by [1],

$$L'(t_i)L''(t_i) > 0 \text{ , all } i$$
$$\|L\|_{i+j} \geq 2^j \|L\|_i \text{ , all } j \geq i$$

with

$$\|L\|_j := \max \{|L'(t_j)|, |L''(t_j)|\} .$$

Suppose now that, by way of contradiction, $\underline{M} \in \mathfrak{N}^- \setminus \{0\}$ satisfies $M'(t_i)M''(t_i) \leq 0$ for some i. Then, again by [1], $M'(t_j)M''(t_j) < 0$ for $j < i$ and

$$\|M\|_{i-j} \geq 2^j \|M\|_i \text{ , } j=1,2,3,\ldots,$$

showing that then $M'(t_j)$ and $M''(t_j)$ both would increase exponentially as $j \longrightarrow -\infty$, while M itself stays bounded since $\underline{M} \in \mathfrak{N}^-$. This would imply that Δt_j decreases exponentially as $j \longrightarrow -\infty$ and then, given that $2^{-j}\|L\|_i \geq \|L\|_{i-j}$ for $j=1,2,\ldots$, $\underline{L}$ would surely also be in $\mathfrak{N}^-$ contradicting the fact that $m^- = 1$.

One proves analogously that

(9) $\underline{L} \in \mathfrak{N}^+ \setminus \{0\}$ implies $L'(t_i)L''(t_i) < 0$, all i.

(8) and (9) imply (see [1]) that

(10) for all $\underline{L} \in \mathfrak{N}^* \setminus \{0\}$, $L''(t_i)L''(t_{i+1}) < 0$, all i.

Next, we claim that

(11) for all $\underline{L} \in \mathfrak{N}^* \setminus \{0\}$, $\underline{L}_j \, L''(t_{j+2}) < 0$, all j.

For this, recall (e.g., from[2;p.270]) that $f = \Sigma \alpha_j N_{j,4}$ implies

$$\text{(12)} \quad \alpha_j = \sum_{r<4} (-)^{3-r} \psi^{(3-r)}(\tau) f^{(r)}(\tau)$$

for any $\tau \in (t_j, t_{j+4})$, with

$$\psi(x) := (t_{j+1}-x)(t_{j+2}-x)(t_{j+3}-x)/3! .$$

Since $L = \Sigma \, \underline{L}_j \, N_{j,4}$, we then get

$$\underline{L}_j = \psi''(t_{j+1})L'(t_{j+1}) - \psi'(t_{j+1})L''(t_{j+1})$$

while $\psi''(t_{j+1}) > 0 > \psi'(t_{j+1})$. Therefore, with (8),

$$\underline{L}_j L''(t_{j+1}) = \psi''(t_{j+1})L'(t_{j+1})L''(t_{j+1}) - \psi'(t_{j+1})[L''(t_{j+1})]^2 > 0$$

if $\underline{L} \in \mathfrak{N}^- \setminus \{0\}$, and (10) now finishes the proof of (11) for this case. The case $\underline{L} \in \mathfrak{N}^+ \setminus \{0\}$ uses (12) with $\tau = t_{j+3}$ instead.

Finally, let now $\underline{L}$ be one of the columns of A^{-1},

$$\underline{L} = A^{-1}(\cdot, i)$$

say. Since $m^- = m^+ = 1$, we have

$$(13) \quad \underline{L}(j) = \begin{cases} \underline{L}^-(j), & \text{for } j < i+2 \\ \underline{L}^+(j), & \text{for } j > i \end{cases}$$

with neither $\underline{L}^-$ nor $\underline{L}^+$ just 0 . Consequently, by (10) and (11), $\underline{L}$ alternates (strictly) and, as to the sense of that alternation, we have from (12) (with $j = i-2$) that

$$\begin{aligned} A^{-1}(i-2,i) &= \psi''(t_{i-1})L'(t_{i-1}) - \psi'(t_{i-1})L''(t_{i-1}) \\ &= \psi''(t_{i+1})L'(t_{i+1}) - \psi'(t_{i+1})L''(t_{i+1}) \end{aligned}$$

while, from (13), (8) and (9),

$$\psi''(t_{i-1}) > 0 > \psi'(t_{i-1}) \;, \quad L'(t_{i-1})L''(t_{i-1}) > 0$$

$$\psi''(t_{i+1}),\ \psi'(t_{i+1}) < 0 \;, \quad L'(t_{i+1})L''(t_{i+1}) < 0$$

We conclude that

$$\operatorname{sign} A^{-1}(i-2,i) = \operatorname{sign} L''(t_{i-1}) = \operatorname{sign} L''(t_{i+1}) .$$

But this implies that $A^{-1}(i-2,i) > 0$, since the contrary assumption would give $L'(t_i^-) > 0 > L'(t_i^+)$ (since $L(t_i) = 1$), an impossibility.

In conclusion, if $m^- = m^+ = 1$, then

$$(14) \quad (-)^{i+j}A^{-1}(i,j) > 0 \text{ , all } i,j,$$

and, in particular, the solution of the linear system $A\underline{\alpha} = ((-)^i)$ does satisfy the condition $(-)^i\alpha_i > 0$, all i, needed in the argument for the finiteness of the local mesh ratio.

Note that we proved (14) here without recourse to finite sections. Even in this simple case, I still do not know whether the matrix A has a main diagonal (though I don't think it would be very hard to prove).

REFERENCES

1. Birkhoff, G. and C. de Boor, Error bounds for spline interpolation, J.Math.Mech. 13 (1964), 827-836.

2. Boor, C. de, The quasi-interpolant as a tool in elementary polynomial spline theory, in "Approximation Theory", G. G. Lorentz ed., Academic Press, 1973, 269-276.

3. Boor, C. de, On bounding spline interpolation, J.Approximation Theory 14 (1975), 191-203.

4. Boor, C. de, Odd-degree spline interpolation at a biinfinite knot sequence, in "Approximation Theory, Bonn 1976", R. Schaback & K. Scherer eds., Lecture Notes Math. 556, Springer, Heidelberg, 1976, 30-53.

5. Demko, S., Inverses of band matrices and local convergence of spline projectors, SIAM J.Numer.Anal. 14 (1977), 616-619.

6. Micchelli, C.A., Infinite spline interpolation, in Proc. Conf.Approximation Theory, Siegen, Germany, 1979, G. Meinardus ed.

7. Boor, C. de, Dichotomies for band matrices, submitted to SIAM J.Numer.Anal.

C. de Boor
Mathematics Research Center
610 Walnut Street
Madison, WI 53706

Supported in part by the United States Army under Contract No. DAAG29-75-C-0024 . Some of the work was done while the author was a very contented and grateful guest of the SFB 72 at the University of Bonn, Germany in the summer of 1979.

RATIONAL APPROXIMATION AND EXOTIC LIPSCHITZ SPACES

Yu. A. Brudnyi

We discuss the relationship between the structural properties of a function and its' degree of approximation by rational functions.

Let $r_n(f;p)$ denote the error of best approximation to f in $L_p[0,1]$, $1 \leq p \leq \infty$, by rational functions of degree less than or equal to n. We are interested in the complete description of the intrinsic properties of the class of functions

$$(1)\quad R_p^\alpha = \{f:\ r_n(f;p) = 0(n^{-\alpha}),\ n \to \infty\}\ ,\ \alpha > 0.$$

In analogy to the classical results of S. Bernstein, Ch. de la Valee Poussin, D. Jackson, and A. Zygmund, it may be expected that the classical Lipschitz spaces Λ_p^α should play an essential role in the description of the spaces R_p^α. Recall that Λ_p^α is defined as the space of all functions f for which the semi-norm

$$(2)\quad |f|_{\Lambda_p^\alpha} = \sup_{h>0} \{h^{-\alpha}\ \|\Delta_h^k f\|_{L_p[0,1-kh]}\}\ ,\ k = [\alpha] + 1\ ,$$

is finite.

It turns out, as can be seen from the articles [7] and [11] and the report of V. Popov at this conference, that Λ_p^α is a proper subset of R_p. Nevertheless, it is possible to give quite a bit of information about the space R_p^α by using the spaces Λ_p^α for $0 < p$. Here, we define Λ_p^α to be the space of those functions for which the semi-norm (2) is finite when $p < 1$, as well. It is important to point out that although there are several different but equivalent possible definitions of the spaces Λ_p^α when $1 \leq p \leq \infty$, these

ISBN 0-12-213650-0

lead to different function spaces when $p < 1$. We shall see that these "exotic" Lipschitz spaces appear naturally and in some sense are inevitable for the description of R_p^α.

In order to formulate our main result, let

$$\Lambda_{p-0}^\alpha = \bigcap_{q<p} \Lambda_q^\alpha , \quad \Lambda_{p+0}^\alpha = \bigcup_{q>p} \Lambda_q^\alpha .$$

THEOREM 1. *If* $q = (\alpha + 1/p)^{-1}$, *then*

$$\Lambda_{q+0}^\alpha \subset R_p^\alpha \subset \Lambda_{q-0}^\alpha . \tag{3}$$

The right hand inclusion may be strengthened. For this purpose we need the spaces $\Lambda_{p\infty}^\alpha$ which are defined by replacing the L_p norm in (2) by the Lorentz $L_{p\infty}$ norm. With this, it is now possible to replace the right hand side of (3) by $\Lambda_{q\infty}^\alpha$. The left hand inclusion does not admit a strengthening of this type.

The theorem may be extended to the more general situation dealing with the class of functions

$$R_p^\omega = \{f \in L_p: \ r_n(f,p) = 0(\omega(n^{-1})), \ n \to \infty\} .$$

Here, ω is any increasing function which tends to 0 as $t \downarrow 0$ and satisfies

$$\omega(2t) \leqq \omega(t) , \quad 0 < t .$$

In this case, we have the following two results.

THEOREM 2. *If* $q = (\alpha + 1/p)^{-1}$, *then*

$$r_n(f,p) \leqq \frac{\gamma}{n^\alpha} \left\{ \sum_{j \geqq n} j^{-1} [j^\alpha \omega_k(f,j^{-1})_{L_q}]^q \right\}^{1/q}$$

for every $n > k$.

Here, $\gamma = \gamma(\alpha,p,q)$ *and* $\omega_k(f;t)_{L_q} = \sup_{h \leqq t} \|\Delta_h^k f\|_{L_q(0,1-kh)}$.

In particular, if $\omega_k(f;t)_{L_q} = 0(t^\alpha \omega(t))$ for some $q > (\alpha + 1/p)^{-1}$ and a function ω, as above, then we have

COROLLARY 1. $r_n(f;p) = 0(n^{-\alpha}\omega(n^{-1}))$.

The following result is inverse to Theorem 2.

THEOREM 3. *If* $q = (\alpha + 1/p)^{-1}$ *and* $q^* = \min(1,q)$, *then*

$$\omega_k(f;\tfrac{1}{n})_{L_q} \leq \frac{\gamma}{n^\alpha}\left\{\sum_{j\leq n} j^{-1}[j^\alpha r_j(f;p)]^{q^*}\right\}^{1/q^*}$$

for every n.

The particular case $q = \lambda = 1$, $p = \infty$ of Theorem 3 has been obtained by E. P. Dolzenko [4] (see also [13]).

I would like to point out three corollaries of these general results.

THEOREM 4. *If* $f \in W_q^s(0,1)$, $1 \leq q \leq \infty$, *then*

$$r_n(f;p) \leq \gamma n^{-s}\omega_k(f;n^{-1})_{L_q} .$$

Here, $\gamma = \gamma(k,s,q)$ *and* $p = \infty$ *when* $s \geq 2$ *or* $s = 1$, $q > 1$; *in the exceptional case,* $s = q = 1$, *we can choose an arbitrary* $p < \infty$.

This theorem may be considered as an analogue of the second Jackson theorem. In the particular case $s \geq 2$, $p = \infty$ and $r = q = 1$, V. A. Popov and F. Szabados [10] have obtained a weaker result with the additional factor $(\log n)^{2s+2}$ on the right side and a constant which depends on $\|f'\|_\infty$. In the exceptional case, we also have

THEOREM 5. *If* $f \in W_1^1(0,1)$ *and* $\omega_2(f';t)_{L_1} = O(t^\alpha)$, $0 < \alpha \leq 1$, *then*

$$r_n(f;\infty) = O(n^{-1-\alpha}) .$$

In particular, if $\omega_2(f^{(s)};t)_L = O(t)$, $s \geq 1$, *then*

$$r_n(f;\infty) = O(n^{-s-1}) . \tag{4}$$

The last statement strengthens a result of V. A. Popov's [9]. He obtained the estimate (4) under the more strict condition Var $f^{(s)} < \infty$. I should like to note that the remarkable method of Popov plays a part in the proof of Theorem 2.

THEOREM 6. *If the function* f *is in* R_p^α, *then it can be changed on a set of arbitrarily small measure so that the resulting function* F *has continuous derivatives of order less than* α *and for bigger derivatives we have the inequality*

$$(5)\quad |\Delta_h^2 F^{(s)}(x)| \leq \gamma h^{\alpha-s}(\log \tfrac{1}{h})^{\alpha+1/p+\varepsilon} .$$

Here, $\varepsilon > 0$ *is arbitrarily small and* $\gamma = \gamma(\alpha,p,\varepsilon) \to \infty$ *as* $\varepsilon \to 0$.

In the case when α is noninteger, the second difference in the left side of (5) may be replaced by the first one. The weaker statement was established earlier by A. A. Gonchar [9] with another method in the case $p = \infty$. The results of Theorem 4 cannot be improved, in fact, there exists a function f from R_p^α for which the statement is false for $\varepsilon = 0$.

In order to establish the complete description of the class R_p^α introduced in Theorem 1, we shall give the corresponding result for the similar problem for the class

$$(6)\quad S_p^\alpha = \{f; s_n^k(f;p) = 0(n^{-\alpha}), n \to \infty\} .$$

Here, $s_n^k(f;p)$ is the best approximation of f in $L_p(0,1)$ metric by splines of degree $k - 1$ with n free knots, $k = [\alpha] + 1$.

THEOREM 7. *If* $q = (\alpha + 1/p)^{-1}$, *then*

$$(7)\quad B_q^\alpha \subset S_p^\alpha \subset \Lambda_{q\infty}^\alpha .$$

Here, the left space in (7) *is defined by the Besov semi-norm*

$$|f|_{B_q^\alpha} = \left\{ \int_0^1 \left(\frac{\| \Delta_h^k f \|_{L_q}}{h^\alpha} \right)^q \frac{dh}{h} \right\}^{1/q}, \quad \alpha < k.$$

It is clear that

$$\Lambda_{q+0}^\alpha \subset B_q^\alpha$$

so that (7) is more exact than (3). If one stays in the scale of Lipschitz spaces then no further improvement of the result is possible. However, the complete description of the class S_p^α obtained in my article [2] is connected with a scale of function spaces of quite another kind. By the way, the shortest method of proving Theorem 7 is based on the embedding theorems for the mentioned spaces and Lipschitz spaces.

The proof of the main theorem is very long and complicated. In it, the combination of methods of rational approximation and of the local polynomial approximation plays an important part. For example, the proof of the left inclusion of (3) is based on the embedding theorems for the class V_{pq}^α which is defined by the variation of local approximation (the exact definition is given in my article [1]). The machinery which is developed in the works of G. Freud [9], V. A. Popov [6], and myself [3], permits an improvement of this embedding which, in turn, gives the left hand side of (3).

For the proof of the right hand side of (3), the classical method of S. Bernstein is applied. In this case, a new inequality for rational functions plays the main role. (It is interesting to compare this inequality with the inequality of Dolzenko for the derivatives of rational functions [5], [12]).

THEOREM 8. <u>If</u> R <u>is</u> <u>a</u> <u>rational</u> <u>function</u> <u>of</u> <u>degree</u> n <u>and</u> $q = (\alpha + 1/p)^{-1}$, <u>then</u>

$$|R|_{\Lambda_q^\alpha} \leq \gamma n^\alpha \|R\|_{L_p} ,$$

where $\gamma = \gamma(\alpha,p)$.

Finally, I should like to mention that in the proof of Theorem 7 an important part is played by a Lebesque type theorem on the differential properties of functions from the class V_{pq}^α (see [1], p. 120), as well as its extension to the case when α is integer. The fact that Theorem 7 cannot be improved follows from the example constructed by K. F. Oskolkov [8].

REFERENCES

1. Brudnyi, Yu. A., Spaces defined by means of local approximations, Trans. Moscow Math. Soc., 24 (1971), 74-139.

2. Brudnyi, Yu. A., Spline approximation and functions of bounded variation, Sov. Math. Doklady, 15 (1974), 518-521.

3. Brudnyi, Yu. A., Some non-linear methods of best approximation in: Theory of Approximation of Functions, Proceeding of the Kalouga Conference, 1975, M. Nauda, 14 (1977).

4. Dolzenko, E. P., Some estimates concerning algebraic hyperplanes and derivatives of rational functions, Dokl. Akad. Nauk, 139 (1961), 1287-1290.

5. Dolzenko, E. P., Uniform approximation by rational (algebraic and trigonometric) functions and their global functional theoretic properties, Doklady Akad. Nauk, 166 (1966), 526-529.

6. Freud, G., Über die Approximation reeler Funktionen durch gebrochenen Functionen, Acta. Math. Acad. Sci. Hung., 17 (1966), 313-324.

7. Gonchar, A. A., The degree of approximation by rational functions and properties of functions, Proceeding of the International Math. Conf., Mir (1968), 329.

8. Oskolkov, K. J., Uniform modulus of continuity of summable functions on a set of positive measure.

9. Popov, V., On rational approximation of functions from the class V_r, C. R. Acad. Bulg. Sci., 29 (1976), 791-799.

10. Popov, V. and J. Szabados, A remark on the rational approximation of functions, C. R. Acad. Bulg. Sci., 28 (1975), 1303-1306.

11. Reddy, A. R., Recent advances in Chebyshev rational approximation on finite and infinite intervals, J. Approx. Th., 22 (1978), 59-84.

12. Sevastianov, E. A., Some estimates for derivatives of rational functions in intergral metrics, Mat. Zametki, 13 (1973), 499-510.

13. Sevastianov, E. A., The uniform approximation by monotone functions and some applications in ϕ variations and Fourier series, Dokl. Akad. Nauk USSR, 217 (1974), 27-30.

Yu. A. Brudnyi
Yaroslavl State University
USSR

SOME MATHEMATICAL PROPERTIES OF EMPIRICAL GIBBS FUNCTIONS

H.G. Burchard[1] and R.A. Waldo[2]

Mathematical properties of a model of phase equilibrium used by chemical engineers are investigated.

§ 1 INTRODUCTION

In this paper we develop some of the mathematical properties of a model for the phase equilibrium problem in chemical engineering, [1,2,3,5].

Due to the high dimensions involved, efficient algorithms must be expected to contain some heuristic elements. But, there is hope that a mathematical approach to formulating and analyzing the problem might turn up results that lead to improved algorithms.

We do not begin to discuss algorithms in this paper. However, we deal with questions directly relevant to computation. An example is a result on stability under perturbation of the data.

A statement in general terms of the mathematical problem is easy to give. Given is a convex function $\hat{G}$ defined on the $(c-1)$-dimensional simplex S . Also given is $x \in S$. Find a (maximal) face F of the epigraph of $\hat{G}$ to which $(x,\hat{G}(x))$ belongs, and find the set of extreme points extr(F) of F .

The physical interpretation is that x represents a vector of mole fractions of a mixture of c chemical components (species) that decomposes into equilibrium phases, the unknown extr(F), at a given P and T . The function $\hat{G}$ is the Gibbs free energy function at (P,T). Chemical reactions may be taken into account by adding stoichiometric equations but we do not do so here.

ISBN 0-12-213650-0

One difficulty is that $\hat{G}$ is not given explicitely. Instead one is given the <u>primitive</u> Gibbs function $G(x,P,T)$ [2]. Taking the convex hull

$$\text{conv}\ \{(x,g)\ :\ x \in S,\ g \geqq G(x,P,T)\}$$

of the epigraph of $G(\cdot,P,T)$ one obtains the epigraph of $\hat{G}(\cdot,P,T)$. A further difficulty lies in the multivaluedness and singularities of G. Fortunately, the definition of G can be modified so that $G(\cdot,P,T)$ is at least single-valued and continuous on S.

In § 2 we describe the minimum problem connected with forming $\hat{G}$ from G. In § 3 we analyze one concrete model of a primitive Gibbs function based on the Soave-Redlich-Kwong equation of state. In § 4 we state in full detail a theorem of stability of the solutions of the minimum problem under small perturbations of the data.

§ 2 MINIMUM PROBLEM AND DUAL

We present this material not in the usual "theorem-proof" form since basically the material is well-understood mathematically. We have frequent recourse to Rockafellar's excellent account of convex sets and functions [6].

The basic data is a real-valued function $G(x,P,T)$, the <u>primitive Gibbs free energy function</u>, defined and continuous for $x \in R^{c+} = \{x \in R^c\ :\ x \geqq o\}$, $P > o$, $T_1 \geqq T \geqq o$. Also,

$$G(\alpha x,P,T) = \alpha G(x,P,T), \qquad \alpha \geqq o\ . \tag{2.1}$$

Hence G is determind by its values on the (c-1)-dimensional simplex $S = \{x \in R^{c+}\ :\ \eta^t x = 1\}$, $\eta = (1,\ldots,1)^t$. We shall see below that G may initially be multiple valued.

The <u>phase equilibrium problem</u> is to <u>minimize for a given</u> (x^o,P^o,T^o) <u>the expression</u>

$$\sum_{i=1}^{\phi} c_i\ G(x^i,P^o,T^o)\ , \tag{2.2}$$

over all $x^i \in R^{c+}$, and all real $c_i > o$ such that $x^o = \sum_{i=1}^{\phi} c_i x^i$,

$\sum_{i=1}^{\phi} c_i=1$. Here ϕ may be any positive integer. Any set of x^i that minimizes (2.2) is a set of equilibrium phases. By (2.1) it is easy to see that it suffices to consider $x^o, x^i \in S$, as we mostly do from here on.

Let $\Gamma(P,T)$ be the graph of $G(\cdot,P,T)$ on S, and $\hat{\Gamma}(P,T) = \text{conv}\,\Gamma(P,T)$. Than $\Gamma(P,T)$ is compact and hence so is $\hat{\Gamma}(P,T)$ [6 , Thm 17.2]. By Caratheodory's Theorem [6 , Thm 17.1] we see that the minimum in (2.2) is always attained with $\phi \leq c$, and its value is the value of the <u>dissipated Gibbs free energy function</u>

(2.3) $$\hat{G}(x,P,T) = \min\{g : (x,g) \in \hat{\Gamma}(P,T)\}.$$

at (x^o,P^o,T^o).

The following properties of $\hat{G}$ can be shown using [6 , Thms 5.3, 10.3 and 10.7] and the continuity of G on compact S :

(2.4) (a) $\hat{G}(\cdot,P,T)$ is convex for each (P,T),

(b) $\hat{G}$ is continuous on $S \times (o,\infty) \times [o,T_1]$.

Next, we consider $x^o \in$ rel int (S), the relative interior of S, i.e. $x^o=(x_1^o,\ldots,x_c^o)$, and $x_1^o>o,\ldots,x_c^o>o$. This is not a serious restriction since a zero component simply means that a certain species does not occur in the mixture at all. Then by [6 , Thm 23.4] the subdifferential $\partial_x \hat{G}(x^o,P^o,T^o) \neq \emptyset$. Using (2.1) this can be expressed as follows: $\exists \mu \in R^c$,

(2.5) (a) $\mu^t y \leq \hat{G}(y,P^o,T^o) \leq G(y,P^o,T^o), \quad \forall y \in R^{c+}$,

(b) $\mu^t x^o = \hat{G}(x^o,P^o,T^o)$.

If (2.5) holds, μ is an <u>equilibrium chemical potential</u>. From this we can see that for $x \in$ rel int S

(2.6) $$\hat{G}(x,P,T) = \max\{\mu^t x : \mu^t y \leq G(y,P,T), \ \forall y \in R^{c+}\}.$$

Therefore μ can be viewed as the solution of a maximum problem <u>dual to the phase equilibrium problem</u>. For arbitrary $x \in S$ we obtain (2.6), but with "sup" instead of "max", by (2.4) and [6 , Thm 12.1].

It is tempting to suggest solving one of these problems approximately by discretizing and using the Simplex Algorithm. However, in practical cases $c\approx 50$ leads to upward of 2^{50} variables in the primal so that carrying out an individual Simplex Step would lead to astronomical computation times.

It should be said that the duality mentioned is nothing but Young-Fenchel duality, with

$$\hat{G}(\cdot,P,T) = G(\cdot,P,T)^{**} ,$$

where * indicates Young-Fenchel conjugation, [6 , p 104]. However, this was well unterstood by Gibbs [2].

We note the usual orthogonality relation for a pair of solutions $\{x^1,\ldots,x^\phi\}$ and μ of the two dual problems which here takes the following form.

$$\mu^t x^i = G(x^i,P^o,T^o) = \hat{G}(x^i,P^o,T^o), \quad i=1,\ldots,\phi . \tag{2.7}$$

Notice that (2.5) and (2.7) imply that $G(\cdot,P^o,T^o)$ is subdifferentiable at x^i. If $x^i \in \text{rel int } (S)$ and $G(\cdot,P^o,T^o)$ is differentiable at x^i there follows at once that

$$\mu = \mu(x^i,P^o,T^o), \quad i=1,\ldots,\phi , \tag{2.8}$$

where the right hand side is the <u>chemical potential</u>

$$\mu(x,P,T) = \nabla_x G(x,P,T) .$$

Incidentally, differentiating w.r. to α in (2.1) we obtain Euler's differential equation

$$G(x,P,T) = \mu(x,P,T)^t x .$$

<u>Remark</u>. In principle (2.8) could posses a continuum of solutions x^i , so that equilibrium phases need not be discrete. However, it is easy to see that under the mildest of conditions of transversality, the phase equilibrium cannot be stable if solutions with $\phi>c$ occur. Below we obtain sufficient conditions for stability when $\phi\leq c$ (§ 4).

In § 3 we discuss a specific class of functions $G(x,P,T)$ with $\frac{\partial}{\partial x_i} G(x,P,T,) = -\infty$ whenever $x_i = o$. It follows by [6 , pp 216 f] that $\partial_x G(x,P,T)=\emptyset$. Therefore for such a model there holds

(2.9) $x^i \in$ rel int (S) if $x^o \in$ rel int (S).

In these cases we can show that $G(\cdot,P_o,T_o)$ is differentiable at x^i, so we have (2.8), cf. § 3, 6.

A more geometric formulation of the phase equilibrium problem can be given in terms of faces of the epigraph E(P,T) of $\hat{G}(\cdot,P,T)$,

$$E(P,T) = \{(x,g) : x\in S, \quad g\geq\hat{G}(x,P,T)\}.$$

It is more convenient, however, to work with the projections of such faces into S. Hence we say that a subset F of S is a *face$_o$* of E(P,T) iff $\bar{F}=\{(x,g) : x\in F, g=\hat{G}(x,P,T)\}$ is a face of E(P,T) in the usual sense [6 , p 162].Similarly, we say that $F\subset S$ is an *exposed* face$_o$ of E(P,T) iff $\bar{F}$ is a face of E(P,T) exposed by a non-vertical hyperplane. This means in effect that there is an affine function H on S such that $H\leq\hat{G}(\cdot,P,T)$ on S and $F=\{x\in S : H(x) = \hat{G}(x,P,T)\}$.

Not every face$_o$ need be exposed.

LEMMA 1. *A maximal face$_o$ of E(P,T) which meets rel int (S) is exposed.*

This follows from [6 , Cor. 18.1.2 and Thm 11.6].

COROLLARY 2. *For any face$_o$ F of E(P,T) the map $x\to(x,\hat{G}(x,P,T))$ of F onto $\bar{F}$ is an affine isomorphism.*

COROLLARY 3. *If F is a face$_o$ of E(P,T) and $x\in$ extr (F), then $\hat{G}(x,P,T) = G(x,P,T)$.*

Proof: By 2., $(x,\hat{G}(x,P,T))$ is an extreme point of $\bar{F}$, hence of $\hat{\Gamma}(P,T)$, hence $(x,\hat{G}(x,P,T)) \in \Gamma(P,T)$. QED.

An affine function H on S can be written uniquely as $H(x)=\mu^t x$, $\mu\in R^c$. Hence (2.5), (2.7) express that $x^o,x^1,\ldots,x^\phi$ belong to the same exposed face$_o$ of E(P,T). The condition

$$G(x^i,P^o,T^o) = \hat{G}(x^i,P^o,T^o)$$

also occurs. For this, 3. provides a sufficient condition in terms of the geometry of a face. Also note that $x^o\in$ rel int (S) is sufficient for an exposed face$_o$ containing x^o to exist.

We summarize some of these considerations.

THEOREM 4. *Given* (x^o, P^o, T^o) *with* $x^o \in$ rel int (S), *equilibrium phases* $x^1, \ldots, x^\phi$ *in* S *and equilibrium chemical potential* μ *in* R^c *always exist, with* $\phi \leq c$, *and are characterized by the following conditions*:

(2.10) $\exists$ *a face*$_o$ F of $E(P^o, T^o)$, *exposed by* $H(x) = \mu^t x$;

(2.11) $x^o \in \text{conv}\,(x^1, \ldots, x^\phi) \subset F$;

(2.12) $\hat{G}(x^i, P^o, T^o) = G(x^i, P^o, T^o), \quad i = 1, \ldots, \phi.$

Without loss of generality we may assume that F is a maximal face.

§ 3 EMPIRICAL GIBBS FUNCTIONS BASED ON THE SOAVE-REDLICH-KWONG EQUATION OF STATE

This equation [7] is related to van der Waal's equation (c=1). For c>1 no rigorous foundation in statistical mechanics exists for such equations. For more on c=1, cf. [3]. Also, see [4] for a further contribution for c>1.

The Soave equation of state has the form

$$(3.1) \qquad P = f(x,V,T) = \frac{RT\eta^t x}{V-b} - \frac{a}{V(V+b)}, \quad V>b,$$

$$a = x^t W(T) x, \quad b = \beta^t x, \quad 0 \leq T \leq T_1 .$$

Here $x \in R^{c+}$ is the vector of mole fractions, $\eta = (1, \ldots, 1)^t$, W(T) is a positive-definite c×c-matrix, continuously differentiable in T in the interval $[0, T_1]$, β a constant positive vector. W(T) and β are material dependent, with W(T) reflecting attempts to model pair interactions. V is volume. Note that f is homogeneous of degree 0 in (x,V).

We refer to the thermodynamics literature for the full derivation of the following formula [5] :

$$(3.2)\qquad G(x,P,T) = \int_{\infty}^{V}\left[\frac{RT\eta^{t}x}{v} - f(x,v,T)\right] dv + RTx^{t}\log\left(\frac{RT}{V}x\right) + PV + (u^{o}-Ts^{o})^{t}x \ ,$$

$$\text{provided} \quad P=f(x,V,T) \ .$$

The difficulty, to be discussed, is that (3.1) may have up to three roots V for given (x,P,T), and hence up to three possible values of G(x,P,T) are defined. There is at least one root for P>o. It is easy to show that the integral converges in the proper manner, so that each _sheet_ of G as defined by (3.2) is smooth (use the implicit function theorem). Singularities occur only due to the logarithmic term when one of the components of x vanishes, i.e. not for x∈ rel int (S), cf. § 2, or on _sheet boundaries_. Incidentally, the integrand vanishes for an ideal gas, so the remaining terms are obtained by ideal gas considerations. The improper integral reflects the thermodynamic limit (proper phase equilibrium can be expected to exist only over infinite volumes). Also, one checks that (2.1) holds, for each sheet, hence it suffices to consider (3.2) for x∈S.

We are initially unable to apply the full theory of § 2. There is no difficulty in formulating the minimum problem (2.2) for multisheeted functions, but discontinuities at sheet boundaries could conceivably be so severe that little might remain of the convexity theory that would be useful. Fortunately, in the above model matters turn out be more favorable. We modify the definition of G.

DEFINITION 5. _Let_ G(x,P,T) _denote the minimum of the values that occur in (3.2)._

This change in definition clearly has no effect on possible solutions to the problem of minimizing (2.2) when G is initially multi-valued.

THEOREM 6. _Let_ G(x,P,T) _be defined as in 5. Then_ G _is a continuous function for_ $x \in R^{c+}$, $P>o$, $o \le T \le T_1$, _positively homogeneous of degree_ 1 _in_ x. _Furthermore,_ G _is smooth_ (C^{∞}) _in_ x _except for_ x _on the relative boundary of_ S _or at sheet inter-_

<u>section points</u>. <u>Whenever</u> $x \in$ rel int (S) <u>and</u> $G(x,P,T)=\hat{G}(x,P,T)$ <u>then</u> $\nabla_x G(x,P,T)$ <u>exists</u>.

REMARKS 7.

<u>7.1</u> The presence of the logarithmic term (not compensated by any other infinity) forces $\frac{\partial}{\partial x_i} G(x,P,T) = -\infty$ whenever $x_i = o$. In § 2 we noted that this implies (2.9).

<u>7.2</u> The term "sheet intersection points" in <u>6</u>. refers mainly to the situation when (3.1) has three distinct roots and two of these happen to give the same value in (3.1). However, in <u>6</u>. we require the closure of this set in $S \times (o,\infty) \times [o,T_1]$. Hence a sheet intersection point can be a sheet boundary point, too. More on this below.

<u>7.3</u> Formal differentiation of (3.2) gives, in accordance with thermodynamics,

$$(3.3) \qquad \frac{\partial}{\partial P} G(x,P,T) = V \ .$$

In fact, (3.2) is merely the result of inverting (3.1) to get (3.3), followed by integration, using proper boundary values. These are available for $P \to o$, $V \to \infty$ where ideal gas behavior is approached. Unfortunately, however, (3.1) is not monotone non-increasing in V, as thermodynamics would require unless metastable states are permitted. The failure of (3.1) to be monotone in V is remedied by <u>Maxwell's construction</u>. It so happens that this gives the same result as our modified definition <u>5</u>., or rather this is the deeper reason behind it. To avoid misunderstandings: <u>Only for</u> c=1 <u>does Maxwell's construction yield the correct value</u> $\hat{G}(x,P,T)$ for all (x,P,T). For c>1, it is necessary to carry out the procedures of § 2 before obtaining $\hat{G}$. We next sketch part of the proof of <u>6</u>. where Maxwell's construction is fully explained.

<u>Proof of 6.: Continuity</u>. Each sheet is smooth in its interior and continuous up to the boundary, as a careful examination with the aid of the implicit function theorem shows. There

remains to prove continuity across sheet boundaries under definition 5. We sketch the proof, which is via Maxwell's construction. In fact this makes G concave in P, with jump discontinuities allowed in (3.3). Hence continuity in P is immediate. Joint continuity therefore follows from continuity in (x,T) by [6 , Thm 10.7]. We now prove continuity in x only, the dependence on T being more trivial. Suppose first (3.1) has three roots

$b<V_o<V_1<V_2$

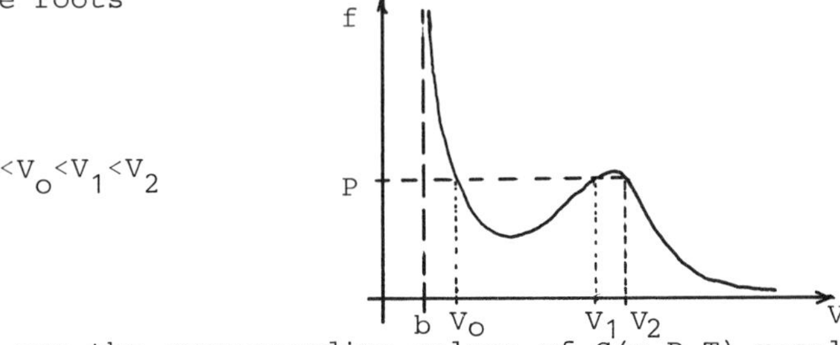

and G_o, G_1, G_2 are the corresponding values of G(x,P,T) resulting from (3.2). One easily calculates from (3.2)

$$(3.4) \qquad G_i - G_1 = \int_{V_1}^{V_i} [P - f(x,v,T)]\, dv .$$

It is apparent from this that

$$(3.5) \qquad G_i \le G_1 , \quad i=0,2; \quad \text{strict inequality holds unless } V_i = V_1 .$$

Accordingly, by definition 5. we find that

$$(3.6) \qquad G(x,P,T) \in \{G_o, G_2\} < G_1$$

unless $G_o=G_1=G_2$, i.e., $V_o=V_1=V_2$. Also (x,P,T) _is a sheet intersection point iff_ $G_2=G_o$.

REMARK 8. Pause in Proof.

By (3.4) $G_2=G_o$ occurs iff

$$(3.7) \qquad \int_{V_1}^{V_2} [P-f(x,v,T)]dv = \int_{V_o}^{V_1} [f(x,v,T) - P]\, dv .$$

There is a unique value $P=P_{Maxw.}(x,T)$ (unless $f(x,\cdot,T)$ already is monotone) for which (3.6) holds. Thus sheet intersection points occur iff $P=P_{Maxw.}(x,T)$. Now Maxwell's construction is to invert (3.1) by the rule.

(3.8) $$V=V_o \quad \text{for } P>P_{Maxw.}; \quad V=V_2 \text{ for } P<P_{Maxw.}.$$

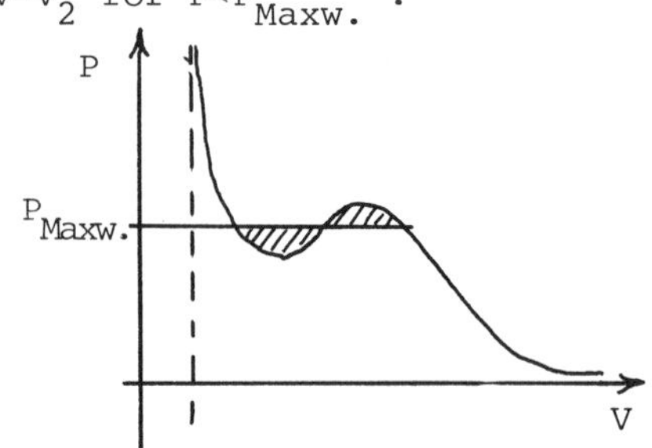

Thus the jump discontinunity in $\frac{\partial}{\partial x} G(x,P,T)=V$ occurs exactly at $P=P_{Maxw.}$. For V_i determined by this value, Maxwell redefined $f(x,V,T)\equiv P_{Maxw.}$ for $V_o \leqq V \leqq V_1$.

Proof of 6.: Continuity, continued. Sheet boundaries can only occur if (3.1) has a double root, but individual sheets are continuous, as remarked earlier. However, it is then clear from (3.6) that

$$G(x,P,T) = \min\{G_o,G_2\} < \max\{G_o,G_2\} = G_1$$

at (x,P,T) where a sheet boundary occurs unless $V_o=V_1=V_3$, counting multiplicities. In the former case the lowest sheet is continuous, it does not have a boundary point at (x,P,T), which, however, occurs on the highest sheet. In the latter case, values of $G(x,P,T)$ tend to the triple value from all three sheets. QED.

We merely remark that the remainder of 6. follows by similar arguments, also using subdifferentiability of each sheet. We have not investigated thoroughly the behavior of derivatives of order higher than one at sheet boundaries.

It is clear that considerations such as those leading to 6. are really only the beginning of an effective analysis of the problem that might lead to good algorithms, Yet, little of this sort of work seems to have been done. Concerning the alleged relationship between catastrophy theory and phase equilibrium we wonder if the importance of convexity has been overlooked. However, the unfolding of the several sheets of G at the points where (3.1) has triple roots bears some definite relationship. Work of this sort would require careful study of (3.2). Incidentally, the integration is trivial to carry out. Note that triple root points are not critical points in this model.

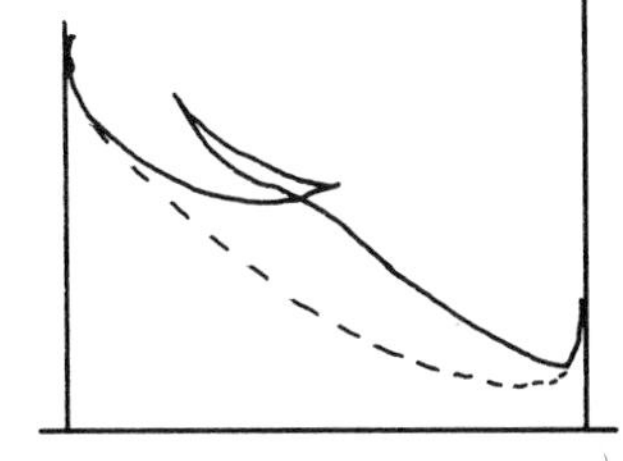

A typical one-dimensional section of $G(\cdot,P,T)$ across the simplex S may now look like this. The graph shows sheet boundary points, sheet intersection and the vertical slopes due to the log at the rel. boundary of S. The dashed line indicates part of the graph of $\hat{G}(\cdot,P,T)$. We drew it so that it shows the influence of parts of $\Gamma(P,T)$ not visible in this section.

As a <u>conjecture</u> we offer that no one dimensional section can show more than one crossing of sheets. This would be suggested by our observations.

The Soave equation was developed for liquid-vapor equilibrium. Qualitatively it gives reasonable results also in other cases. $\phi \geqq 3$ apparently has never been proved to occur. Models with separate correlations for liquid and vapor phases are in use. These apparently do not require significant changes in the basic structure of the analysis in § 2.

§ 4 STABILITY AND LOCAL SOLVABILITY OF EQUILIBRIUM EQUATIONS

The equilibrium equations are given in (2.8)

$$\mu = \mu(x^i, P^o, T^o), \quad i=1,\ldots,\phi \ . \tag{4.1}$$

We seek conditions under which these may be stably inverted and for which this inversion provides a stable solution for the problem of minimizing (2.2). Here stability is w.r. to small perturbations of (x^o, P^o, T^o).

By the homogeneity property (2.1), the Hessean $\nabla_x^2 G(x,P,T)$ is never invertible. However, for $x \in S$ we may consider G as a function of (c-1) variables $x,\ldots,x_{c-1}$, besides P,T. In practice this reduction in the number of variables can be accomplished in many ways. We assume henceforth that this has been done. But now, $\mu=\nabla_x G$ no longer holds, hence (4.1) takes a new form, e.g., we may write it as

$$G(x^i,P^o,T^o) = \hat{G}(x^o,P^o,T^o) + \mu^t(x^i-x^o),\ i=1,\ldots,\phi\ . \tag{4.2}$$

Treating $\hat{g}=\hat{G}(x^o,P^o,T^o)$ as unknowns, too, we have a full set of unknowns in the equilibrium equations

$$B = (x^1,\ldots,x^\phi,\mu,c_1,\ldots,c_\phi,\hat{g})$$

while the data are $A=(x,P,T)$. We have as many scalar equations as scalar unknowns.

$$\begin{aligned}(4.3)\qquad & \nabla_x G(x^i,P^o,T) - \mu = 0\ ,\quad i=1,\ldots,\phi\ ;\\ & \sum_{i=1}^{\phi} c_i x^i - x = 0\ ;\\ & \sum_{i=1}^{\phi} c_i - 1 = 0\ ;\\ & G(x^i,P,T)-\hat{g}-\mu^t(x^i-x)=0\ ,\quad i=1,\ldots,\phi\ .\end{aligned}$$

We collect these into one vector equation $F(A,B)=0$ in $R^{(c-1)\phi+c+\phi}$. We consider this for A near A^o, where a solution is assumed to exist, and apply the implicit function theorem. The Jacobian is easily analyzed under special circumstance stated next.

THEOREM 9. *Suppose we have values* A^o,B^o *such that* $F(A^o,B^o)=0$, where it is assumed that

$$x^o,x^i \in \text{rel int }(S)\ ; \tag{4.4}$$

$$\nabla_x G(x,P,T),\ \nabla_x^2 G(x,P,T)\ \textit{are continuous in neighborhoods of}\ (x^i,P^o,T^o),\quad i=1,\ldots,\phi\ ; \tag{4.5}$$

(4.6) $\nabla_x^2 G(x^i,P^o,T^o) > 0, \quad i=1,\ldots,\phi$;

(4.7) <u>The points</u> $x^1,\ldots,x^\phi$ <u>are affinely independent</u> $(\Rightarrow \phi \leq c)$;

(4.8) $c_i > 0, \quad i=1,\ldots,\phi$.

In the above, x^i, c_i refer to the corresponding components of B^o. Then the Jacobian $\frac{\partial F}{\partial B}(A^o,B^o) \neq 0$, and hence we have smooth maps

$$A=(x,P,T) \rightarrow \left(x^i(A),\ \mu(A),\ c_i(A),\ \hat{g}(A)\right) = \gamma(A) = B$$

defined in a neighborhood $V(A^o)$ of A^o such that $F\left(A,\gamma(A)\right) = 0$ and $\gamma(A^o)=B^o$. In particular, Newton's method for solving (4.3) converges locally.

<u>Remark</u>. One can also obtain results where (4.8) is not required, which have some practical importance in algorithms where A^o is subject to large perturbations, or where (4.8) simply is not satisfied. These are more difficult to prove.

The preceding result answers the question about locally stable solvability of the equilibrium equations (first order contact conditions). Stronger, and global hypotheses are needed to prove local stability of the solutions of minimizing (2.2). We state a result describing what appears to be the typical situation. The proof relies heavily on <u>9</u>.and the analysis of § 2.

THEOREM 10. <u>Given is a point</u> $A^o=(x^o,P^o,T^o)$ <u>such that conditions (4.4,5,6) are met, where</u> $x^1,\ldots,x^\phi$ <u>are given points in</u> S, <u>and furthermore we have</u>

(4.9) $x^o \in \text{rel int }(F)$, where $F = \text{conv}(x^1,\ldots,x^\phi)$ is a maximal face_o of $E(P,T)$;

(4.10) $x^i \in \text{extr }(F), \quad i=1,\ldots,\phi$;

(4.11) $\hat{G}(x,P^o,T^o) < G(x,P^o,T^o), \ \forall x \in F\setminus\{x^1,\ldots,x^\phi\}$;

(4.12) the points $x^1,\ldots,x^\phi$ are affinely independent, in particular $\phi \leq c$.

Then the phase equilibrium problem is stable at A^o in the sense that the assumptions of 9. are satisfied at A^o with a suitable B^o which includes the given x^i, $i=1,\ldots,\phi$, and the map $A \to \gamma(A)$ of 9. supplies the unique solution of the minimum problem for (2.2) for all A in some $V(A^o)$. In particular, ϕ is constant and the equilibrium phases $x^i(A)$ are continuous functions of A in $V(A^o)$, $i=1,\ldots,\phi$.

Remark. The Assumption that F is maximal is actually required. Otherwise there is no guarantee that F is exposed (which we could assume instead, but the remaining assumptions would then actually imply that F is maximal).

Proof of 10. By 1. F is exposed, hence we can find an affine function H(x) such that

(4.13) $H(x)=\hat{G}(x,P^o,T^o), \ \forall x \in F \ ; \ H(x)<\hat{G}(x,P^o,T^o), \ \ \forall x \in S\setminus F.$

Hence by 3. and (4.10)

(4.14) $H(x^i) = \hat{G}(x^i,P^o,T^o) = G(x^i,P^o,T^o), \quad i=1,\ldots,\phi$.

From (4.11,13,14) we have

(4.15) $H(x) \leq G(x,P^o,T^o), \ \ \forall x \in S$ with equality iff $x \in \{x^1,\ldots,x^\phi\}$.

There follows from (4.5,6) that H makes first order contact with $G(\cdot,P^o,T^o)$ at the x^i, i.e.,

(4.16) $H(x) = \hat{G}(x^o,P^o,T^o) + \mu^t(x-x^o), \quad \mu = \nabla_x G(x^i,P^o,T^o),$ $i=1,\ldots,\phi$.

From this we see that the assumptions of 9. are satisfied for the given A^o and

$$B^o = (x^1,\dots,x^\phi, \mu, c_1,\dots,c_\phi, \hat{G}(A^o)),$$

with the c_i determined by (4.9). Therefore, 9. provides us with functions $x^i(A)$, $\mu(A)$, $c_i(A)$, $\hat{g}(A)$ continuous in $V(A^o)$ which satisfy the equilibrium equations. In particular,

$$\mu(A) = \nabla_x G(x^i(A),P,T), \quad A=(x,P,T) \in V(A^o), \quad i=1,\dots,\phi . \tag{4.17}$$

Let us define

$$H(y;A) = \hat{g}(A) + \mu(A)^t(y-x) .$$

By (4.17) and the equilibrium equations the affine function $H(\cdot;A)$ makes first order contact with $G(\cdot,P,T)$ at the points $x^i(A)$. It follows from (4.16) that

$$H = H(\cdot,A^o), \quad \hat{g}(A^o) = \hat{G}(A^o) ,$$

since first order contact at even one point determines an affine function uniquely. We show now that we can choose $V(A^o)$ so small that

$$D(y;A) = G(y,P,T) - H(y;A) \geqq 0 , \quad \forall y \in S , \text{ and } A=(x,P,T) \in V(A^o) \text{ with equality iff } y \in \{x^i(A) : i=1,\dots,\phi\}. \tag{4.18}$$

For $A=A^o$, this is just (4.15). By (4.5,6) we can find $\delta>o$ such that

$$\nabla_x^2 G(y,P,T)>o \quad \text{for } |y-x^i(A^o)|<\tfrac{3}{2}\delta, |(P,T)-(P^o,T^o)|<\delta,$$

and also we may assume that

$$|x^i(A) - x^i(A^o)| < \tfrac{1}{2}\delta \quad \text{if } A\in V(A^o) . \tag{4.19}$$

Then

$$\nabla_x^2 G(y,P,T)>o \quad \text{for } |y-x^i(A)|<\delta , \quad A\in V(A^o) . \tag{4.20}$$

Because of the resulting strict convexity we see that

$$(4.21) \qquad D(y;A) \geqq o \quad \text{for } A \in V(A^o),\ |y-x^i(A)| < \delta\ ,$$
$$\text{with equality only for } y=x^i(A).$$

To complete the proof of (4.18), let

$$\varepsilon = \min\{D(y;A^o)\ :\ |y-x^i(A^o)| \geqq \delta/2,\quad y \in S\}.$$

Then $\varepsilon > o$ because of (4.15). Determine $V(A^o)$ by the usual covering argument of the compact S so that

$$|D(y,A) - D(y,A^o)| \leqq \varepsilon/2 \quad \forall y \in S, \quad \forall A \in V(A^o).$$

Then if $|y-x^i(A)| \geqq \delta$, $A \in V(A^o)$, we have by (4.19)

$$|y-x^i(A^o)| \geqq |y-x^i(A)| - |x^i(A)-x^i(A^o)|$$
$$\geqq \delta - \delta/2 \geqq \delta/2\ ,$$

therefore

$$D(y,A) \geqq D(y,A^o) - \varepsilon/2 \geqq \varepsilon - \varepsilon/2 > 0\ .$$

This proves (4.18) for those cases not covered by (4.21). From this the theorem follows directly by our work in § 2.QED.

REFERENCES

1. Buckingham, M.J., Thermodynamics, chpt. 1 of _Phase Transitions and Critical Phenomena_, Vol. 2, C. Domb and M.S. Green, eds., Academic Press (1972).

2. Gibbs, J.W., _The Collected Works_, Vol. 1, Longmans, Green and Co. (1928).

3. Lebowitz, J.L., Statistical mechanics of equilibrium systems: Some rigorous results, in: _Critical Phenomena_, J. Brey and R.B. Jones, eds., Lect. Notes in Physics 54, Springer (1976), 215-248.

4. Peng, D. and Robinson, D.B., A new two-constant equation of state, Ind. Eng. Chem. Fundamentals _15_ (1976), 59.

5. Prausnitz, J.M., _Molecular Thermodynamics of Fluid-Phase Equilibria_, Prentice-Hall (1969).

6. Rockafellar, R.T., _Convex Analysis_, Princeton University Press (1970).

7. Soave, G., Equilibrium Constants from a modified Redlich-Kwong equation of state, Chem. Eng. Sci. _27_, (1972), 1197.

H.G. Burchard
Dept. of Mathematics
Oklahoma State University
Stillwater, OK. 74074
U.S.A.
and
Institut für Angewandte
Mathematik / SFB 72
Universität Bonn
5300 Bonn
Fed.Rep.of Germany

R.A. Waldo
Phillips Petroleum Company
Bartlesville, OK. 74004
U.S.A.

MULTIVARIATE INTERPOLATION AND THE RADON TRANSFORM, PART II: SOME FURTHER EXAMPLES

A. S. Cavaretta, Jr., Charles A. Micchelli, and A. Sharma

Here we offer three examples of multivariate interpolation operators which are determined by certain corresponding univariate operators. These examples are based on results presented in reference [1].

1. INTRODUCTION

In a recent paper [1] we studied a new method for extending univariate interpolation procedures to higher dimensions. This idea is based on the requirement that the multivariate extension is related to its univariate analog on the class of ridge functions.

A particularly simple instance of this approach occurs in the relationship between Taylor polynomial of degree n in one variable

$$T_n(g|t_0)(t) = g(t_0) + g'(t_0)(t-t_0) + \dots + (1/n!)g^{(n)}(t_0)(t-t_0)^n$$

and its analog in R^k, $\mathcal{T}_n(f|x^0)(x)$, x^0, $x \in R^k$. Specifically, it is easy to verify that

$$\mathcal{T}_n(f|x^0)(x) = T_n(g|\lambda\cdot x^0)(\lambda \cdot x)$$

for the ridge function $f(x) = g(\lambda \cdot x)$.

In its general form our problem has defined for every $(t_1,\dots,t_m) \in R^m$, a distribution $g \to T(g|t_1,\dots,t_m)$ of compact support on $C_0^\infty(R^1)$. We ask the question whether or not this family of distributions can be extended to $C_0^\infty(R^k)$ for all k in the following sense. Given $x^1,\dots,x^m \in R^k$ does

ISBN 0-12-213650-0

there exist a distribution $\mathcal{T}_k(f|x^1,\ldots,x^m)$ of compact support on $C_0^\infty(R^k)$ such that

(1.1) $\mathcal{T}_k(f|x^1,\ldots,x^m) = T(g|\lambda\cdot x^1,\ldots,\lambda\cdot x^m)$

whenever $f(x) = g(\lambda\cdot x)$.

It can be easily shown that not all distributions $T(\cdot|t_1,\ldots,t_m)$ can be extended. However, recently P. Kergin [2] showed that Hermite interpolation $p(t) = H(g|t_0,\ldots,t_n)(t)$ does have an extension. Here $t_0,\ldots,t_n$ denotes the nodes at which the polynomial p (of degree n) agrees with g and its derivatives.

This interesting result and its equivalent formulation in terms of a generalized mean value property for the multivariate extension motivated us to prove in [1] that every complex-poised Hermite-Birkhoff interpolation procedure can be extended. We also proved that the requirement of being complex poised is also necessary for the existence of a multivariate extension in the sense we described above for H-B interpolation.

Moreover, we showed in [1] that extending a distribution $T(\cdot|t_1,\ldots,t_m)$ to R^k for all k is (essentially) equivalent to finding the inverse Radon transform of this distribution. These facts are explained in [1] and we will not require them here. Instead, our objective in this paper is to present several more explicit examples of univariate procedures from approximation theory which can be extended in the sense described above. In each case, we give precise formulas for the multivariate extension and study their generalized mean value properties.

The examples we consider, in the order that they are discussed below, are Abel-Gončarov interpolation, area matching by polynomials, and three-point Lidstone interpolation.

2. MULTIVARIATE ABEL-GONČAROV INTERPOLATION

Our objective here is to give an explicit formula for the multivariate extension of Abel-Gončarov interpolation.

Given n+1 points $t_0, t_1, \ldots, t_n$ not necessarily distinct and $g \in C^n(R^1)$, the Abel-Gončarov interpolating polynomial $p \in \pi_n$ satisfies the following conditions:

$$p^{(j)}(t_j) = g^{(j)}(t_j), \quad j = 0,1,\ldots,n. \tag{2.1}$$

We denote the polynomial in (2.1) by $A_n(g|t_0,\ldots,t_n)(t)$. Theorem 1 of [1] guarantees the existence of a multivariate extension of A_n for any vectors $x^0,\ldots,x^n \in R^k$; we shall denote this multivariate extension by $\mathcal{A}_n$.

As is well known, the fundamental polynomials for the interpolation problem (2.1) are given by

$$\begin{aligned} &p_m(t|t_0,\ldots,t_{m-1}) = \int_{t_0}^{t} \cdots \int_{t_{m-1}}^{s_{m-1}} ds_m \ldots ds_1, \quad m = 1,\ldots,n, \\ &p_0(t) = 1. \end{aligned} \tag{2.2}$$

Thus $p_m^{(j)}(t_j) = \delta_{jm}$, so that

$$A_n(g|t_0,\ldots,t_n)(t) = \sum_{j=0}^{n} p_j(t|t_0,\ldots,t_{j-1})g^{(j)}(t_j). \tag{2.3}$$

The function $p_m(t|t_0,\ldots,t_{m-1})$ is a homogeneous polynomial in the m variables $t-t_0, t_0-t_1, \ldots, t_{m-2}-t_{m-1}$. Indeed, this fact is seen immediately from the representation (2.2) on re-writing it as

$$p_m(t|t_0,\ldots,t_{m-1}) = \int_0^{t-t_0} ds_1 \int_0^{t_0-t_1+s_1} ds_2 \cdots \int_0^{t_{m-2}-t_{m-1}+s_{m-1}} ds_m \tag{2.4}$$

which is obtained from (2.2) by a simple change of variables.

Thus, we may express $p_m(t)$ in the form

$$(2.5)\quad p_m(t|t_0,\dots,t_{m-1}) = \sum_{|\alpha|=m} c_\alpha (t-t_0)^{\alpha_0}\dots(t_{m-2}-t_{m-1})^{\alpha_{m-1}}$$

for appropriate constants c_α with $|\alpha| = \sum_0^{m-1} \alpha_\nu = m$.

We are now in a position to define certain linear differential operators necessary for our formula. Let $x, x^0,\dots,x^m \in R^k$ and define the differential operator

$$(2.6)\quad p_m(D|x^0,\dots,x^m,x) = \sum_{|\alpha|=m} c_\alpha D^{\alpha_0}_{x-x^0} D^{\alpha_1}_{x^0-x^1} \cdots D^{\alpha_{m-1}}_{x^{m-2}-x^{m-1}}$$

where $D_y f$ is the directional derivative of f in the direction y. Then we have

PROPOSITION 1. <u>The multivariate Abel-Gončarov interpolation map is given by</u>

$$(2.7)\quad \mathcal{A}_n f(x) = \sum_{m=0}^{n} p_m(D|x^0,\dots,x^m,x) f(x^m).$$

<u>Proof</u>. First note that $\mathcal{A}_n f(x)$ is a polynomial of total degree n in the variable x and also that we require $f \in C^n(K)$, K = convex hull of $x^0,\dots,x^n$ for $\mathcal{A}_n f$ to be well-defined.

For $f(x) = g(\lambda\cdot x)$ we find that

$$(2.8)\quad \mathcal{A}_n f(x) = A_n(g|\lambda\cdot x^0,\dots,\lambda\cdot x^n)(\lambda\cdot x).$$

Indeed, from (2.6) and (2.5), we have

$$\mathcal{A}_n f(x) = \sum_{m=0}^{n} \sum_{|\alpha|=m} c_\alpha (\lambda\cdot x-\lambda\cdot x^0)^{\alpha_0}\dots(\lambda\cdot x^{m-2}-\lambda\cdot x^{m-1})^{\alpha_{m-1}} g^{(m)}(\lambda\cdot x^m)$$

$$= \sum_{m=0}^{n} p_m(\lambda\cdot x|\lambda\cdot x^0,\dots,\lambda\cdot x^{m-1}) g^{(m)}(\lambda\cdot x^m)$$

$$= A_n(g|\lambda\cdot x^0,\dots,\lambda\cdot x^n)(\lambda\cdot x).$$

Thus, $\mathcal{A}_n$ satisfies our requirement, expressed in general terms by (1.1), which identifies it as the multivariate extension of A_n.

We remark that this formula for A_n is a special case of one given in [1].

Concerning the interpolation properties of $A_n f$, we have

PROPOSITION 2. *Let q_ℓ be any constant coefficient homogeneous differential operator of order ℓ, $0 \leq \ell \leq n$ and let K be the convex hull of $x^0,\ldots,x^n \in R^k$. Then the operator $A_n\colon C^n(K) \to \pi_n(R^k)$, given by (2.7), satisfies*

$$q_\ell(A_n f)(x^\ell) = q_\ell(f)(x^\ell) \; . \tag{2.9}$$

Moreover, these interpolation conditions completely determine the polynomial operator A_n.

Proof. Let us start with the uniqueness statement. If there were two such operators, then the difference operator, D, would satisfy $q_\ell(Df)(x^\ell) = 0$, $\ell = 0,1,\ldots,n$ for each f. Since Df is in $\pi_n(R^k)$, $q_n(Df)$ is a constant which vanishes at x^n and so must be zero. Thus $Df \in \pi_{n-1}(R^k)$ and, proceeding similarly, we get $Df = 0$; hence A_n is unique.

To verify that A_n given by (2.7) does satisfy (2.9) we use (2.8). If $f(x) = g(\lambda\cdot x)$, then

$$A_n f(x) - f(x) = A_n(g|\lambda\cdot x^0,\ldots,\lambda\cdot x^n)(\lambda\cdot x) - g(\lambda\cdot x) \; .$$

Hence,

$$q_\ell(A_n f - f)(x^\ell) = q_\ell(\lambda)\{(A_n^{(\ell)}g)(\lambda\cdot x^\ell) - g^{(\ell)}(\lambda\cdot x^\ell)\} = 0 \; . \tag{2.10}$$

Since the functions $g(\lambda\cdot x)$, $\lambda \in R^k$, $g \in C^n(R)$ are dense in $C^n(K)$, (2.10) follows for all $f \in C^n(K)$, and this proves (2.9).

3. AREA MATCHING BY POLYNOMIALS

For a given function g, let p be a polynomial of degree less than or equal to n defined by the requirement that

$$\int_{t_i}^{t_{i+1}} p(t)\,dt = \int_{t_i}^{t_{i+1}} g(t)\,dt,\ i = 0,1,\dots,n,$$

where the points $t_0 < t_1 < \dots < t_{n+1}$ are fixed in advance. Thus, $p(t) = M(g|t_0,\dots,t_{n+1})(t)$ has the same integral as g over each of the intervals $[t_i,t_{i+1}]$, $i = 0,1,\dots,n$. In this section we will give an explicit formula for the multivariate version of this mapping.

To this end, we observe that

$$M(g|t_0,\dots,t_{n+1})(t) = H'(h|t_0,\dots,t_{n+1})(t) \tag{3.1}$$

where $h(t) = \int_0^t g(\sigma)\,d\sigma$ and $H(h|t_0,\dots,t_{n+1})(t)$ is Hermite interpolation to h at $t_0,\dots,t_{n+1}$. We write H in its Newton divided difference form

$$H(h|t_0,\dots,t_{n+1})(t) = \sum_{j=0}^{n+1} (t-t_0)\dots(t-t_{j-1})[t_0,\dots,t_j]h$$

which by the Hermite-Genocchi formula gives the representation

$$H(h|t_0,\dots,t_{n+1})(t) = $$
$$= \sum_{j=0}^{n+1} (t-t_0)\cdots(t-t_{j-1})\int_{S^j} g^{(j-1)}(v_0t_0+\cdots+v_jt_j)\,dv_1\cdots dv_j$$

where $S^j = \{(v_0,\dots v_j) \mid \sum_{\ell=0}^{j} v_\ell = 1,\ v_\ell \geq 0\}$.

Thus from (3.1)

$$M(g|t_0,\dots t_{n+1})(t) = $$
$$\sum_{j=0}^{n} \frac{d}{dt}((t-t_0)\dots(t-t_j))\int_{S^{j+1}} g^{(j)}(v_0t_0+\dots+v_{j+1}t_{j+1})\ dv_1\dots dv_{j+1}.$$

Now, for every set of vectors $x^0,x^1,\dots,x^n \in R^k$, we define differential operators for $0 \leq j \leq n$ by

$$m_0(D|x^0,x) = \text{Identity}$$

$$m_1(D|x^0,x^1,x) = D_{x-x^0} + D_{x-x^1}$$

$$\vdots$$

$$m_j(D|x^0,\ldots,x^j,x) = \sum_{\ell=0}^{j} D_{x-x^0}\cdots D_{x-x^{\ell-1}} D_{x-x^{\ell+1}}\cdots D_{x-x^j}$$

and we also define the linear functional

$$\int_{[x^0,\ldots,x^n]} f = \int_{S^n} f(v_0x^0 + \ldots + v_nx^n)\,dv_1\ldots dv_n \,.$$

Then we have

PROPOSITION 3. <u>The multivariate extension</u> M: $C^n(K) \to \pi_n(R^k)$, K = <u>convex hull of</u> $x^0,\ldots,x^n$, <u>of the map</u> (3.1) <u>is given by</u>

$$M(f|x^0,\ldots,x^{n+1})(x) = \sum_{j=0}^{n} \int_{[x^0,\ldots,x^{j+1}]} m_j(D|x^0,\ldots,x^j,x)f.$$

<u>Proof</u>. For $f(x) = g(\lambda\cdot x)$

$$M(f|x^0,\ldots,x^{n+1})(x) =$$

$$\sum_{j=0}^{n} (t-\lambda\cdot x^0)\cdots(t-\lambda\cdot x^j))\cdot(\lambda\cdot x)\,[\lambda\cdot x^0,\ldots,\lambda\cdot x^{j+1}]g^{(-1)}$$

$$= M(g|\lambda\cdot x^0,\ldots,\lambda\cdot x^{n+1})(\lambda\cdot x).$$

PROPOSITION 4. <u>Let</u> q_ℓ <u>be any homogeneous differential operator of order</u> ℓ, $0 \leq \ell \leq n$ <u>and</u> $\{x^{i_0},\ldots,x^{i_{\ell+1}}\}$ <u>any subset of</u> $\{x^0,\ldots,x^n\}$. <u>Then</u>

$$\int_{[x^{i_0},\ldots,x^{i_{\ell+1}}]} q_\ell(Mf) = \int_{[x^{i_0},\ldots,x^{i_{\ell+1}}]} q_\ell(f)\,.$$

<u>Proof</u>. As before, it is sufficient to prove this condition for functions of the form $f(x) = g(\lambda\cdot x)$. In this case, we have

$$\begin{aligned}\int_{[x^{i_0},\dots,x^{i_{\ell+1}}]} q_\ell(Mf) &= q_\ell(\lambda)[\lambda\cdot x^{i_0},\dots,\lambda\cdot x^{i_{\ell+1}}]M^{(-1)}(g|\lambda\cdot x^{i_0},\dots,\lambda\cdot x^{i_{\ell+1}}) \\ &= q_\ell(\lambda)[\lambda\cdot x^{i_0},\dots,\lambda\cdot x^{i_{\ell+1}}]g^{(-1)} \\ &= \int_{[x^{i_0},\dots,x^{i_{\ell+1}}]} q_\ell(f).\end{aligned}$$

4. LIDSTONE TYPE INTERPOLATION ON m POINTS

In the univariate case, multipoint Lidstone interpolation was worked out in detail by H. Poritsky [4]. In this section, we follow his lead and provide a multivariate extension of this procedure.

In general, the incidence matrix of this interpolation scheme is an iteration of the incidence matrix of Lagrange interpolation on m nodes. The case $m = 2$ is the usual Lidstone case. Below, we describe the case $m = 3$ which has all the salient features of the general situation. All methods extend easily for larger values of m. Our goal is to give a reasonably explicit formula for the multivariate operator.

To fix our ideas, we take three distinct points, t_0, t_1, t_2. If $g(t)$ is sufficiently differentiable, we set

$$L_n(g|t_0,t_1,t_2)(t) = \sum_{m=0}^{n} g^{(3m)}(t_0)A_m(t|t_0,t_1,t_2) + \sum_{m=0}^{n} g^{(3m)}(t_1)B_m(t|t_0,t_1,t_2) + \sum_{m=0}^{n} g^{(3m)}(t_2)C_m(t|t_0,t_1,t_2), \tag{4.1}$$

where $A_m(t)$, $B_m(t)$, $C_m(t)$ are the fundamental polynomials of interpolation for the matrix

$$E = \begin{pmatrix} 1 & 0 & 0 & 1 & & 1 & 0 & 0 \\ 1 & 0 & 0 & 1 & \cdots & 1 & 0 & 0 \\ 1 & 0 & 0 & 1 & & 1 & 0 & 0 \end{pmatrix}.$$

Note that L_0 means Lagrange interpolation in the nodes t_0, t_1, t_2. Poritsky has determined the generating functions for the polynomials $A_n(t)$, $B_n(t)$ and $C_n(t)$ by using the so-called Olivier functions $S_1(t)$ and $S_2(t)$ given by

$$\text{(4.2)} \quad \begin{aligned} S_1(t) &= \tfrac{1}{3}(e^t + \omega^2 e^{\omega t} + \omega e^{\omega^2 t}) = \sum_{n=0}^{\infty} \frac{t^{3n+1}}{(3n+1)!} \\ S_2(t) &= \tfrac{1}{3}(e^t + \omega e^{\omega t} + \omega^2 e^{\omega^2 t}) = \sum_{n=0}^{\infty} \frac{t^{3n+2}}{(3n+2)!} \end{aligned}$$

where $\omega^3 = 1$. Following Poritsky, we define

$$\text{(4.3)} \quad N(z|t_0,t_1,t_2) = \begin{vmatrix} S_1(z(t_1 - t_0)) & S_1(z(t_2 - t_0)) \\ S_2(z(t_1 - t_0)) & S_2(z(t_2 - t_0)) \end{vmatrix}.$$

Poritsky has shown that

$$\text{(4.4)} \quad \sum_{n=0}^{\infty} A_n(t|t_0,t_1,t_2) z^{3n} = \frac{N(z|t,t_1,t_2)}{N(z|t_0,t_1,t_2)}.$$

This may easily be seen by observing first that $A_n(t_0) = \delta_{0n}$, $A_n(t_1) = A_n(t_2) = 0$ and that $A_n'''(t) = A_{n-1}(t)$, $n = 1,2,\ldots$. The latter equation follows from the fact

$$\left(\frac{\partial^3}{\partial t^3} - z^3\right) N(z|t,t_1,t_2) = 0 ,$$

because, acording to (4.2) and (4.3), $N(z|t,t_1,t_2)$ is the sum of terms whose typical form is $e^{z\omega^{\ell} t}$, $\ell = 0,1,2$. The generating functions for B_n and C_n are similar since

$$\text{(4.5)} \quad B_n(t) = A_n(t|t_1,t_0,t_2) , \; C_n(t) = A_n(t|t_2,t_0,t_1) .$$

Integral functions of the kind given by (4.3) were studied by G. Polya [3], who proved the existence of an infinite number of roots for them and also investigated the distribution of these roots at infinity.

We shall need the polynomials

$$\begin{aligned} P_n(t) &= A_n(t) + B_n(t) + C_n(t) \\ Q_n(t) &= (t_1 - t_0)B_n(t) + (t_2 - t_0)C_n(t) \qquad (4.6) \\ R_n(t) &= (t_2 - t_0)(t_2 - t_1)C_n(t) . \end{aligned}$$

Using these polynomials, we may express $L_n(g)$ in the form

$$\begin{aligned} L_n(g|t_0,t_1,t_2)(t) = &\sum_{m=0}^{n} g^{(3m)}(t_0)P_m(t|t_0,t_1,t_2) + \\ &\sum_{m=0}^{n} [t_0,t_1]g^{(3m)}Q_m(t|t_0,t_1,t_2) + \qquad (4.7) \\ &\sum_{m=0}^{n} [t_0,t_1,t_2]g^{(3m)}R_m(t|t_0,t_1,t_2) . \end{aligned}$$

Concerning these functions, we have the following

LEMMA 1. *The polynomials* $P_n(t|t_0,t_1,t_2)$, $Q_n(t|t_0,t_1,t_2)$ *and* $R_n(t|t_0,t_1,t_2)$ *are homogeneous polynomials in the differences* $t - t_0, t - t_1, t - t_2$ *of degree* 3n, 3n+1 *and* 3n+2, *respectively*.

Proof. First let us note that

$$\begin{vmatrix} S_1(zx_1) & S_1(zx_2) \\ S_2(zx_1) & S_2(zx_2) \end{vmatrix} = z^3 x_1 x_2 (x_2 - x_1) \sum_{n=0}^{\infty} z^{3n} E_n(x_1,x_2)$$

where $E_n(x_1,x_2)$ is a symmetric homogeneous polynomial of degree 3n with $E_0 = 1/2$. Thus, using (4.4), (4.5) and (4.6), we have

$$\left(\sum_{n=0}^{\infty} R_n(t)z^{3n}\right)\left(\sum_{n=0}^{\infty} E_n(t_0 - t_2, t_1 - t_2)z^{3n}\right)$$

$$= (t_0 - t)(t_1 - t)\sum_{n=0}^{\infty} E_n(t_1 - t, t_0 - t)z^{3n} .$$

Multiplying the power series on the left and comparing powers of z on both sides in this identity, it follows recursively that $R_n(t|t_0,t_1,t_2)$ is a homogeneous polynomial of degree 3n+2 in $t - t_0, t - t_1, t - t_2$.

We will write $R_n(t) = R(t-t_0, t-t_1, t-t_2)$ which is, according to (4.5) and (4.6), equal to

$$(t_2-t_0)(t_2-t_1)A_n(t|t_2,t_0,t_1) .$$

Now it is an easy matter to see that

$$Q_n(t) = \frac{R(t-t_0,t-t_2,t-t_1) - R(t-t_0,t-t_1,t-t_2)}{t_1-t_2},$$

from which it then follows that Q_n is a homogeneous polynomial of degree 3n+1. As before, we write $Q(t-t_0,t-t_1,t-t_2)$ for this polynomial. Thus,

$$Q(t-t_0,t-t_1,t-t_2) = (t_1-t_0)A_n(t|t_1,t_0,t_2) +$$
$$(t_2-t_0)A_n(t|t_2,t_0,t_1) ,$$

and it immediately follows that

$$P_n(t) = \frac{Q(t-t_0,t-t_1,t-t_2) - Q(t-t_1,t-t_0,t-t_2)}{t_1-t_0},$$

which is also obviously a homogeneous polynomial of degree 3n. This completes the proof of the lemma.

On the basis of the above lemma, we can now define the differential operators $P_n(D|x^0,x^1,x^2,x)$, $Q_n(D|x^0,x^1,x^2,x)$ and $R_n(D|x^0,x^1,x^2,x)$. The differential operators are obtained by replacing $t-t_0$ by D_{x-x^0}, $t-t_1$ by D_{x-x^1} and $t-t_2$ by D_{x-x^2} in the expansions of P_n, Q_n and R_n in powers of $t-t_0$, $t-t_1$ and $t-t_2$. This procedure of defining differential operators has been used several times earlier. Thus, we have

PROPOSITION 5. <u>The multivariate three-point Lidstone interpolation map is given by</u>

$$L_n(f|x^0,x^1,x^2)(x) = \sum_{m=0}^{n} (P_m(D|x)f(x^0) + \int_{[x^0,x^1]} Q_m(D|x)f +$$
$$\int_{[x^0,x^1,x^2]} R_m(D|x)f) .$$

Proof. The proof of this result follows arguments similar to those given in Propositions 1 and 3 and so we omit the details.

We also have the following recursion formula for this multivariate map.

PROPOSITION 6. *The multivariate operator* L_n *given above satisfies the recursion formula*

$$L_n f(x) = L_0 f(x) + \int_{[x,x^0,x^1,x^2]} L_{n-1}(D_{x-x^0} D_{x-x^1} D_{x-x^2} f)$$

where

$$L_0 f(x) = f(x^0) + \int_{[x^0,x^1]} D_{x-x^0} f + \int_{[x^0,x^1,x^2]} D_{x-x^0} D_{x-x^1} f .$$

Proof. In the univariate case, we have $L_n^{(3)} g = L_{n-1} g^{(3)}$. Hence, using the Newton form for interpolation at t_0, t_1, t_2 with remainder

$$g(t) = g(t_0) + (t - t_0)[t_0,t_1]g + (t - t_0)(t - t_1)[t_0,t_1,t_2]g + (t - t_0)(t - t_1)(t - t_2)[t,t_0,t_1,t_2]g ,$$

we obtain

$$L_n g = g(t_0) + (t - t_0)[t_0,t_1]g + (t - t_0)(t - t_1)[t_0,t_1,t_2]g + (t - t_0)(t - t_1)(t - t_2)[t,t_0,t_1,t_2]L_{n-1}^{(-3)} g^{(3)} .$$

Now, by the usual argument with ridge functions, the multivariate version follows directly.

Finally, concerning the interpolation properties of $L_n f$, we have

PROPOSITION 7. *Let* q_ℓ *be any homogeneous differential operator with constant coefficients. Then* L_n *is the unique linear map from* $C^{(3n+2)}(K)$, $K =$ *convex hull of* x^0, x^1, x^2, *into* $\pi_{3n-1}(R^k)$ *which satisfies the following conditions:*

(i) If $\ell = 3n$, then $q_\ell(L_n f)(x^i) = q_\ell(f)(x^i)$, $i = 0,1,2$.

(ii) If $\ell = 3n+1$, then for each pair, x^i, x^j, $i \neq j$,

$$\int_{[x^i,x^j]} q_\ell(L_n f) = \int_{[x^i,x^j]} q_\ell(f) .$$

(iii) If $\ell = 3n+2$, then

$$\int_{[x^0,x^1,x^2]} q_\ell(L_n f) = \int_{[x^0,x^1,x^2]} q_\ell(f) .$$

Proof. The proof follows the type of argument used in Propositions 2 and 4 and so we omit the details.

REFERENCES

1. Cavaretta, A. S., Charles A. Micchelli and A. Sharma, Multivariate interpolation and the Radon transform, IBM Research Report, 1980.

2. Kergin, Paul, A natural interpolation of C^K functions, to appear J. Approximation Theory.

3. Polya, G., Bemerkungen zur Interpolation und zur Naherungstheorie der Balken-Biegung, Z. Angew. Math. Mech., 11 (1931), 445-449.

4. Poritsky, H., On certain polynomial and other approximations to analytic functions, Trans. Amer. Math. Soc., 34 (1932), 274-331.

A. S. Cavaretta, Jr.
Department of Mathematics
Kent State University
Kent, OH 44242

Charles A. Micchelli
IBM Research Center
PO Box 218
Yorktown Heights, NY 10598

A. Sharma
Department of Mathematics
University of Alberta
Edmonton, Alberta
CNANDA T6G 2G1

LACUNARY TRIGONOMETRIC INTERPOLATION ON EQUIDISTANT NODES

A. S. Cavaretta, Jr., A. Sharma and R. S. Varga

Dedicated to Professor A. Zygmund on his 79th birthday, December 26, 1979

This paper studies the interpolation at equidistant nodes of a given function and certain of its derivatives by trigonometric polynomials. Essentially, unique interpolation is possible only if even derivative values and odd derivative values are used in equal quantities (cf. Theorems 1 and 2). In addition, explicit forms for the fundamental polynomials are derived.

1. INTRODUCTION

We shall say that an interpolation problem is <u>regular</u> on some given nodes if it is uniquely solvable on those nodes. Recently, we have shown [1] that the problem of $(0,m_1,\ldots,m_q)$-interpolation by algebraic polynomials is regular on the n-th roots of unity with some natural growth conditions on the m_i's. Our object here is to solve the problem of regularity of $(0, m_1, \ldots, m_q)$-interpolation by trigonometric polynomials on the equidistant nodes $X := \{x_k\}_0^{n-1}$, where $x_k := \frac{2k\pi}{n}$, $k = 0, 1, \ldots, n-1$.

For our problem, it is convenient to distinguish between the following two classes of trigonometric polynomials: For m a positive integer,

$$(1.1)\quad \mathcal{T}_m := \{T(x) = a_0 + \sum_{\nu=1}^{m} (a_\nu \cos \nu x + b_\nu \sin \nu x) : \ a_\nu, b_\nu \text{ are complex numbers}\},$$

$$(1.2)\quad \mathcal{T}_{m,\varepsilon} := \{T(x) = a_0 + \sum_{\nu=1}^{m-1} (a_\nu \cos \nu x + b_\nu \sin \nu x) + a_m \cos (mx - \tfrac{\pi\varepsilon}{2}) : \ a_\nu, b_\nu \text{ are complex numbers}\},$$

where $\varepsilon = 0$ or 1. For m a positive integer, the inclusion $\mathcal{T}_{m,\varepsilon} \subseteq \mathcal{T}_m \subseteq \mathcal{T}_{m+1,\varepsilon}$ evidently holds.

ISBN 0-12-213650-0

We propose to solve the following problems:

PROBLEM A. *Given* q *positive integers* $m_1, m_2, \ldots, m_q$ *and the nodes* $X = \{\frac{2k\pi}{n}\}_0^{n-1}$, $n \geq 1$, *determine conditions on the* m_i's *which will guarantee the existence of a unique trigonometric polynomial* T(x) *of the appropriate type and order* (depending on n and q) *such that*

$$(1.3) \quad T^{(m_\nu)}(x_k) = a_{k\nu}; \; k = 0, 1, \ldots, n-1; \; \nu = 0, 1, \ldots, q, \quad \text{with } m_0 := 0,$$

for any given data $\{a_{k\nu}\}$.

For example, it is easy to see that if $q \geq 2$, then all the m_i's cannot be even, nor can they be all odd if the interpolation problem A is regular. For, if all the m_i's were even, the functions sin nx and the identically zero function would both satisfy all the conditions of (1.3) with zero data, thereby contradicting the uniqueness of interpolation. Similarly, if all the m_i's are odd, the function 1-cos nx satisfies (1.3) with zero data.

PROBLEM B. *If Problem A has a unique solution, find an explicit form of the interpolatory polynomial. In other words, find the explicit form of the associated fundamental polynomials, defined by*

$$(1.4) \quad \rho_{i,m_\nu}^{(m_j)}(x_k) := \delta_{j,\nu}\delta_{k,i}; \; i, k = 0, 1, \ldots, n-1; \; j, \nu = 0, 1, \ldots, q.$$

In order to solve these problems, we shall denote by o_q and e_q the number of odd m_i's and even m_i's, respectively, in $\{m_1, m_2, \ldots, m_q\}$, and, with $[[\tau]]$ denoting the integer part of the real number τ, we shall use throughout the notation

$$(1.5) \quad M := [[(nq + 1)/2]] .$$

With this notation, we shall establish the following two theorems:

THEOREM 1. *If* $n = 2r + 1$ *is odd, the problem of* $(0, m_1, \ldots, m_q)$-*interpolation by trigonometric polynomials is regular on* $X = \{\frac{2k\pi}{n}\}_0^{n-1}$, *precisely when*

$$(1.6)\quad \begin{cases} o_q - e_q = 0 & \text{if } q = 2p, \\ \qquad\quad = \pm 1 & \text{if } q = 2p + 1. \end{cases}$$

Moreover, when $q = 2p$, *interpolation is within the class* $\mathcal{T}_{M+r}$, *while if* $q = 2p + 1$, *interpolation is within the class* $\mathcal{T}_{M+r,\varepsilon}$ *where* $\varepsilon = 1$ *or* 0, *according as* $o_q - e_q = +1$ *or* -1, *respectively*.

THEOREM 2. *If* $n = 2r$ *is even, the problem of* $(0,m_1,\cdots,m_q)$-*trigonometric interpolation is regular on* X *precisely when*

$$(1.7)\quad \begin{cases} o_q - e_q = 0 & \text{if } q = 2p, \\ \qquad\quad = 1 & \text{if } q = 2p + 1. \end{cases}$$

Moreover, the interpolation is within the class $\mathcal{T}_{M+r,\varepsilon}$ *where* $\varepsilon = 0$ *or* 1, *according as* q *is even or odd, respectively*.

Only special cases of Theorems 1 and 2 are known in the literature. The simplest case of Lagrange interpolation, i.e., $q = 0$, on X, as well as the Hermite interpolation case of $q = 1$ and $m_1 = 1$ on X, can be found in Zygmund [11, p. 1 and p. 23]. In [5], Kiš settled the (0,2) case on X, the first lacunary case so settled. The results of Kiš were subsequently extended by Sharma and Varma [7, 8], Varma [9], Čuprigin [2], and Zeel' [10]. The results of these authors are all particular cases of Theorems 1 and 2.

Certain determinants, analogous to Vandermonde determinants but bearing on our lacunary problem, occur in our proofs, and a discussion of these determinants is relegated to §2. In §3, we prove Theorem 1 by establishing two lemmas which formalize an inductive proof of Theorem 1. The proof of Theorem 2 is along completely similar lines, and will not, for reasons of brevity, be given. Problem B is finally addressed in §4.

It is worth remarking that our work here has been recently extended by A. Sharma, P. W. Smith, and J. Tzimbalario [6].

2. SOME DETERMINANTS

In order to facilitate matters later, we show in this

section that certain determinants are nonzero. We begin with the following known result:

LEMMA 1 (cf. Gantmacher [3, p. 99]). *Let* $m_1 < m_2 < \cdots < m_q$ *be distinct real numbers, and let* $t_1 < t_2 < \cdots < t_q$ *be positive numbers. Then, the determinant*

$$(2.1)\quad \begin{vmatrix} t_1^{m_1} & t_1^{m_2} & \cdots & t_1^{m_q} \\ t_2^{m_1} & t_2^{m_2} & \cdots & t_2^{m_q} \\ \vdots & & & \vdots \\ t_q^{m_1} & t_q^{m_2} & \cdots & t_q^{m_q} \end{vmatrix}$$

is positive.

Next, we need an analogous result, which is formulated as

LEMMA 2. *Let* $m_1 < m_2 < \cdots < m_k$ *be distinct positive even integers, and let* $m_{k+1} < m_{k+2} < \cdots < m_q$ *be distinct positive odd integers, where* $0 \le k < q$. *For any* $q+1$ *positive numbers* $t_1 < t_2 < \cdots < t_{q+1}$, *the following determinant,*

$$(2.2)\quad B := \begin{pmatrix} 1 & (-t_1)^{m_1} & \cdots & (-t_1)^{m_k} & (-t_1)^{m_{k+1}} & \cdots & (-t_1)^{m_q} \\ 1 & t_2^{m_1} & \cdots & t_2^{m_k} & t_2^{m_{k+1}} & \cdots & t_2^{m_q} \\ 1 & (-t_3)^{m_1} & \cdots & (-t_3)^{m_k} & (-t_3)^{m_{k+1}} & \cdots & (-t_3)^{m_q} \\ \vdots & \vdots & & \vdots & \vdots & & \vdots \\ 1 & ((-1)^{q+1}t_{q+1})^{m_1} & \cdots & ((-1)^{q+1}t_{q+1})^{m_k} & ((-1)^{q+1}t_{q+1})^{m_{k+1}} & \cdots & ((-1)^{q+1}t_{q+1})^{m_q} \end{pmatrix}$$

is nonzero. More precisely,

$$(2.3)\quad \operatorname{sgn} B = (-1)^{(q-k)(q+k-1)/2}.$$

In addition, the determinant in (2.2), *in which the* $(j+1)$-st *column is replaced by* $\{t_1^{m_j}, (-t_2)^{m_j}, t_3^{m_j}, \cdots, ((-1)^q t_{q+1})^{m_j}\}^T$ *for each* $j=1, 2, \cdots, q$, *is similarly nonzero, with sign* $(-1)^{(q-k)(q+k+1)/2}$.

Proof. We shall use the Laplace expansion of B of (2.2) in terms of the first k+1 columns of B. Then (cf. Karlin [4, p. 6],

$$(2.4)\qquad B = \Sigma(-1)^{i_1+i_2+\dots+i_{k+1}+\frac{(k+1)(k+2)}{2}} B\begin{pmatrix} i_1,\dots,i_{k+1} \\ 1,\dots,k+1 \end{pmatrix} \times B\begin{pmatrix} i_1',\dots,i_{q-k}' \\ k+2,\dots,q+1 \end{pmatrix},$$

where $B\begin{pmatrix} i_1,\dots,i_{k+1} \\ 1,\dots,k+1 \end{pmatrix}$ is the determinant of the rows $i_1,\dots,$ i_{k+1} and columns 1, 2, ..., k+1 of B, and $B\begin{pmatrix} i_1',\dots,i_{q-k}' \\ k+2,\dots,q+1 \end{pmatrix}$ is the determinant formed from the complementary rows $i_1',\dots,i_{q-k}'$ and complementary columns k+2, ..., q+1 of B. Here, the sum (2.4) is taken over all integers i_1, ..., i_{k+1} for which $1 \le i_1 < i_2 < \dots < i_{k+1} \le q+1$. By definition,

$$(2.5)\qquad i_1+i_2+\dots+i_{k+1}+i_1'+i_2'+\dots+i_{q-k}' = 1+2+\dots+q+1 = (q+1)(q+2)/2.$$

Now, applying Lemma 1, $\operatorname{sgn} B\begin{pmatrix} i_1,\dots,i_{k+1} \\ 1,\dots,k+1 \end{pmatrix} = 1$ for all $1 \le i_1 < \dots < i_{k+1} \le q+1$. Next, note that the j-th row of $B\begin{pmatrix} i_1',\dots,i_{q-k}' \\ k+2,\dots,q+1 \end{pmatrix}$ contains all negative (positive) entries if i_j' is odd (even). Then, multiplying each element of row i_ν' by $(-1)^{i_\nu'}$ in the determinant $B\begin{pmatrix} i_1',\dots,i_{q-k}' \\ k+2,\dots,q+1 \end{pmatrix}$ for each $\nu = 1, 2, \dots, q-k$, gives a determinant which is positive from Lemma 1. Consequently,

$$\operatorname{sgn} B\begin{pmatrix} i_1',\dots,i_{q-k}' \\ k+2,\dots,q+1 \end{pmatrix} = (-1)^{\sum_{\nu=1}^{q-k} i_\nu'}.$$

Thus, each term of the sum (2.4) is nonzero and has the same sign, namely,

$$(-1)^{\{\sum_{\nu=1}^{k+1} i_\nu + (k+1)(k+2)/2 + \sum_{\nu=1}^{q-k} i_\nu'\}}$$

which, using (2.5), is equivalent to the desired result of (2.3). The proof for the alternate form of the determinant B is completely similar. ∎

Now, with $m_1, m_2, \cdots, m_q$ any distinct, not necessarily ordered, positive integers, and with (1.5), let

$$(2.6)\quad \Delta_{M-j,q} := \begin{vmatrix} 1 & (M-j)^{m_1} & \cdots & (M-j)^{m_q} \\ 1 & (M-j-n)^{m_1} & & (M-j-n)^{m_q} \\ \vdots & \vdots & & \vdots \\ 1 & (M-j-qn)^{m_1} & \cdots & (M-j-qn)^{m_q} \end{vmatrix}.$$

Similarly, let $\Delta^*_{M-j,q}$ denote the determinant obtained by deleting the last row and last column in $\Delta_{M-j,q}$. Then, as applications of Lemma 2, we have

LEMMA 3. *For* $n \geq 3$ *and* $q = 2p$,

$$(2.7)\quad \Delta_{M-j,q} \neq 0 \text{ and } \Delta^*_{M-j,q} \neq 0 \text{ for all } j=1,2,\cdots,[\![(n-1)/2]\!].$$

Similarly, for $n = 2r \geq 4$ *and* $q = 2p+1$,

$$(2.8)\quad \Delta_{M-j,q} \neq 0 \text{ and } \Delta^*_{M-j,q} \neq 0 \text{ for all } j=1,2,\cdots,[\![(n-1)/2]\!],$$

while for $n=2r+1 \geq 3$ *and* $q = 2p+1$,

$$(2.9)\quad \Delta_{M-1-j,q} \neq 0 \text{ and } \Delta^*_{M-j-1,q} \neq 0 \text{ for all } j=0,1,\cdots,[\![(n-1)/2]\!].$$

Proof. To establish (2.7), assume $n \geq 3$ and $q = 2p$, so that (cf. (1.5)) $M = np$. Then, the numbers $\{M-j-\lambda n\}_{\lambda=0}^{q}$ form a strictly decreasing sequence, of which the first p terms are positive and the remaining terms negative, for any j with $1 \leq j \leq n-1$. These numbers can be arranged, in order of increasing absolute values, as follows:

$$-j,\ n-j,\ -n-j,\ 2n-j,\ \cdots,\ -pn-j.$$

Setting $t_1, t_2, \cdots, t_{q+1}$ to be their successive absolute values, i.e.,

$$t_{2\ell+1} = \ell n+j \text{ for } \ell=0,1,\cdots,p;\ t_{2\ell} = \ell n-j \text{ for } \ell=1,2,\cdots,p,$$

then, with the hypothesis that $1 \leq j \leq [\![(n-1)/2]\!]$, it follows that $0 < t_1 < t_2 < \cdots < t_{q+1}$. Now, the determinant $\Delta_{M-j,q}$ of (2.6) in this case, is, after a suitable interchange of rows and columns, just the determinant B of (2.2) which is nonzero from Lemma 2. Thus, $\Delta_{M-j,q} \neq 0$ and similarly

$\Delta^*_{M-j,q} \neq 0$, establishing (2.7). Establishing (2.8) and (2.9) is similar, except that the latter part of Lemma 2 is applied in each case. ∎

Another determinant, which closely resembles that of (2.6) when j=0, must also be considered. It is the occurrence of this type of determinant which is responsible for the conditions on o_q-e_q in (1.6) and (1.7) in Theorems 1 and 2. With $m_1, m_2, \cdots, m_q$ distinct positive integers and with (1.5), define

$$\Phi_{M,q} := \begin{pmatrix} 1 & M^{m_1}\cos\tfrac{1}{2}\pi m_1 & M^{m_2}\cos\tfrac{1}{2}\pi m_2 & \cdots & M^{m_q}\cos\tfrac{1}{2}\pi m_q \\ \vdots & \vdots & \vdots & \vdots & \vdots \\ 1 & (M-pn)^{m_1}\cos\tfrac{1}{2}\pi m_1 & (M-pn)^{m_2}\cos\tfrac{1}{2}\pi m_2 & \cdots & (M-pn)^{m_q}\cos\tfrac{1}{2}\pi m_q \\ 0 & [M-(p+1)n]^{m_1}\sin\tfrac{1}{2}\pi m_1 & [M-(p+1)n]^{m_2}\sin\tfrac{1}{2}\pi m_2 & \cdots & [M-(p+1)n]^{m_q}\sin\tfrac{1}{2}\pi m_q \\ \vdots & \vdots & \vdots & \vdots & \vdots \\ 0 & (M-qn)^{m_1}\sin\tfrac{1}{2}\pi m_1 & (M-qn)^{m_2}\sin\tfrac{1}{2}\pi m_2 & \cdots & (M-qn)^{m_q}\sin\tfrac{1}{2}\pi m_q \end{pmatrix} \tag{2.10}$$

Similarly, let $\Phi^*_{M,q}$ denote the determinant obtained by deleting the last row and last column of $\Phi_{M,q}$. We next establish

LEMMA 4. <u>Let</u> $m_1, m_2, \cdots, m_p$ <u>be distinct positive even integers, and let</u> $m_{p+1}, m_{p+2}, \cdots, m_q$ <u>be distinct positive odd integers. For</u> $q = 2p+1$ <u>and</u> $n > 1$, <u>then</u>

$$\Phi_{M,q} \neq 0 \text{ and } \Phi^*_{M,q} \neq 0. \tag{2.11}$$

<u>Further, if</u> $n = 2r$ <u>and if</u> $\Phi_{M+r-n,q}$ <u>and</u> $\Phi^*_{M+r-n,q}$ <u>are defined by (2.10) with M replaced by</u> $M+r-n$, <u>we similarly have that</u>

$$\Phi_{M+r-n,q} \neq 0 \text{ and } \Phi^*_{M+r-n,q} \neq 0. \tag{2.12}$$

<u>Proof</u>. First, note that $\Phi_{M,q}$ of (2.10) is a determinant of order $q+1 = 2p+2$. From the hypotheses on the positive integers $\{m_i\}_{i=1}^q$, it follows that the matrix in (2.10) reduces to block-diagonal form, so that

$$\Phi_{M,q} = \pm V_1 \cdot V_2,$$

where V_1 and V_2 are generalized Vandermonde determinants of order p+1, given by

$$V_1 := \begin{pmatrix} 1 & M^{m_1} & \cdots & M^{m_p} \\ 1 & (M-n)^{m_1} & \cdots & (M-n)^{m_p} \\ \vdots & & & \vdots \\ 1 & (M-pn)^{m_1} & \cdots & (M-pn)^{m_p} \end{pmatrix} \quad V_2 := \begin{pmatrix} [M-(p+1)n]^{m_{p+1}} & \cdots & [M-(p+1)n]^{m_q} \\ [M-(p+2)n]^{m_{p+1}} & \cdots & [M-(p+2)n]^{m_q} \\ \vdots & & \vdots \\ [M-qn]^{m_{p+1}} & \cdots & [M-qn]^{m_q} \end{pmatrix}$$

Applying Lemma 1 with n > 1, then $V_1 \neq 0$ and $V_2 \neq 0$, whence $\Phi_{M,q} \neq 0$. Since $\Phi^*_{M,q}$ has the same structure as that of $\Phi_{M,q}$, then $\Phi^*_{M,q} = V_1 V_2^*$, where V_2^* is obtained from V_2 by omitting its last row and last column. From Lemma 1 again, $\Phi^*_{M,q} \neq 0$. In the same way, (2.12) follows. ■

3. THE CASE n ODD: n = 2r+1

Theorem 1 is an easy consequence of the following two lemmas. Here, $m_1, m_2, \cdots, m_q$ are again distinct, not necessarily ordered, positive integers.

LEMMA 5. *If* $n = 2r+1$, $q = 2p+1$, *and if* $(0, m_1, \cdots, m_{q-1})$-*interpolation is regular on* X *with respect to* $\mathcal{T}_{M-1}$, *then* $(0, m_1, \cdots, m_q)$-*interpolation is regular on* X *with respect to* $\mathcal{T}_{M+r,\varepsilon}$, *provided that* $\varepsilon=+1$ *if* m_q *is odd, and* $\varepsilon=0$ *if* m_q *is even*.

LEMMA 6. *If* $n = 2r+1$, $q = 2p$, *and if* $(0, m_1, \cdots, m_{q-1})$-*interpolation is regular on* X *with respect to* $\mathcal{T}_{M,\varepsilon}$ *where* ε *is as in Theorem* 1, *then* $(0, m_1, \cdots, m_q)$-*interpolation is regular on* X *with respect to* $\mathcal{T}_{M+r}$, *provided that* $o_q - e_q = 0$.

We first outline the proof of Theorem 1, assuming the validity of Lemmas 5 and 6, and then we shall turn to the proofs of these lemmas. Since Lagrange interpolation, corresponding to the case q = 0, is evidently regular on X with respect to $\mathcal{T}_r$, we may apply Lemma 5 to deduce that $(0, m_1)$-interpolation is regular on X with respect to $\mathcal{T}_{2r+1,\varepsilon}$, where $\varepsilon = 1$ if m_1 is odd and $\varepsilon = 0$ if m_1 is even. Now, applying Lemma 6 with q = 2, we deduce that $(0, m_1, m_2)$-interpolation is regular on X with respect to $\mathcal{T}_{3r+1}$, provided that $o_2-e_2=0$, i.e., m_1 and m_2 are of different parity. Thus, by repeated alternate use of Lemmas 5 and 6, we see that $(0,m_1,\cdots,m_q)$-interpolation is regular on X as described in Theorem 1, provided that m_1, m_2, $\cdots$, m_q satisfy (1.6).

Before we turn to the proofs of Lemmas 5 and 6, we observe that in each of these lemmas, it is assumed that $(0, m_1, \cdots, m_{q-1})$-interpolation is regular on X with respect to an appropriate $\mathcal{T}_s$ or $\mathcal{T}_{s,\varepsilon}$. This implies that there exists a linear operator $\mathcal{L}_n$ of the form

$$(3.1) \quad \mathcal{L}_n(x;\ f) := \sum_{\nu=0}^{q-1} \sum_{k=0}^{n-1} f^{(m_\nu)}(x_k)\rho_{k,m_\nu}(x), \qquad (m_0 := 0),$$

mapping any sufficiently differentiable function of period 2π into a trigonometric polynomial, where the $\rho_{k,m_\nu}(x)$ are the associated fundamental trigonometric polynomials of $(0, m_1, \cdots, m_{q-1})$-interpolation on X, i.e., (cf. (1.4)),

$$(3.2) \quad \rho_{k,m_\nu}^{(m_j)}(x_i) = \delta_{j,\nu}\delta_{k,i};\ i,\ k=0,1,\cdots,n-1;\ j,\nu=0,1,\cdots,q-1.$$

<u>Proof of Lemma 5</u>. As the problem in (1.3) is linear, we consider instead of (1.3) the associated homogeneous system defined by

$$(3.3) \quad T^{(m_\nu)}(x_k) = 0,\ k = 0,1,\cdots,n-1;\ \nu = 0,1,\cdots,n-1,$$

where $T(x) \in \mathcal{T}_{M+r,\varepsilon}$. The lemma is proved if we can show that $T(x) \equiv 0$.

First, for any $T(x) \in \mathcal{T}_{M+r,\varepsilon}$, we can write $T(x)$ as the sum

$$(3.4) \quad T(x) = A(x) + B_{M,r}(x),$$

where $A(x) \in \mathcal{T}_{M-1}$, and where

$$(3.5) \quad B_{M,r}(x) = \sum_{j=0}^{r-1} \{a_j \cos(M+j)x + b_j \sin(M+j)x\} + c_r \cos[(M+r)x - \tfrac{\pi\varepsilon}{2}],$$

with $\varepsilon = 1$ or 0, according as m_q is odd or even. Now, by the regularity hypothesis, the linear operator $\mathcal{L}_n$ of (3.1) is a projection on $\mathcal{T}_{M-1}$, and (3.3) gives $\mathcal{L}_n(x;\ T(x)) \equiv 0$. Thus, applying $\mathcal{L}_n$ to (3.4) gives

$$A(x) = -\mathcal{L}_n(x;\ B_{M,r}(x)),$$

so that from (3.4),

$$(3.6)\quad T(x) = B_{M,r}(x) - \mathcal{L}_n(x; B_{M,r}) := \sum_{j=0}^{r-1}\{a_j\omega_j(x) + b_j\mu_j(x)\} + c_r\omega_{r,\varepsilon}(x),$$

where we set

$$(3.7)\quad \begin{cases} \omega_j(x) := \cos(M+j)x - \mathcal{L}_n(x;\ \cos(M+j)x),\ j = 0,1,\cdots,r-1, \\ \mu_j(x) := \sin(M+j)x - \mathcal{L}_n(x;\ \sin(M+j)x),\ j = 0,1,\cdots,r-1, \\ \omega_{r,\varepsilon}(x) := \cos[(M+r)x - \frac{\pi\varepsilon}{2}] - \mathcal{L}_n(x;\ \cos[(M+r)x - \frac{\pi\varepsilon}{2}]). \end{cases}$$

We claim that

$$(3.8)\quad \begin{cases} \omega_j^{(m_q)}(x_k) = \Delta_{M-j-1,q}\cos[(r-j)x_k + \frac{\pi m_q}{2}]/\Delta^*_{M-j-1,q}, \\ \qquad\qquad j = 0,1,\cdots,r-1, \\ \mu_j^{(m_q)}(x_k) = \Delta_{M-j-1,q}\ \sin[(r-j)x_k + \frac{\pi m_q}{2}]/\Delta^*_{M-j-1,q}, \\ \qquad\qquad j = 0,1,\cdots,r-1, \\ \omega_{r,\varepsilon}^{(m_q)}(x_k) = \Delta_{M-r-1,q}\ \cos[(m_q-\varepsilon)\frac{\pi}{2}]/\Delta^*_{M-r-1,q}, \end{cases}$$

for $k = 0, 1, \cdots, n-1$, where $\Delta_{M-j-1,q}$ and $\Delta^*_{M-j-1,q}$ are defined in §2 (cf. (2.6)). We shall establish (3.8) later in this section, and we continue with the preceding argument.

Applying the remaining conditions of (3.5), namely

$$T^{(m_q)}(x_k) = 0,\quad k = 0, 1, \cdots, n-1,$$

to the representation of (3.6), we see from (3.8) that the sum

$$(3.9)\quad \sum_{j=0}^{r-1}[\Delta_{M-j-1,q}\{a_j\ \cos[(r-j)x + \frac{\pi m_q}{2}] + b_j\ \sin[(r-j)x + \frac{\pi m_q}{2}]\}/\Delta^*_{M-j-1,q}] + c_r\Delta_{M-r-1,q}\cos[(m_q-\varepsilon)\frac{\pi}{2}]\ /\Delta^*_{M-r-1,q},$$

which is a trigonometric polynomial of degree r, vanishes at all n nodes of X. Since $n = 2r+1$, this means that this sum must vanish identically. Since, by (2.9) of Corollary 3, $\Delta_{M-j-1,q} \neq 0$ for all $j = 0, 1, \ldots, r-1$, and since $m_q - \varepsilon$ is, (cf. Theorem 1) by definition, even, it follows that all the coefficients a_j, b_j, and c_r of (3.9) are all zero. But then, from (3.6), $T(x) \equiv 0$. ∎

Proof of (3.8). To calculate the a_j's of (3.5), we observe from the definition of M that, since n and q are both odd, $\cos(M+j)x = \cos[M-j-1-qn]x$, so that from (3.1), we have

$$(3.10)\qquad \mathcal{L}_n(x;\ \cos(M+j)x) = \sum_{\nu=0}^{q-1} (M-j-1-qn)^{m_\nu} I_{j,\nu}(x),$$

where

$$(3.11)\qquad I_{j,\nu}(x) := \sum_{k=0}^{n-1} \cos[(M-j-1)x_k + \frac{\pi m_\nu}{2}]\cdot \rho_{k,m_\nu}(x),$$
$$j = 0, 1, \ldots, r-1.$$

Next, we further observe that, as $2M = nq+1$, then

$$-M < M-j-1-\lambda n < M, \text{ for } j = 0, 1, \ldots, r;\ \lambda = 0, 1, \ldots, q-1,$$

so that by the reproducing character of the operator $\mathcal{L}_n$ on $\mathcal{T}_{M-1}$, we further have the identities

$$(3.12)\qquad \cos(M-j-1-\lambda n)x = \sum_{\nu=0}^{q-1} (M-j-1-\lambda n)^{m_\nu} I_{j,\nu}(x);$$
$$j = 0, 1, \ldots, r;\ \lambda = 0, 1, \ldots, q-1.$$

Combining (3.12) for the cases $j = 0, 1, \ldots, r-1$ with (3.10), this implies that the following determinants are zero:

$$(3.13)\qquad \begin{vmatrix} 1 & (M-j-1)^{m_1} & \cdots & (M-j-1)^{m_{q-1}} & \cos(M-j-1)x \\ 1 & (M-j-1-n)^{m_1} & \cdots & (M-j-1-n)^{m_{q-1}} & \cos(M-j-1-n)x \\ \vdots & \vdots & & \vdots & \vdots \\ 1 & [M-j-1-(q-1)n]^{m_1} & & [M-j-1-(q-1)n]^{m_{q-1}} & \cos[M-j-1-(q-1)n]x \\ 1 & [M-j-1-qn]^{m_1} & \cdots & [M-j-1-qn]^{m_{q-1}} & \mathcal{L}_n(x;\ \cos(M+j)x) \end{vmatrix}$$
$$= 0,\ j=0,1,\ldots,r-1.$$

Expanding this determinant in terms of its last column and noting that the cofactor of $\mathcal{L}_n(x;\ \cos(M+j)x)$ is just $\Delta^*_{M-j-1,q}$ which is nonzero from (2.9) of Corollary 3, it easily follows from the definition of $\omega_j(x)$ in (3.7) that

$$\omega_j(x) = \Delta/\Delta^*_{M-j-1,q}, \quad j=0,1,\ldots,r-1,$$

where is the determinant obtained by putting $\cos(M-j-1-qn)x$ in the determinant of (3.13) in place of $\mathcal{L}_n(x;\cos(M+j)x)$. Then, since all columns of Δ, except the last, are independent of x, we see that differentiating $\omega_j(x)$ m_q-times and putting x_k for x gives, with the definition of $\Delta_{M-j-1,q}$ in (2.6), the desired first equation of (3.8). The second and third equations of (3.8) are similarly deduced, the derivation of the third relation in (3.8) making use of the fact that (3.12) is valid for $j = r$. ∎

In a similar fashion, we give the

Proof of Lemma 6. Here, $n = 2r+1$ and $q = 2p$, so that $M=np$. Again, we consider the homogeneous system defined by

$$(3.14) \quad T^{(m_\nu)}(x_k) = 0,\ \nu = 0, 1, \cdots, q;\ k = 0, 1, \cdots, n-1,$$

where $T(x) \in \mathcal{T}_{M+r}$. By hypothesis, the linear operator $\mathcal{L}_n$ of the form (3.1) exists and maps f into $\mathcal{T}_{M,\varepsilon}$, where each $\rho_{k,m_\nu}(x)$ of (3.1) is now in $\mathcal{T}_{M,\varepsilon}$. Also, since we require that $o_q-e_q = 0$, we may suppose that $m_1,\cdots,m_p$ have the same parity as ε, while $m_{p+1},\cdots,m_{2p}$ have the opposite parity to ε.

For $T(x) \in \mathcal{T}_{M+r}$, we write $T(x) = A(x) + B_{M,r}(x)$, where $A(x) \in \mathcal{T}_{M,\varepsilon}$, and where

$$B_{M,r}(x)=b_0 \sin(Mx-\tfrac{\pi\varepsilon}{2}) + \sum_{j=1}^{r}\{a_j \cos(M+j)x+b_j \sin(M+j)x\}.$$

Following the method of proof of Lemma 5, we see again that the conditions $\nu=0,1,\cdots,q-1$; $k=0,1,\cdots,n-1$ of (3.14) imply that

$$(3.15) \quad T(x) = B_{M,r}(x)-\mathcal{L}_n(x;B_{M,r}(x)) = \sum_{j=1}^{r}\{a_j\omega_j(x) + b_j\mu_j(x)\} + b_0\mu_{0,\varepsilon}(x),$$

where

$$(3.16) \quad \begin{cases} \omega_j(x) = \cos(M+j)x - \mathcal{L}_n(x;\cos(M+j)x),\ j=1,2,\cdots,r, \\ \mu_j(x) = \sin(M+j)x - \mathcal{L}_n(x;\sin(M+j)x),\ j=1,2,\cdots,r. \\ \mu_{0,\varepsilon}(x) = \sin(Mx-\tfrac{\pi\varepsilon}{2}) - \mathcal{L}_n(x;\sin(Mx - \tfrac{\pi\varepsilon}{2})). \end{cases}$$

We shall show that

$$(3.17)\quad \begin{cases} \omega_j^{(m_q)}(x_k)=\Delta_{M-j,q}\cos(jx_k-\frac{\pi m_q}{2})/\Delta^*_{M-j,q}, & j=1,2,\cdots,r \\ \mu_j^{(m_q)}(x_k)=\Delta_{M-j,q}\sin(jx_k-\frac{\pi m_q}{2})/\Delta^*_{M-j,q}, & j=1,2,\cdots,r, \\ \mu_{0,\varepsilon}^{(m_q)}(x_k)=-\Phi_{M,q}\sin[(m_q+\varepsilon)\frac{\pi}{2}]/\Phi^*_{M,q}, & \end{cases}$$

where $\Delta_{M-j,q}$ is given in (2.6), $\Phi_{M,q}$ is given in (2.10), and $\Delta^*_{M-j,q}$ and $\Phi^*_{M,q}$ are obtained from $\Delta_{M-1,q}$ and $\Phi_{M,q}$, respectively, by omitting their last row and last column.

We shall establish (3.17) later in this section, and we continue with the preceding argument. Applying the remaining conditions of (3.14), namely

$$T^{(m_q)}(x_k) = 0, \qquad k = 0, 1, \cdots, n-1,$$

to the representation of (3.15), we deduce from (3.17) that the sum

$$(3.18)\quad \begin{aligned} &-b_0\Phi_{M,q}\sin[(m_q+\varepsilon)\tfrac{\pi}{2}]/\Phi^*_{M,q}+\sum_{j=1}^{r}[\Delta_{M-j,q}\{a_j\cos(jx-\tfrac{\pi m_q}{2})+ \\ &\qquad b_j\sin(jx-\tfrac{\pi m_q}{2})\}/\Delta^*_{M-j,q}], \end{aligned}$$

which is a trigonometric polynomial of degree r, vanishes at all n = 2r+1 nodes of X, and thus vanishes identically. By Lemmas 3 and 4 (cf. (2.8) and (2.11)), the various determinants which appear in (3.18), i.e., $\Delta_{M-j,q}$, $\Delta^*_{M-j,q}$, $\Phi_{M,q}$, and $\Phi^*_{M,q}$, are all nonzero. In addition, the hypothesis of Lemma 6 and the condition $o_q-e_q = 0$, together imply that $m_q+\varepsilon$ is odd, so that $\sin(m_q+\varepsilon)\frac{\pi}{2}$ is nonzero. Thus, as the trigonometric polynomial in (3.18) vanishes identically, then the a_j's $(j=1,2,\cdots,r)$ and the b_j's $(j=0,1,\cdots,r)$ are all zero, whence $T(x) \equiv 0$ from (3.15). ∎

<u>Proof of (3.17)</u>. Since q = 2p, then 2M = nq, so that $\cos(M+j)x = \cos(M-j-qn)x$ for all $j=1,2,\cdots,r$. Thus, with (3.1), we can write

$$\mathscr{L}_n(x;\cos(M+j)x) = \sum_{\nu=0}^{q-1}(M-j-qn)^{m_\nu}\,\hat{I}_{j,\nu}(x),$$

where

$$\hat{I}_{j,\nu}(x) := \sum_{k=0}^{n-1} \cos[(M-j)x_k+\frac{\pi m_\nu}{2}]\cdot\rho_{k,m_\nu}(x), \quad j=1,2,\cdots,r.$$

Since $\mathcal{L}_n$ reproduces the functions $\cos(M-j-\lambda n)x$ and $\sin(M-j-\lambda n)x$ for $\lambda=0,\cdots,q-1$, we can use the resulting identities similar to (3.12) to obtain $\omega_j(x)$ as $(\Delta^*_{M-1,q})^{-1}$ times a determinant, defined by adjoining the following extra row to $\Delta^*_{M-j,q}$, namely,

$$\{1,\ (M-j-qn)^{m_1},(M-j-qn)^{m_2},\cdots(M-j-qn)^{m_{q-1}},\ \cos(M-j-qn)x\},$$

and the following extra column:

$$\{\cos(M-j)x,\ \cos(M-j-n)x,\ \cdots,\ \cos(M-j-qn)x\}^T.$$

Then, on differentiating m_q-times this resulting expression for $\omega_j(x)$ and putting x equal to x_k, the first relation of (3.17) is obtained. The proof of the second relation in (3.17) is similar and is omitted.

In order to derive the last formula in (3.17), we observe that

$$(3.19)\quad \mathcal{L}_n(x;\ \sin(Mx-\frac{\pi\varepsilon}{2})) = \sum_{\nu=0}^{q-1} M^{m_\nu}\sin(m_\nu-\varepsilon)\frac{\pi}{2}\cdot J_\nu(x),$$

where $J_\nu(x) := \sum_{k=0}^{n-1} \rho_{k,m_\nu}(x)$. We also note the identities

$$(3.20)\quad \begin{cases} \cos(Mx-\frac{\pi\varepsilon}{2}) = \sum_{\nu=0}^{q-1} M^{m_\nu}\cos(m_\nu-\varepsilon)\frac{\pi}{2}\cdot J_\nu(x), \\ \cos(M-\lambda n)x = \sum_{\nu=0}^{q-1} (M-\lambda n)^{m_\nu}\cos\frac{\pi m_\nu}{2}\cdot J_\nu(x), \quad \lambda=1,2,\cdots,p, \\ \sin(M-\lambda n)x = \sum_{\nu=0}^{q-1} (M-\lambda n)^{m_\nu}\sin\frac{\pi m_\nu}{2}\cdot J_\nu(x), \quad \lambda=p+1,\cdots,2p-1. \end{cases}$$

Since $m_1,\ \cdots,\ m_p$ have the same parity as ε and since $m_{p+1},\cdots,m_q$ have the opposite parity, we shall treat the case when $\varepsilon = 0$ and $m_1,\cdots,m_p$ are even; the other case is analogously treated.

Thus, we assume that $\varepsilon = 0$, and that $m_1,\cdots,m_p$ are even. Now, the q identities given in (3.20) are linearly independent since the associated coefficient matrix is exactly $\Phi^*_{M,q}$ of Lemma 4 (cf. (2.10)), and is thus nonzero. On combining the q+1 equations (3.19) and (3.20), and setting $\mu_{0,0}(x) = \sin Mx - \mathcal{L}_n(x; \sin Mx)$, we obtain

$$\mu_{0,0}(x) =$$

$$\frac{-1}{\Phi^*_{M,q}} \begin{vmatrix} 1 & M^{m_1} \cos \tfrac{1}{2}\pi m_1 & \cdots & M^{m_{q-1}} \cos \tfrac{1}{2}\pi m_{q-1} & \cos Mx \\ 1 & (M-n)^{m_1} \cos \tfrac{1}{2}\pi m_1 & \cdots & (M-n)^{m_{q-1}} \cos \tfrac{1}{2}\pi m_{q-1} & \cos(M-n)x \\ \vdots & \vdots & & \vdots & \vdots \\ 1 & (M-pn)^{m_1} \cos \tfrac{1}{2}\pi m_1 & \cdots & (M-pn)^{m_{q-1}} \cos \tfrac{1}{2}\pi m_{q-1} & \cos(M-pn)x \\ 0 & (M-(p+1)n)^{m_1} \sin \tfrac{1}{2}\pi m_1 & \cdots & (M-(p+1)n)^{m_{q-1}} \sin \tfrac{1}{2}\pi m_{q-1} & \sin(M-(p+1)n)x \\ \vdots & \vdots & & \vdots & \vdots \\ 0 & (M-qn)^{m_1} \sin \tfrac{1}{2}\pi m_1 & \cdots & (M-qn)^{m_{q-1}} \sin \tfrac{1}{2}\pi m_{q-1} & \sin(M-qn)x \end{vmatrix}$$

From this, it is easy to see that

$$\mu_{0,0}^{(m_q)}(x_k) = \Phi_{M,q} \sin \frac{\pi m_q}{2} / \Phi^*_{M,q}, \quad k = 0, 1, \cdots, n-1.$$

Similarly, if $\varepsilon = 1$ and $m_1, \cdots, m_p$ are odd, we have

$$\mu_{0,0}^{(m_q)}(x_k) = \Phi_{M,q} \cos \frac{\pi m_q}{2} / \Phi^*_{M,q}, \quad k = 0, 1, \cdots, n-1,$$

which completes the proof of (3.17). ∎

4. EXPLICIT FORMS FOR THE FUNDAMENTAL POLYNOMIALS

The determinant $\Delta_{M-1,q}$ of (2.6), which occurred in the solution of Problem A, can also be effectively used to answer Problem B. We assume that the conditions of Theorem 1 or 2 are satisfied, so that the $(0,m_1,\cdots,m_q)$-interpolation problem is regular on the nodes $x_k = \frac{2k\pi}{n}$. Then, the linear system given by (1.4) has a unique solution $\rho_{i,m_\nu}(x)$ within the proper trigonometric class as defined by our Theorems.

As our nodes x_i are equidistant, it is clear that

$$\rho_{i,m_\nu}(x) = \rho_{0,m_\nu}(x-x_i) \text{ for } i = 1, \cdots, n-1.$$

It is therefore sufficient to give explicit forms for the function $\rho_{0,m_\nu}(x)$ which is defined by the linear system

$$(4.1)\quad \begin{cases} \rho_{0,m_\nu}^{(m_j)}(x_k) = 0, & k = 0, 1, \cdots, n-1 \text{ and } j \neq \nu, \\ \rho_{0,m_\nu}^{(m_\nu)}(x_k) = \delta_{0,k}, & k = 0, 1, \cdots, n-1. \end{cases}$$

Using the determinants $\Delta_{M-j,q}$ as given in (2.6), we define certain trigonometric polynomials $N_{M-j}(x;\, m_\nu, q)$ which, for each $\nu = 0,1,\cdots,q$, are obtained from $\Delta_{M-j,q}$ by replacing its $(\nu+1)$-st column by

$$\{\cos[(M-j)x-\frac{\pi m_\nu}{2}],\ \cos[(M-j-n)x-\frac{\pi m_\nu}{2}],\cdots,\ \cos[(M-j-qn)x-\frac{\pi m_\nu}{2}]\}^T.$$

With these polynomials, we have

THEOREM 3. *Let* $n = 2r+1$, *and let* $m_1, m_2, \cdots, m_q$ *be positive integers which satisfy condition* (1.6) *of Theorem* 1. *Then, for* $\nu = 0,1,\cdots,q$, *we have*

$$(4.2)\quad \rho_{0,m_\nu}(x) = \begin{cases} \dfrac{1}{n}\dfrac{N_M(x;\, m_\nu,\, q)}{\Delta_{M,q}} + \dfrac{2}{n}\sum_{j=1}^{r}\dfrac{N_{M-j}(x;\, m_\nu,\, q)}{\Delta_{M-j,q}} & \text{for } q = 2p, \\ \dfrac{1}{n}\dfrac{N_{M-r-1}(x;\, m_\nu, q)}{\Delta_{M-r-1,q}} + \dfrac{2}{n}\sum_{j=1}^{r}\dfrac{N_{M-j}(x;\, m_\nu,\, q)}{\Delta_{M-j,q}} & \text{for } q = 2p+1. \end{cases}$$

Proof. The proof follows easily by observing that $\rho_{0,m_\nu}(x)$, as given by (4.2), satisfies (4.1). Indeed, we observe that

$$N_{M-j}^{(m_\nu)}(x_k;\, m_\nu,\, q) = \cos[(M-j)x_k]\cdot\Delta_{M-j,q}$$

and so, if $q = 2p$, then $M = pn$ and it follows that

$$\rho_{0,m_\nu}^{(m_\nu)}(x_k) = \frac{1}{n}[1+2\sum_{j=1}^{r}\cos j\, x_k] = \begin{cases} 0 & \text{if } k \neq 0; \\ 1 & \text{if } k = 0, \end{cases}$$

as desired. If $q = 2p+1$, then $M = pn + m + 1$, so that

$$\rho_{0,m_\nu}^{(m_\nu)}(x_k) = \frac{1}{n}[1 + 2 \sum_{j=1}^{r} \cos(r + 1 - j)x_k]$$

$$= \begin{cases} 0 & \text{if } k \neq 0; \\ 1 & \text{if } k = 0. \end{cases}$$

Thus, Theorem 3 is verified directly. The case when $n = 2r$ is exactly similar, and we can give

THEOREM 4. *If* $n = 2r$ *and if* $m_1, \cdots, m_q$ *are positive integers which satisfy condition* (1.7) *of Theorem* 2, *then for* $\nu = 0, 1, \cdots, q-1$,

$$\rho_{0,m_\nu}(x) = \frac{1}{n}\frac{N_M(x;\, m_\nu,\, q)}{\Delta_{M,q}} + \frac{2}{n}\sum_{j=1}^{r-1}\frac{N_{M-j}(x;\, m_\nu,\, q)}{\Delta_{M-r,q}} + \frac{1}{n}\frac{N_{M-r}(x;\, m_\nu,\, q)}{\Delta_{M-r,q}}. \tag{4.3}$$

5. CONCLUSION

A natural problem which now arises is the problem of *convergence*; more precisely, as the number of nodes is allowed to increase, how well do our interpolating trigonometric polynomials approximate the function we interpolate?

We fix q positive integers $m_1, \cdots, m_q$ which satisfy condition (1.6) or (1.7), according as n is odd or even. Then, according to our theorems, there are defined linear operators $\mathcal{L}_n(x;\, f)$, given by

$$\mathcal{L}_n(x;\, f) = \sum_{k=0}^{n-1} f(x_k)\rho_{k,0}(x) + \sum_{\nu=1}^{q}\sum_{k=0}^{n-1} \beta_{k,\nu}\rho_{k,m_\nu}(x) \tag{5.1}$$

where $\rho_{k,0}$ and ρ_{k,m_ν} are the fundamental polynomials given in §5, and the $\{\beta_{k,\nu}\}_{k=0,\ \nu=1}^{n-1,\ q}$ are certain given numbers. If f is continuous and periodic with period 2π, we are interested in finding conditions on the $\beta_{k,\nu}$ which will ensure that

$$\lim_{n\to\infty} \mathcal{L}_n(x;\, f) = f(x).$$

We hope to return to this problem.

REFERENCES

1 Cavaretta, Jr., A. S., A. Sharma, and R. S. Varga, Hermite-Birkhoff interpolation in the n-th roots of unity, Trans. Amer. Math. Soc. (to appear).

2 Čuprigin, O. A., On trigonometric (0, 1, 2, ···, r-2, r)-interpolation, (Russian), Vesci Akad. Navuk BSSR Ser. Fiz.-Mat. Navuk 1965, no. 1, 129-131; MR 33, #486.

3 Gantmacher, F. R., The Theory of Matrices, Volume II, Chelsea Publishing Co., New York, 1959.

4 Karlin, Samuel, Total Positivity, Stanford University Press, Stanford, California, 1968.

5 Kiš, O., On trigonometric interpolation, (Russian), Acta. Math. Acad. Sci. Hung. 11(1960), 256-276.

6 Sharma, A., P. W. Smith, and J. Tzimbalario, Polynomial interpolation in roots of unity with applications, (to appear).

7 Sharma, A. and A. K. Varma, Trigonometric interpolation, Duke Math. J. 32(1965), 341-358.

8 Sharma, A. and A. K. Varma, Trigonometric interpolation: (0, 2, 3) case, Ann. Polon. Math. 21(1968), 51-58.

9 Varma, A. K., Trigonometric interpolation, J. Math. Anal. Appl. 28(1969), 652-659.

10 Zeel', E. O., Trigonometric (0, p, q)-interpolation, (Russian), Izv. Vysš. Učebn. Zaved. Matematika 1970, no. 3(94), 27-35.

11 Zygmund, A., Trigonometric Series, Volume II, second edition, Cambridge, 1959.

A. S. Cavaretta Jr.
Dept. of Math.
Kent State Univ.
Kent, OH 44242

A. Sharma*
Dept. of Math.
Univ. of Alberta
Edmonton, Alberta
CANADA T6G 2G1

R. S. Varga**
Dept. of Math.
Kent State Univ.
Kent, OH 44242

*The work done by this author was performed while the author was visiting at KSU and on leave from the University of Alberta. Research supported in part by the Canadian NRC.

**Research supported in part by the Air Force Office of Scientific Research and by the Department of Energy.

MONOTONE APPROXIMATION BY SPLINE FUNCTIONS

C. K. Chui, P. W. Smith, and J. D. Ward

The purpose of this paper is to establish estimates for the approximation of monotone nondecreasing functions by monotone nondecreasing splines with knots governed by a diffeomorphism. The general L_p, $1 \leqq p \leqq \infty$, setting is considered.

1. INTRODUCTION

The problem of degree of monotone approximation by polynomials was studied as early as 1968 by Lorentz and Zeller [10]. Since then, several interesting results have been obtained. Perhaps, the most complete result in this direction is that of DeVore [7], where uniform approximation was considered. An analogous result in approximation by monotone spline functions with equally spaced knots was also obtained by DeVore [6]. In [4], these authors used fairly elementary techniques to prove DeVore's result in [6] and at the same time extended the L_∞ result to L_p, $1 \leqq p \leqq \infty$.

In this paper, we will consider the problem of monotone L_p approximation by spline functions with knot sequence governed by a diffeomorphism. More precisely, let $t \in C^1[0,1]$ satisfy the conditions $t(0) = 0$, $t(1) = 1$ and

(1.1) $\quad \eta := \min\{t'(s): \quad 0 \leqq s \leqq 1\} > 0.$

We will consider the approximation of functions from the space $S(k,N)$ of all splines of order k and with knot sequence

(1.2) $\quad \underline{T}_N: \quad 0 = t_0 < \ldots < t_N = 1,$

with $t_i = t(i/N)$, $i = 0,\ldots,N$. Following [4], if A is a collection of functions defined on the unit interval [0,1], then A* will denote the subcollection of functions in A which are nondecreasing on [0,1]. That is, $f \in A^*$ if and only if $f \in A$ and $f\uparrow$. Hence, $S^*(k,N)$ is the collection

ISBN 0-12-213650-0

of splines s in $S(k,N)$ with $s\uparrow$. For each integer j, $j \geqq 0$ and $1 \leqq p \leqq \infty$, let $L_p^0[0,1] = L_p[0,1]$ and $L_p^j[0,1]$ be the space of functions which are j-fold integrals of $L_p[0,1]$ functions. If $f \in L_p^{k*}[0,1]$, that is, $f \in L_p^k[0,1]$ and $f\uparrow$, we will study the $L_p[0,1]$ distance of f from $S^*(k,N)$, denoted by

$$(1.3) \quad E^*_{N,p}(f,k) = \inf\{\|f - s\|_p : \quad s \in S^*(k,N)\}$$

where $\|\cdot\|_p$ denotes the usual $L_p[0,1]$ norm $\|\cdot\|_{L_p[0,1]}$. We will also use the notation

$$(1.4) \quad M_{p,k}(f) = \|(t')^{k+\frac{1}{p}} f^{(k)} \circ t\|_p .$$

The following result will be established.

THEOREM 1.1. <u>Let</u> k <u>be a positive integer</u>. <u>There exists a positive constant</u> C, <u>depending only on</u> k, <u>such that if</u> $f \in L_p^{k*}[0,1]$, <u>then</u>

$$(1.5) \quad E^*_{N,p}(f,k) \leqq CN^{-k}M_{p,k}(f)$$

<u>for</u> $1 \leqq p \leqq \infty$ <u>and all</u> $N \geqq N_0(\eta)$.

In addition, the following notation will be used throughout the paper:

(a) π_k: = the space of all polynomials with degree $< k$,

(b) $\pi(k,\underline{\tau}_N)$: = the space of all functions $f \in C[0,1]$ such that $f|_{[t_{i-1},t_i]} \in \pi_k$, $i = 1,\ldots,N$,

(c) $\omega(g;\delta)$: = $\sup\{|g(s) - g(t)|: \quad 0 \leqq s,\ t \leqq 1,\ |s - t| \leqq \delta\}$,

(d) $D^*_{N,p}(f,k)$: = $\inf\{\|f - g\|_p: \quad g \in \pi^*(k,\underline{\tau}_N)\}$, and

(e) $P_k f$ denotes the polynomial in π_k such that $(P_k f - f)(\frac{i}{k-1}) = 0$, $i = 0,\ldots,k-1$.

2. MONOTONE APPROXIMATION BY PIECEWISE POLYNOMIALS

The main result in this section is the following

THEOREM 2.1. *There exists a positive constant* C, *depending only on* k, *such that for every* p, $1 \leqq p \leqq \infty$, *and every* $f \in L_p^{k*}[0,1]$,

$$D^*_{N,p}(f,k) \leqq CN^{-k}M_{p,k}(f) \tag{2.1}$$

where $N = 1,2,\ldots$.

We need the following three lemmas.

LEMMA 2.1. *There exists a positive constant* C_1, *depending only on* k, *such that*

$$\|(f - P_k f)^{(i)}\|_p \leqq C_1 \|f^{(k)}\|_p \tag{2.2}$$

for $i = 0,\ldots,k-1$ *and every* $f \in L_p^k[0,1]$, $1 \leqq p \leqq \infty$. *If, in addition,* $f \in L_p^{k*}[0,1]$, *then*

$$-\min[\min_{0 \leqq x \leqq 1} (P_k f)'(x), 0] \leqq C_1 \|f^{(k)}\|_p . \tag{2.3}$$

To prove this lemma, we consider the functional

$$Eg = (g - P_k g)^{(i)}(x) ,$$

so that

$$(f - P_k(f))^{(i)}(x) = \int_0^1 E[(\cdot - t)_+^{k-1}/(k-1)!]f^{(k)}(t)dt .$$

Hence, $|(f - P_k(f))^{(i)}(x)| \leqq C_1 \|f^{(k)}\|_p$ by Holder's inequality $i = 0,\ldots,k-1$. This holds for all $x \in [0,1]$, and (2.2) follows. If, in addition, $f' \geqq 0$, we apply the above inequality with $i = 1$ to yield

$$-\min[(P_k f)'(x), 0] \leqq |(f - P_k f)'(x)| \leqq C_1 \|f^{(k)}\|_p$$

for all $x \in [0,1]$. By taking the maximum over x, we have (2.3). This completes the proof of the lemma.

We will now study monotone approximation and establish the following

LEMMA 2.2. *There exists a positive constant* C_2, *depending only on* k, *such that for every* $f \in L_p^{k*}[0,1]$, *there is a* $g \in \pi_k^*$ *satisfying* $g(0) = f(0)$, $g(1) = f(1)$, *and*

$$\|(f-g)^{(i)}\|_p \leqq C_2 \|f^{(k)}\|_p \tag{2.4}$$

for all $i = 0,\ldots,k-1$ *and* $1 \leqq p \leqq \infty$.

Let $f \in L_p^{k*}[0,1]$. Without loss of generality, we may assume that $f(0) = 0$. Set

$$g(x) = [P_k(f)(x) + d_f x]\, \frac{f(1)}{f(1) + d_f}$$

where

$$d_f = -\min[\min_{0 \leqq x \leqq 1} (P_k f)'(x), 0]\,.$$

It is clear that

$$g(0) = 0 = f(0)$$

$$g(1) = [(P_k f)(1) + d_f]\, \frac{f(1)}{f(1) + d_f} = f(1)\,.$$

Also,

$$g'(x) = [(P_k f)'(x) + d_f]\, \frac{f(1)}{f(1) + d_f} \geqq 0$$

so that $g \in \pi_k^*$. To prove (2.4), we note that

$$\|(f-g)^{(i)}\|_p \leqq \|(f - P_k f)^{(i)}\|_p + \|(g - P_k f)^{(i)}\|_p$$
$$\leqq \|(f - P_k f)^{(i)}\|_p\, \frac{f(1)}{f(1) + d_f} + \frac{d_f f(1)}{f(1) + d_f}\,,$$

where the last term on the right-hand side can be dropped if $i \geqq 2$. By applying Lemma 2.1, we have

$$\|(f-g)^{(i)}\|_p \leqq \|(f - P_k f)^{(i)}\|_p + d_f + \|(P_k f)^{(i)}\|_p\, \frac{d_f}{f(1) + d_f}$$

$$\leqq 2C_1 \|f^{(k)}\|_p + \|(P_k f)^{(i)}\|_p\, \frac{d_f}{f(1) + d_f}\,. \tag{2.5}$$

To estimate the last term on the right, we observe that all $L_p[0,1]$ norms are equivalent on π_k, so that by the Markov inequality, there is a positive constant C', depending only on k, such that

$$\|(P_k f)^{(i)}\|_p \leqq C'\|P_k f\| .$$

Choose $x_0 \in [a,b]$ such that $|P_k f(x_0)| = \|P_k f\|_\infty$. If $P_k f(x_0) = \|P_k f\|_\infty$, then by the Mean Value Theorem, there is an $a \in [0,1]$ such that

$$f(1) - \|P_k f\| = (P_k f)(1) - (P_k f)(x_0)$$

$$= (P_k f)'(a)(1 - x_0) \geqq -d_f(1 - x_0) \geqq -d_f$$

since $d_f \geqq 0$. Hence,

$$\|(P_k)^{(i)}\|_p \leqq C'(f(1) + d_f)$$

and from (2.5), we have

$$\|(f - g)^{(i)}\|_p \leqq 2C_1\|f^{(k)}\|_p + C'd_f .$$

By applying (2.3), we have (2.4) with $C_2 = 2C_1 + C'$. Similarly, if $(P_k f)(x_0) = -\|P_k f\|_\infty$, then again by the Mean Value Theorem, there is a constant $b \in [0,1]$ such that

$$\|P_k f\|_\infty = -(P_k f)(x_0) = (P_k f)(0) - (P_k f)(x_0)$$

$$= (P_k f)'(b)(0 - x_0) = -x_0(P_k f)'(b) \leqq d_f x_0 \leqq d_f ,$$

and from (2.5), it follows that

$$\|(f - g)^{(i)}\|_p \leqq 2C_1\|f^{(k)}\|_p + C'd_f^2/(f(1) + d_f)$$

$$\leqq 2C_1\|f^{(k)}\|_p + C'd_f .$$

Again, by applying (2.3), we have (2.4) with $C_2 = 2C_1 + C'$. This completes the proof of the lemma.

The next lemma is a generalization of the Mean Value Theorem for integrals.

LEMMA 2.3. Let $1 \leq p \leq \infty$, $g \in L^p[0,1]$ and $h \in C[0,1]$. Then there exists a $\xi_p \in [0,1]$ such that

$$|h(\xi_p)|\,\|g\|_p = \|hg\|_p .$$

This result follows trivially from the Intermediate Value Theorem since it can be shown that

$$\min_{0 \leq x \leq 1} |h(x)|\,\|g\|_p \leq \|hg\|_p \leq \max_{0 \leq x \leq 1} |h(x)|\,\|g\|_p .$$

We are now ready to establish Theorem 2.1. Let $f \in L_p^{k*}[0,1]$ and define $f_i(s) = f(t_{i-1} + s(t_i - t_{i-1}))$, $0 \leq s \leq 1$, $i = 1,\ldots,N$. For each i, by applying Lemma 2.2, there is a polynomial $q_i \in \pi_k^*$ such that $g_i(0) = f_k(0)$, $g_i(1) = f_i(1)$, and

$$\|f_i - g_i\|_p \leq C_2 \|f_i^{(k)}\|_p . \tag{2.6}$$

Let h be defined on $[0,1]$ by

$$h(x) = g_i\left[\frac{x - t_{i-1}}{t_i - t_{i-1}}\right], \quad t_{i-1} \leq x \leq t_i ,$$

$i = 1,\ldots,N$. Hence, $h \in \pi^*(k,\tau_N)$.

We first consider $p = \infty$. In this case, we have, by (2.5),

$$\begin{aligned} \|f - h\|_\infty &\leq \max_{1 \leq i \leq N} \|f_i - g_i\|_\infty \leq C_2 \max_{1 \leq i \leq N} \|f_i^{(k)}\|_\infty \\ &\leq C_2 \max_{1 \leq i \leq N} (t_i - t_{i-1})^k \|f^{(k)}\|_{L_\infty[t_{i-1},t_i]} . \end{aligned} \tag{2.7}$$

Hence, by the Mean Value Theorem, we obtain

$$\begin{aligned} N^k D_{N,\infty}^*(f,k) &\leq N^k \|f - h\|_\infty \\ &\leq C_2 \max_{1 \leq i \leq N} \left[\frac{t(\frac{i}{N}) - t(\frac{i-1}{N})}{1/N}\right] \|f^{(k)} \circ t\|_{L_\infty[\frac{i-1}{N},\frac{i}{N}]} \\ &\leq C_2 \max_{1 \leq i \leq N} |t'(\xi_i)|^k \|f^{(k)} \circ t\|_{L_\infty[\frac{i-1}{N},\frac{i}{N}]} \end{aligned}$$

for some $\xi_i \in [\frac{i-1}{N},\frac{i}{N}]$. But Lemma 2.3 gives

$$\|(t')^k f^{(k)} \circ t\|_{L_\infty[\frac{i-1}{N},\frac{i}{N}]} = |t'(\zeta_i)|^k \|f^{(k)} \circ t\|_{L_\infty[\frac{i-1}{N},\frac{i}{N}]}$$

where $\zeta_i \in [\frac{i-1}{N},\frac{i}{N}]$. Since $\omega(t'^k;\frac{1}{N}) \leqq \eta^k$ for all sufficiently large N, say for all $N \geqq N_0$, we have

$$\left| |t'(\xi_i)|^k \|f^{(k)} \circ t\|_{L_\infty[\frac{i-1}{N},\frac{1}{N}]} - \|(t')^k f^{(k)} \circ t\|_{L_\infty[\frac{i-1}{N},\frac{i}{N}]} \right|$$

$$= \left| (t'^k(\xi_i) - t'^k(\zeta_i)) \|f^{(k)} \circ t\|_{L_\infty[\frac{i-1}{N},\frac{i}{N}]} \right|$$

$$\leqq \omega(t'^k;\tfrac{1}{N}) \|f^{(k)} \circ t\|_{L_\infty[\frac{i-1}{N},\frac{i}{N}]}$$

$$\leqq \eta^k \|f^{(k)} \circ t\|_{L_\infty[\frac{i-1}{N},\frac{i}{N}]} \leqq |t'(\zeta_i)|^k \|f^{(k)} \circ t\|_{L_\infty[\frac{i-1}{N},\frac{i}{N}]}$$

$$= \|(t')^k f^{(k)} \circ t\|_{L_\infty[\frac{i-1}{N},\frac{i}{N}]}$$

for all $N \geqq N_0$. Here, we have used the fact that $\eta = \min t'$. Hence,

$$N^k D^*_{N,\infty}(f,k) \leqq 2C_2 \max_{1 \leqq i \leqq N} \|(t')^k f^{(k)} \circ t\|_{L_\infty[\frac{i-1}{N},\frac{i}{N}]} \leqq 2C_2 M_{\infty,k}(f)$$

for all $N \geqq N_0$, where N_0 only depends on t. This proves the theorem for $p = \infty$.

Now, let $1 \leqq p < \infty$. By (2.5), we have

$$\|f - h\|_p^p = \sum_{i=1}^{N} \int_{t_{i-1}}^{t_i} |f(x) - h(x)|^p dx$$

$$(2.8) \qquad = \sum_{i=1}^{N} \int_{t_{i-1}}^{t_i} \left| f_i\left[\frac{x - t_{i-1}}{t_i - t_{i-1}}\right] - g_i\left[\frac{x - t_{i-1}}{t_i - t_{i-1}}\right] \right|^p dx$$

$$= \sum_{i=1}^{N} (t_i - t_{i-1}) \int_0^1 |f_i(s) - g_i(s)|^p ds$$

$$\leqq C_2^p \sum_{i=1}^{N} (t_i - t_{i-1}) \| f_i^{(k)} \|_p^p$$

$$= C_2^p \sum_{i=1}^{N} (t_i - t_{i-1})^{kp+1} \int_0^1 | f^{(k)} (t_{i-1} + s(t_i - t_{i-1})) |^p ds$$

$$= C_2^p \sum_{i=1}^{N} (t_i - t_{i-1})^{kp} \int_{t_{i-1}}^{t_i} | f^{(k)} (x) |^p dx .$$

Hence, by the Mean Value Theorem, we obtain

$$[N^k D^*_{N,p}(f,k)]^p \leqq N^{kp} \| f - h \|_p^p$$

$$\leqq C_2^p \sum_{i=1}^{N} \left[\frac{t(\frac{i}{N}) - t(\frac{i-1}{N})}{1/N} \right]^{kp} \int_{t(\frac{i-1}{N})}^{t(\frac{i}{N})} | f^{(k)}(x)|^p dx$$

$$= C_2^p \sum_{i=1}^{N} (t'(\xi_i))^{kp} \int_{\frac{i-1}{N}}^{i/N} | f^{(k)} \circ t(s) |^p t'(s) ds$$

for some $\xi_i \in (\frac{i-1}{N}, \frac{i}{N})$. By Lemma 2.3, it also follows that

$$\| (t')^{k+\frac{1}{p}} f^{(k)} \circ t \|^p_{L_p[\frac{i-1}{N}, \frac{i}{N}]} = \int_{\frac{i-1}{N}}^{i/N} [t'(s)]^{kp+1} | f^{(k)} \circ t(s)|^p ds$$

$$= \int_{\frac{i-1}{N}}^{i/N} [(t'(s)]^{kp} | f^{(k)} \circ t(s) |^p t'(s) ds$$

$$= [t'(\zeta_i)]^{kp} \int_{\frac{i-1}{N}}^{i/N} | f^{(k)} \circ t(s) |^p t'(s) ds .$$

By the same argument as in the $p = \infty$ case, there exists an N_1, depending only on t, such that for all $N \geqq N_1$, we have

$$[N^k D^*_{N,p}(f,k)]^p \leq 2C_2^p \sum_{i=1}^{N} \|(t')^{k+\frac{1}{p}} f^{(k)} \circ t\|^p_{L_p[\frac{i-1}{N},\frac{i}{N}]}$$

$$= 2C_2^p \|(t')^{k+\frac{1}{p}} f^{(k)} \circ t\|_p^p = 2C_2^p M_{p,k}^p(f) .$$

This completes the proof of the theorem.

3. MONOTONE APPROXIMATION BY SPLINES

In this section, we will approximate by monotone splines. We first need some notation. Let $\underline{t}$: $\{t_i\}_{i=-\infty}^{k-1}$ and $\underline{s} = \{s_i\}_{i=0}^{\infty}$ where $|t_i - 1| \leq \varepsilon \leq 1/4$ for $i \leq k-1$, $|s_i + k - 1 - i| \leq \varepsilon \leq 1/4$ for $i \geq k$ and $\varepsilon > 0$ is to be chosen later. The reader should be aware that all knot sequences depend on ε in this section. We further assume that $s_{i+k-1} = t_i$ for $i = 1,\ldots,k-1$. For $-\infty < i \leq -1$, let $N_i = N_{i,k,\underline{t}}$ and for $0 \leq i < \infty$ let $N_i = N_{i,k,\underline{s}}$ be the normalized B-splines with knots at $\underline{t}$ and $\underline{s}$ as indicated by the third subscript (cf [1]). Also, let X be the spline subspace spanned by $\{N_i\}_{i=-\infty}^{\infty}$ and Y the subspace spanned by $\{N_i\}_{i=-\infty}^{-1} \cup \{N_i\}_{i=k-1}^{\infty}$. The knot sequence for Y will be denoted by $\underline{z} = \{z_j\}$ with $z_0 = 0$. We define a "smoothing operator" T mapping X to Y as follows:

$$T\left\{\sum_{i=-\infty}^{\infty} a_i N_i\right\} := \sum_{i=-\infty}^{-1} a_i N_i + \sum_{i=k+1}^{\infty} a_i N_i .$$

Set $E = I - T$ where I is the identity map. Note that for any $s \in X$, Es has finite support and hence is an element of $L_p^{k-1}[(-\infty,0) \cup (0,\infty)]$, $1 \leq p \leq \infty$. In addition, the following estimate is not difficult to prove (see [1,2] for the uniform case).

LEMMA 3.1. *There is a positive constant* C_3 *depending only on* k, *such that for any* $s \in X$,

$$(3.1) \quad \|(Es)^{(j)}\|_{L_p(R)} \leq C_3 \sum_{i=0}^{k-2} |s^{(i)}(0^+) - s^{(i)}(0^-)|$$

for $0 \leq j \leq k-1$ and $1 \leq p \leq \infty$.

Here, by $(Es)^{(j)}$ we mean the function defined pointwise by taking j derivatives. Since Y is the kernel of E, we may assume that $s^{(i)}(0^-) = 0$ for $i = 0,\ldots,k-1$ by simply adding a suitable polynomial of degree $\leqq k-1$ to s. Thus, if $s = \Sigma\alpha_i N_i$, then

$$\|(Es)^{(j)}\|_{L_p(R)} = \left\|\sum_{i=0}^{k-2} \alpha_i N_i^{(j)}\right\|_{L_p(R)}$$

$$\leqq \left[\sum_{i=0}^{k-2} |\alpha_i|\right] \max_{0\leqq i\leqq k-2} \|N_i^{(j)}\|_{L_p(R)}$$

$$\leqq C_3 \sum_{i=0}^{k-2} |s^{(i)}(0^+) - s^{(i)}(0^-)|$$

for all j, $0 \leqq j \leqq k-1$, and for some constant depending only on k. Here we have used the fact that $\|N_i^{(j)}\|_{L_p(R)}$ are uniformly bounded independent of the knot locations provided $\varepsilon < 1/4$. This completes the proof of Lemma 2.1.

The next lemma may be found in [4].

LEMMA 3.2. *Let k be a positive integer and $1 \leqq p \leqq \infty$. There exist positive numbers δ_1 and δ_2, depending only on k, such that if q is a polynomial of degree $< k$ with $q(0) = 0$ and $\|q\|_p \geqq 1$ there is an interval $I \subset [0,1]$ of length δ_1, such that*

$$\min\{|q'(x)|: \quad x \in I\} \geqq \delta_2 .$$

Now let C(R) be the space of all continuous functions on the real line R. As mentioned in Section 1, the asterisk indicates subcollections of nondecreasing functions. Hence, C*(R) will denote the collection of nondecreasing functions which are continuous on R and Y* will denote the nondecreasing spline functions in Y. We have the following result on monotone approximation of piecewise polynomials by monotone splines.

LEMMA 3.3. *There is a positive constant C_4, depending only on k, and on $\varepsilon_0 > 0$ so that for all positive $\varepsilon < \varepsilon_0$ and for every $f \in C^*(R)$ whose restrictions to $(-\infty,0)$ and $(0,\infty)$ are*

$(k-1)^{st}$ *degree polynomials, there is an* $s \in Y^*$ *such that* $s = f$ *on* $(-\infty, z_{-4k^2})$ *and* (z_{4k^2}, ∞), *and*

$$\text{(3.2)} \quad \|s - f\|_{L_p(R)} \leqq C_4 \sum_{i=1}^{k-2} |f^{(i)}(0^+) - f^{(i)}(0^-)| .$$

Before beginning the proof of this result, we state one last lemma whose proof will be deferred to the end of this section.

LEMMA 3.4. *Let* k *be a given positive integer and let* $f \in C^*(R)$ *be as above, considering* f *as an element of* X. *Then if* $f' = \Sigma \alpha_i N_{i,k-1}$ *the sequence* $\{\alpha_i\}_{i=1}^{4k^2}$ *can have at most* $(k-2)$ *sign changes. Similarly, the sequence* $\{\alpha\}_{i=-4k^2}^{-1}$ *can have at most* $(k-2)$ *sign changes.*

We now prove Lemma 3.3. Let $d_2 = z_{4k^2}$, $d_1 = z_{-4k^2}$ and let $F = \{f \in C^*(R): \; f|_{(-\infty,0)}, f|_{(0,\infty)} \in \pi_{k-1}$, $\sum_{i=1}^{k-1} |f^{(i)}(0^+) - f^{(i)}(0^-)| \leqq 1\}$. It is sufficient to prove, for $\varepsilon_0 > 0$ small enough and every $f \in F$, there is an $s \in Y^*$ so that $s = f$ on $(-\infty, d_1) \cup (d_2, \infty)$ satisfying

$$\|s - f\|_{L_p(R)} \leqq C'$$

where C' is an absolute constant depending only on k.

We divide the proof into two cases: (i) $\|f\|_{L_p[d_1,d_2]} \leqq \alpha$ and (ii) $\|f\|_{L_p[d_1,d_2]} > \alpha$ where $\alpha > 0$ is to be determined later.

Case (i): Suppose $f \in F$ and $\|f\|_{L_p[d_1,d_2]} \leqq \alpha$. Then $f \in X$ and hence $f = \sum_{-\infty}^{\infty} \alpha_i N_i$ and $g := (Tf)' = \Sigma \beta_i N_{i,k-1,\underline{z}}$ where we recall that $\{z_i\}$ is a knot sequence containing all the knots for elements of Y with $z_0 = 0$. Since $Tf = f$ except on $[z_0, z_{k-1}]$, we conclude, by Lemma 3.4., that $(\beta_1, \ldots, \beta_{4k^2})$ can have at most (k-2) sign changes (unless $f' \equiv 0$ on the set which can be easily handled). Similarly, $(\beta_{-4k^2},\ldots,\beta_{-1})$ has the same property. Since $f' \geqq 0$, we see that for every segment $(\beta_{\ell+1},\ldots,\beta_{\ell+k-1})$ for $\ell > 0$ or $\ell \leqq -k$ there is an i with $\ell+1 \leqq i \leqq \ell+k-1$ so that $\beta_i \geqq 0$. Thus, there must be an ℓ^* satisfying $0 \leqq \ell^* \leqq 4k^2 - 2(k-1)$ so that all the coefficients $(\beta_{\ell^*+1},\ldots,\beta_{\ell^*+2k-2})$ are nonnegative. The proof is by contradiction since otherwise we could force more than (k-2) weak sign changes in $(\beta_1,\ldots,\beta_{4k^2})$. Similarly, there is a sequence $(\beta_{\ell_*},\ldots,\beta_{\ell_*+2k-3})$ of nonnegative coefficients with $-4k^2 \leqq \ell_* \leqq -2(k-1)$. Now let

$$h = \sum_{i=\ell_*+k-1}^{\ell^*+k-1} \beta_i N_{i,k-1,\underline{z}} .$$

Then $0 \leqq h \leqq g$ on $(-\infty, z_{\ell^*+2k-2}) \cup (z_{\ell^*+k},\infty)$ and $h = g = (Tf)'$ on $(z_{\ell_*+2k-2}, z_{\ell^*+k-1})$. Since f is nondecreasing and $Tf = f$ except on $[z_0, z_{k-1}]$, we have

$$\int_{-\infty}^{\infty} h \geqq \int_{z_{\ell_*+2(k-1)}}^{z_{\ell^*+k-1}} (Tf)' \geqq 0 .$$

Since $\int_{-\infty}^{\infty} N_{0,k-1,\underline{z}}$ is uniformly bounded above and below by a positive number due to the assumption on $\underline{z}$, we set

$$\gamma = \left[\int_{-\infty}^{\infty} N_{0,k-1,\underline{z}}\right]^{-1} \int_{-\infty}^{\infty} h(x)\,dx$$

and finally define $s \in Y$ to be

$$s(x) = f(d_1) + \int_{d_1}^{x} (g - h + \gamma N_{0,k-1,\underline{z}}) .$$

Clearly, $s \in Y^*$ and $s = f$ on $(-\infty, d_1) \cup (d_2,\infty)$ and by applying Lemma 3.1 we conclude that $\|h - N_{0,k-1,\underline{z}}\|_{L_p[d_1,d_2]}$ and $\|s - f\|_{L_p[d_1,d_2]}$ are bounded by some constant which depends only on k and α.

Case (ii): Suppose now that $f \in F$ and $\|f\|_{L_p[d_1,d_2]} > \alpha$. Without loss of generality, we may assume that $f(0) = 0$. It is intuitively clear that f' must be large on substantial sets if α is very large. This allows us to modify Tf in order to guarantee that it is nondecreasing and satisfies (3.2). Indeed, by using Lemma 3.2, we can choose α so large that if $\|f\|_{L_p[d_1,d_2]} \geqq \alpha$ then there is an integer ℓ such that $[z_\ell, z_{\ell+k-1}] \subset [d_1,d_2]$ and $f'(x) \geqq \beta(\int N_{\ell,k-1,\underline{z}})^{-1} := \gamma_2$ on $[t_\ell, t_{\ell+k-1}]$ where

$$\beta = C_3 \int_{-\infty}^{\infty} \sum_{i=2-k}^{k-2} N_{i,k-1,\underline{z}} \, .$$

Here, C_3 is the positive constant in Lemma 3.1. We note that α is chosen independent of f in F. Now set

$$\begin{aligned} s(x) &= (Tf)(x) + \int_{-\infty}^{x} [C_3 \sum_{i=2-k}^{k-2} N_{k,k-1,\underline{z}} - \gamma_2 N_{\ell,k-1,\underline{z}}]: \\ &= (Tf)(x) + M(x) \, . \end{aligned}$$

Then $s \in Y^*$ and

$$\|f - s\|_{L_p[d_1,d_2]} \leqq \|s - Tf\|_{L_p[d_1,d_2]} + \|M\|_{L_p[d_1,d_2]}$$

and both terms on the right are uniformly bounded for all $f \in F$ with $\|f\|_{L_p[d_1,d_2]} \geqq \alpha$. Thus, we have completed the proof of Lemma 3.3.

Let us now sketch a proof of Lemma 3.4. We are indebted to Carl de Boor for this proof which significantly simplifies our original presentation. Suppose $f = \sum_i a_i N_{i,k}$ and that $f^{(j)} = \sum_i a_i^j N_{i,k-j}$ where $a_i^0 := a_i$ and the a_i^j are scaled differences of the a_i^{j-1} (cf [1,2]). Thus, if we denote by $S^-\{a_i^j\}$ the number of sign changes in the sequence $\{a_i^j\}_{i=-\infty}$, then a discrete Rolle's theorem yields

$$S^-\{a_i^j\} + 1 \geqq S^-\{a_i^{j-1}\} \, .$$

Now if $f \in \pi_k$ then $f^{(k-1)}$ is a constant and hence $S^-\{a_i^{k-1}\} = 0$. It follows in this case that

$$(k-1) \geqq S^-\{a_i^0\} .$$

This last statement completes the proof of Lemma 3.4.

We need another lemma to relate the error estimates in (3.2) to $\|f^{(k)}\|_p$. As an application of Lemma 2.2, we note that for every $f \in L_p^{k*}[-1,1]$, there is a nondecreasing continuous function g on $[-1,1]$ such that g interpolates f at -1 and 1, the restrictions of g on $[-1,0]$ and $[0,1]$ are polynomials of degree $\leqq k$ and such that

$$(3.3) \quad \|(f-g)^{(i)}\|_{L_p[0,1]} \leqq C\|f^{(k)}\|_{L_p[-1,0]}$$

and

$$(3.4) \quad \|(f-g)^{(i)}\|_{L_p[0,1]} \leqq C\|f^{(k)}\|_{L_p[0,1]} .$$

Let δ^i be the continuous linear functional defined on $L_p^k[0,1]$ and $L_p^k[0,1]$ by $\delta^i f = f^{(i)}(0)$, $0 \leqq i \leqq k-1$. Here we can use the norms

$$\|f\|_{L_p^k[-1,0]} = \sum_{i=0}^{k} \|f^{(i)}\|_{L_p[-1,0]}$$

and

$$\|f\|_{L_p^k[0,1]} = \sum_{i=0}^{k-1} \|f^{(i)}\|_{L_p[0,1]}$$

on $L_p^k[-1,0]$ and $L_p^k[0,1]$, respectively. The following lemma is proved in a fashion similar to Lemma 4.4 in [4].

LEMMA 3.5. *Let $1 \leqq p \leqq \infty$ and $k \geqq 2$. There is a positive constant C_5 depending only on k, such that for every $f \in L_p^{k*}[-1,1]$ and for every $g \in C^*[-1,1]$ whose restrictions on $[-1,0]$ and $[0,1]$ are polynomials of degree $< k$, and which satisfies (3.3) and (3.4), the inequality*

$$\sum_{i=1}^{k-1} |g^{(i)}(0^+) - g^{(i)}(0^-)| \leqq C_5 \| f^{(k)} \|_{L_p[-1,1]}$$

<u>is satisfied for all</u> $1 \leqq p \leqq \infty$.

An obvious modification of this lemma by superposition and simple affine transformation extends the result to splines whose knots are generated by a distribution function t satisfying (1.1).

We are now ready to prove the main result of this paper, namely Theorem 1.1. Let $f \in L_p^{k*}[0,1]$ and suppose that $N > 3d$ where $d = 4k^2$ is defined in the proof of Lemma 3.3. For each N let us choose a coarser mesh (u_ℓ) from the mesh $\{t(i/N)\}_{i=0}^N$ as follows: Set

$$\{u_\ell = t_{3d\ell}\}, \quad \ell = 0,\ldots,M+1$$

where M is the integer part of $(N/3d - 1)$ and $u_{M+1} := 1$. Let us choose $g_N \in C^*[0,1]$ to be a piecewise polynomial with break points at the $\{u_\ell\}_{i=1}^M$ only and satisfying

$$(3.5) \quad \| f - g_N \|_p \leqq CN^{-k} M_{p,k}(f)$$

as in (2.1). We remark that the constant in (3.5) can be chosen to be a fixed multiple of the constant in (2.1).

We now wish to use the ideas of this section to smooth g_N to a spline s while preserving the monotonicity. Let

$$(3.6) \quad \Delta_\ell = t'(\tfrac{3d\ell}{N})/N, \quad \ell = 1,\ldots,M$$

and set for each $\ell = 1,\ldots,M$

$$(3.7) \quad \hat{g}_{\ell,N}(\tau) := g_N(u_\ell + \tau\Delta_\ell) .$$

Now it is easy to see that $(t_{i+1} - t_i)/\Delta_\ell = (t(\frac{i+1}{N}) - t(\frac{i}{N}))/\Delta_\ell$ is nearly 1 for i near $3d\ell$ and large N. Thus, the mapping $t \to (t - u_\ell)/\Delta_\ell$ maps the knots near t_{3d} onto (nearly) the integers and maps $[u_{\ell-1}, u_{\ell+1}]$ onto (nearly) $[-3d,3d]$. Thus, we may apply Lemma 3.3 (for large N) to obtain an $\hat{s}_\ell$ on $[-3d,3d]$ with knots

$$\tau_j = (t_j - u_\ell)/\Delta_\ell, \quad j \text{ near } 3d\ell ,$$

satisfying (3.2). Now $\hat{s}_\ell = \hat{g}_{\ell,N}$ on

$$(3.8)\quad [(u_{\ell-1} - u_\ell)/\Delta_\ell,\ (u_{\ell+1} - u_\ell)/\Delta_\ell]/[\alpha_\ell,\beta_\ell]$$

where $\alpha_\ell = (t_{3d\ell-d} - u_\ell)/\Delta_\ell$ and $\beta_\ell = (t_{3d\ell+d} - u_\ell)/\Delta_\ell$. That is, on the outer 2/3 of the interval [-3d,3d], $\hat{s}_\ell$ is identically $\hat{g}_{\ell,N}$. Thus, we may unambiguously define

$$(3.9)\quad s(t) = \hat{s}_\ell([t - u_\ell]/\Delta_\ell) \text{ if } t \in \left[\frac{u_\ell + u_{\ell-1}}{2}, \frac{u_\ell + u_{\ell+1}}{2}\right], \qquad \ell = 1,\dots,M$$

and for $t \in [0,\frac{1}{2}u_1] \cup [(1 - u_M)/2]$ set $s(t) = g_N(t)$. Thus, we see that s is an increasing spline of order k with knot sequence $\{t(i/N)\}$.

From Lemma 3.3, we know that

$$\|\hat{s} - \hat{g}_{\ell,N}\|_{L_p[-3d,3d]} \leq C_4 \sum_{i=1}^{k-1} |\hat{g}_{\ell,N}^{(i)}(0^+) - \hat{g}_{\ell,N}^{(i)}(0^-)| .$$

Further, we see that $\|f - s\|_p \leq \|f - g_N\|_p + \|g_N - s\|_p$ so that we only need to estimate the second term on the right as the first term satisfies the correct estimate by (3.5). Let us first argue the case $p = \infty$. Using Lemmas 3.3 and 3.5, we see that

$$\begin{aligned}\|s - g_N\|_\infty &= \max_{1\leq\ell\leq M} \|s - g_{\ell,N}\|_{L_\infty[-3d,3d]}\\ &\leq C_4 \max_{1\leq\ell\leq M} \sum_{i=1}^{k-1} |\hat{g}_{\ell,N}^{(i)}(0^+) - \hat{g}_{\ell,N}^{(i)}(0^-)|\\ &\leq C' \max_\ell \|(\tfrac{d}{d\tau})^k f(u_\ell + \tau\Delta\ell)\|_{L_\infty[-3d,3d]}\\ &\leq C'\Delta_{\ell^*}^k \|f^{(k)}\|_{L_\infty[u_{\ell^*-1},u_{\ell^*+1}]}\end{aligned}$$

for some

$$\ell^* = C'N^{-k}[t'(\tfrac{3d\ell^*}{N})]^k \|f^{(k)}\|_{L_\infty[u_{\ell^*-1},u_{\ell^*+1}]} .$$

Notice for the second inequality we have assumed that the $\hat{g}_{\ell,N}$ satisfy (2.4) as well. Now one can argue as in section 2 to complete the proof.

For $1 < p < \infty$ we compute

$$\|s - g_N\|_p^p \leqq \sum_{j=1}^{M} \|s - g_N\|_{L_p[u_{j-1},u_{j+1}]}^p$$

$$= \sum_{j=1}^{M} \Delta_j \|\hat{s}_j - \hat{g}_{j,N}\|_{L_p[-3d,3d]}^p$$

$$\leqq C_4^p \sum_{j=1}^{M} \Delta_j \left| \sum_{i=1}^{k-1} |\hat{g}_{j,N}^{(i)}(0^+) - \hat{g}_{j,N}^{(i)}(0^-)| \right|^p$$

$$\leqq C_4^p C_5^p \sum_{j=1}^{M} \Delta_j \left\| \left(\frac{d}{d\tau}\right)^k f(u_j + \tau\Delta_j) \right\|_{L_p[-3d,3d]}^p$$

$$= C_4^p C_5^p \sum_{j=1}^{M} \Delta_j^{kp} \|f^{(k)}\|_{L_p[u_{j-1},u_{j+1}]}^p$$

$$\leqq C_4^p C_5^p 2^{p-1} \sum_{j=1}^{M} N^{-kp} [t'(\tfrac{3dj}{N})]^{kp} \int_{t(\frac{3d(j-1)}{N})}^{t(\frac{3d(j+1)}{N})} |f^{(k)}|^p .$$

The argument may now be completed as in the proof of Theorem 2.1 (see 2.8 and following).

REFERENCES

1. de Boor, C., On calculating with B-splines, J. Approximation Theory, 6 (1972), 50-62.

2. __________, A Practical Guide to Splines, Springer-Verlag Applied Mathematical Sciences, Springer-Verlag, NY, Vol. 27, 1978.

3. __________ and G. J. Fix, Spline approximation by quasi-interpolants, J. Approximation Theory, 8 (1973), 19-45.

4. Chui, C. K, P. W. Smith and J. D. Ward, Degree of L_p approximation by monotone splines, SIAM J. Mat. Anal., to appear.

5. DeVore, R., Degree of Approximation, Approximation Theory II, ed. by G. G. Lorentz, C. K. Chui and L. L. Schumaker, Academic Press, NY, 1976, 117-162.

6. __________, Monotone approximation by splines, SIAM J. Math. Anal., 8 (1977), 891-905.

7. __________, Monotone approximation by polynomials, SIAM J. Math. Anal., 8 (1977), 891-905.

8. Karlin, S., Total Positivity, Stanford University Press, 1968.

9. Lorentz, G. G., Monotone approximation, Inequalities III, ed. O. Shisha, Academic Press, NY, 1972, 201-215.

10. Lorentz, G. G. and K. Zeller, Degree of approximation by monotone polynomials I, J. Approximation Theory, 1 (1968), 501-504.

11. __________, Degree of approximation by monotone polynomials II, J. Approximation Theory, 2 (1969), 265-269.

Charles K. Chui*
Department of Mathematics
Texas A&M University
College Station, TX 77843

Joseph D. Ward*
Department of Mathematics
Texas A&M University
College Station, TX 77843

Philip W. Smith*
Department of Mathematics
Texas A&M University
College Station, TX 77843

* Supported by the U.S. Army Research Office under Grant Number DAAG 29-78-0097.

APPROXIMATION BY SMOOTH MULTIVARIATE SPLINES ON NON-UNIFORM GRIDS

Wolfgang Dahmen

In this note, we construct smooth multivariate splines with respect to certain non-uniform grid configurations and establish local error estimates.

1. INTRODUCTION

The objective of this note is centered around the following facts concerning the approximation by smooth piecewise polynomials of several variables. First, constructing for arbitrary spatial dimension and any degree spline spaces of possibly high global smoothness while preserving the "local nature" of these spline spaces is still a delicate problem. Here, "local nature" means, for instance, that the structure is compatible with local refinements. Furthermore, (as a consequence) error bounds for the approximation by smooth multivariate splines involve typically only a globally maximal mesh size parameter. This does, of course, not provide satisfactory information when dealing with highly varying mesh sizes which is, e.g., a reasonable method to adapt singularities. In this context, the multivariate B-spline (cf [2]) turns out to play an interesting role. In fact, besides further advantageous properties, namely it is a non-negative piecewise polynomial determined only by "knots" and may be stably evaluated by practical recurrence relations, it combines highest possible smoothness with a minimal support (cf [5], [7], [10], [11]). Moreover, its geometric interpretation (cf [2]) gives rise to the construction of certain linear spans of such B-splines which provide optimal "global" approximation rates [6],[8]. Our objective here is to construct for any spatial dimension and arbitrary degree smooth spline spaces of the above type arising from locally refining

ISBN 0-12-213650-0

certain basic "uniform" spline spaces which are introduced and discussed in [8] (cf [7]). Due to the particular structure of the uniform spaces, it turns out that these refinements do not interfere at all even with (for non-trivial splines) highest possible global smoothness. Furthermore, we establish local error estimates for the "refined" configurations which will be used in a forthcoming paper to construct smooth adaptive approximation schemes being essentially equivalent to piecewise polynomial adaptive approximation (cf [3]).

2. LOCAL REFINEMENTS OF UNIFORM CONFIGURATIONS OF KNOT SETS

This section deals with the construction of certain smooth spline spaces exploiting and extending the approach in [8]. To this end, we shall frequently use the following notation. [V] denotes the convex hull of a given set V and if $V \subset \mathbb{R}^n$, $n \geqq s$, $(V)_s$ means the projection of V to $\mathbb{R}^s$. The three integers s, k, n will be consistently interrelated by $k = n - s \geqq 0$ where s and k always refer to the spatial dimension and to the (total) degree of splines or polynomials, respectively.

A crucial role in our construction is played by the multivariate B-spline (cf [2], [5], [10]). As in the univariate case, the multivariate B-spline $M(x|P)$ is determined (up to normalization) for any s by a "knot set", $P = \{x^0,\ldots,x^n\} \subset \mathbb{R}^s$, which satisfies

$$\text{(2.1)} \quad \dim \operatorname{span}([P]) = s .$$

As for a detailed discussion of the properties and various representations of the B-spline, we refer to [6],[8] [10],[11]. Here we emphasize that $M(x|P)$ is a piecewise polynomial of total degree $k = n - s$ which even belongs to $C^{k-1}(\mathbb{R}^s)$ (the space of functions having continuous partial derivatives of order k-1 on $\mathbb{R}^s$) whenever the knots in P are in "general position", i.e., every s+1 points in P are affinely independent (cf [5], [10]). Furthermore, we shall make essential use of a certain geometric interpretation of the

B-spline (cf [2]) which actually may serve as its definition. To describe this, let $\sigma = [\{v^o,\ldots,v^n\}]$ be a nondegenerate n-simplex with vertices v^i. The set

$$P(\sigma) = \{v^0\big|_{R^s},\ldots,v^n\big|_{R^s}\}$$

will always denote the set of projected vertices of the simplex σ. Note that $P(\sigma)$ automatically satisfies (2.1). Defining now for $x \in R^s$,

$$(2.2)\quad M_\sigma(x) = \mathrm{vol}_k(\{u \in \sigma:\ u\big|_{R^s} = x\})\ ,$$

the following relation holds (cf [6], [10])

$$(2.3)\quad M_\sigma(x) = \mathrm{vol}_n M(x|P(\sigma))\ .$$

Clearly, (2.2) and (2.3) imply that $M(x|P)$ is a non-negative function supported on $[P]$. This fact and its high global smoothness make the B-spline attractive for our purposes.

Let us briefly review now the construction of certain "uniform" spline spaces introduced in [8]. As a first ingredient we need the so called "Kuhn's triangulation", $\mathcal{K}_n(Q)$ of the unit n-cube $Q = [0,1]^n$. Denoting by $e^i = (\delta_{ij})_{j=1}^n$ the unit coordinate vectors and by Per_n the group of all permutations of the integers $\{1,\ldots,n\}$, we set for $\pi \in \mathrm{Per}_n$

$$(2.4)\quad v_\pi^0 = 0,\ v_\pi^j = v_\pi^{j-1} + e^{\pi(j)},\ j = 1,\ldots,n$$

and

$$\sigma_\pi = [\{v_\pi^0,\ldots,v_\pi^n\}]\ .$$

Then (cf [1], [9])

$$(2.5)\quad \mathcal{K}_n(Q) = \{\sigma_\pi:\ \pi \in \mathrm{Per}_n\}$$

is the n-dimensional Kuhn's triangulation of Q. For every affine image $q = A(Q)$, we define

$$(2.6)\quad \mathcal{K}_n(q) = \{A(\sigma_\pi):\ \pi \in \mathrm{Per}_n\}\ .$$

Let Q_{ij} be the (n-1)-dimensional rectangle which is spanned by $e^i + e^j$ and e^r, $r = 1,\ldots,n$, $r \neq i,j$ and set for $q = A(Q)$, $q_{ij} = A(Q_{ij})$. The following lemma may easily be derived from (2.4) (cf [1], [9]).

LEMMA 2.1. *For any parallelepiped* $q \subset R^n$, *let* $K_n(q)$ *be defined by* (2.6). *The restriction of* $K_n(q)$ *to any lower dimensional face of* q *or to any "diagonal cube"* q_{ij} *is again a Kuhn's triangulation.*

The Kuhn's triangulation allows us now to associate with any set V of parallelepipeds in R^n the following configuration of knot sets.

(2.7) $P(V) = \{P(\sigma): \quad \sigma \in K_n(q),\ q \in V\}$.

This in turn gives rise to the following space of splines of degree $k = n - s$,

(2.8) $S_k(V,D) = \text{span}\{M(x|P): \quad P \in P(V),\ [P] \cap D \neq \emptyset\}$

where D is a given domain in R^s.

In particular, we shall deal with special collections V of parallelepipeds whose construction involves an $(n \times n)$ matrix of the following type

$$H = \begin{pmatrix} (\delta_{ij})_{i,j=1}^{s} & (c_{ij})_{i=1,j=1}^{s,k} \\ 0 & (\delta_{ij})_{i,j=s+1}^{n} \end{pmatrix}.$$

Parallelepipeds of the form

$$q = aH(Q) + u$$

where $a \in R$, $u \in R^n$, will be briefly called H-cubes. Whenever we shall write $u + x$ for $u \in R^n$ and $x \in R^s$, x will be understood to be raised to an n-vector by setting $x_i = 0$, $i = s+1,\ldots,n$.

It is now readily seen that the collection

(2.10) $H = \{H(Q) + \nu: \quad \nu \in Z^s\}$

of H-cubes forms a partition of $R^s \times [0,1]^k$. This fact is used in [7], [8] to show that the corresponding spline space $S_k(H,D)$ (cf (2.8)) satisfies, for any domain $D \subset R^s$, the inclusion

(2.11) $\Pi_k(D) \subset S_k(H,D)$

where $\Pi_k(D) = \{ \sum_{|\alpha| \leq k} c_\alpha x^\alpha : c_\alpha \in R, x \in D\}$ denotes the space of polynomials of (total) degree less than or equal to k. Here we have used for $\alpha \in Z_+^s$ the familiar multiindex notation i.e., $|\alpha| = \alpha_1 + \ldots + \alpha_s$ and $x^\alpha = x_1^{\alpha_1} \cdot \ldots \cdot x_s^{\alpha_s}$. Note that (2.11) is already sufficient to confirm that $S_k(H,D)$ is "appropriate" for approximation on D (cf [8]). Furthermore, it is also shown in [8] that one even has, in general

$$(2.12) \quad S_k(H,D) \subset C^{k-1}(R^s) .$$

More precisely, the global smoothness of the splines in $S_k(H,D)$ is completely controlled by the matrix H. In fact, one can show that (2.12) holds if and only if H is a so called "dispersive" matrix (cf [8, sec. 4]) which means that any s sums over disjoint sets of the restrictions of columns of H to R^s are linearly independent.

This condition does not interfere with the following restriction which we will always assume to be valid.

$$(2.13) \quad \sum_{j=1}^{k} \max(0, * c_{ij}) < 1/2, \quad i = 1,\ldots,n$$

where "*" is either "+" or "-". This means, roughly speaking, that H does not "tilt" the n-cube $[0,1]^n$ "too much". Indeed, it is easy to see that (2.13) implies the existence of a cube $\tilde{C}$ in R^s such that for $\Omega = [0,1]^s$

$$(2.14) \quad \Omega \subset (H(Q))_s \subset \tilde{C}$$

where $\operatorname{diam}(\tilde{C}) \leq 2\operatorname{diam}(\Omega) = 2\sqrt{s}$. Moreover, setting for any parallelepiped $q \subset R^n$, $\mu(q) = \max\{\operatorname{vol}_k(\{u \in q: u|_{R^s} = x\}): x \in (q)_s\}$ and

$$(2.15) \quad I(q) = \{x \in (q)_s : \operatorname{vol}_k(\{u \in q: u|_{R^s} = x\}) = \mu(q)\},$$

the condition (2.13) assures that

$$(2.16) \quad I(H(Q)) \neq \emptyset .$$

Later we shall refer also to the following consequence of (2.13). Let $q_i \in H$, $i = 1,2$, then

$$(2.17) \quad (q_1)_s \cap (q_2)_s \neq \emptyset \text{ iff } q_1 \text{ and } q_2 \text{ are "neighbors",}$$

where two parallelepipeds are called neighbors if and only if their intersection is a lower dimensional face of at least one of them.

Our motivation for dealing with the spaces $S_k(H,D)$ is the remarkable fact that they can be "refined" locally without interfering with the global smoothness properties at all. In fact, it is pointed out in [8] that such refinements can be obtained simply by subdividing the "higher dimensional" H-cubes in H. Denoting for a given H-cube q and a fixed $m \in N$ by $d_m(q)$, the "elementary (m-type) subdivision" of q into m^n congruent parallelepipeds, we call a partition $\rho(q)$ an "m-type partition" if it is obtained by a finite number of subsequent elementary m-type subdivisions. Since we are mainly interested in approximation with respect to a bounded domain in R^s, we focus for simplicity upon

$$H_0 = \{q \in H: \quad (q)_s \cap \Omega \neq \emptyset\}$$

where $\Omega = [0,1]^s$. By assigning now to every $q \in H_0$ a partition $\rho(q)$ of a fixed m-type, we obtain a "refined" partition

$$(2.18) \qquad R = R(H) = \cup\{\rho(q): \quad q \in H_0\}$$

for $\Omega \times [0,1]^k$. $R(H)$ is then said to be an "m-type refinement" of H_0, writing briefly, $H_0 < R(H)$, i.e., every element in H_0 is the union of parallelepipeds in $R(H)$. The next lemma tells us that an analogous statement holds also for the triangulations which are induced by H_0 and $R(H)$.

LEMMA 2.2. <u>Setting for any m-type partition</u> ρ <u>of an H-cube</u> q, $K_n(q,\rho) = \cup\{K_n(q'): \quad q' \in \rho\}$, <u>then</u> $K_n(q,\rho)$ <u>is a "refinement" of</u> $K_n(q)$, i.e., <u>every element of</u> $K_n(q)$ <u>is the union of elements of</u> $K_n(q,\rho)$.

<u>Proof</u>. Since by definition any m-type partition ρ can be obtained by successive elementary m-type subdivisions, it is sufficient to prove the assertion for $\rho = d_m(q)$. Observing that in view of Lemma 2.1 every (n-1)-face of a simplex in $K_n(q)$ is an element of $K_{n-1}(q^*)$ where q^* is either an (n-1)-face of q or one of its diagonal cubes q_{ij}, the assertion follows now for $\rho = d_m(q)$ easily by induction on n.

Let us collect some immediate consequences in

PROPOSITION 2.1. (a) The following inclusions hold for any $H_0 < R(H) < R'(H)$ and any domain $D \subseteq \Omega$.

$$(2.19)\quad S_k(H_0,D) \subseteq S_k(R(H),D) \subseteq S_k(R'(H),D)$$

In particular, one still has

$$\Pi_k(\Omega) \subseteq S_k(R(H),\Omega)\ .$$

(b) For any m-type refinement $R(H)$, (2.18)

$$S_k(R(H),\Omega) \subseteq C^{k-1}(R^s)$$

holds if and only if the matrix H is dispersive (cf (2.12), [8, sec. 4]).

Proof. (a) follows immediately from (2.2), (2.3), (2.8) and Lemma 2.2.

As for (b), note that the construction (2.18) of $R(H)$ clearly affirms that the knot sets in $P(R(H))$ differ from those in $P(H_0)$ only by dilation. In particular, the knots in every $P \in P(R(H))$ are in general position if and only if this is true for the knot sets belonging to the uniform configurations $P(H_0)$. Hence, the B-splines generating $S_k(R(H),\Omega)$ have always exactly the same smoothness properties as those spanning $S_k(H_0,\Omega)$, i.e., in particular,

$$(2.20)\quad S_k(R(H),\Omega) \subset C^{k-1}(R^s) \text{ iff } S_k(H_0,\Omega) \subseteq C^{k-1}(R^s)$$

which in turn is equivalent to H being dispersive (cf (2.12), [8, sec. 4]).

It will be useful to introduce the following terminology (cf [3]). An H-cube q_0 is called an "ancestor" of an H-cube q if $q \in \rho(q_0)$ for some (m-type) partition ρ of q_0. We denote then by $\hat{R}$ the set of all ancestors of the elements of R. In particular, we write $\hat{q}$ for the "parent" of q, i.e., $q \in d_m(\hat{q})$.

Furthermore, let us consider certain subsets of $\hat{R}$ called "layers". For any $q \in R$ the layer L_q is the set of all H-cubes in $\hat{R}$ whose last k components of corresponding vertices coincide with those of q. Hence all elements of a layer L_q have the same "size" as q and L_q induces via (2.7) locally (up to dilation) the same uniform configuration of knot sets as H_0.

Let us mention now a simple observation for the case $s = k = 1$, i.e., $\Omega = [0,1]$, $Q = [0,1]^2$. Assume that $u,v \in \mathbb{Z}$ and consider

$$H = \begin{pmatrix} 1 & u/v \\ 0 & 1 \end{pmatrix}.$$

Let for some $\ell \in \mathbb{N}$, $\ell \geqq 2$, L_1, L_2 be two layers in $d_{\ell v}(H(Q))$ which are separated from each other by v-1 other layers. It is readily seen that L_1 and L_2 are shifted by H with respect to each other by u length units 1/v. Hence their projections coincide on I(H(Q)), i.e., (cf (2.7), (2.15))

$$(2.21)\qquad P(L_1)\Big|_{I(H(Q))} = P(L_2)\Big|_{I(H(Q))} .$$

In order to exploit an analogous effect also in the general case, $s \geqq 1$, $k \geqq 0$. We assume now that H has only rational entries $c_{ij} = u_{ij}/v_{ij}$, $u_{ij}, v_{ij} \in \mathbb{Z}$ and let m(H) be the absolute value of the smallest common multiple of the v_{ij}. We formally set $m(H) = \infty$ if H has also irrational entries.

LEMMA 2.3. <u>Let $m = m(H)$ be finite and $R(H)$ be any m-type refinement of H_0 (cf (2.18)). Furthermore, assume that for some $q_0 \in \hat{R}(H)$, $d_m(q_0) \subset R(H)$ and that the elements of $d_m(q_0)$ have "locally" minimal size in $R(H)$, i.e., those $q' \in R(H)$ for which $(q')_s \cap I(q_0) \neq \emptyset$ have larger or equal size as $q \in d_m(q_0)$. Then the following local equality holds</u>

$$S_k(d_m(q_0), I(q_0))\Big|_{I(q_0)} = S_k(R(H), I(q_0))\Big|_{I(q_0)}$$

<u>where</u> $S_k(V,D)\big|_D = \{S\big|_D\colon\ S \in S_k(V,D)\}$.

<u>Proof</u>. Since the elements of $d_m(q_0)$ have locally minimal size, the structure of an m-type refinement implies that all those H-cubes $q' \in R = R(H)$ which "overlap" q_0, i.e., which satisfy $(q')_s \cap I(q_0) \neq \emptyset$ can be partitioned into H-cubes

having the same size as the H-cubes in $d_m(q_0)$. Let us denote the corresponding refinement of R by R'. From (2.19), we conclude that

$$(2.22)\quad S_k(R, I(q_0))\big|_{I(q_0)} \subset S_k(R', I(q_0))\big|_{I(q_0)} .$$

Observing the definition of $m = m(H)$ and applying the same reasoning which lead us to (2.21) to each of the last k coordinate directions in R^{s+k}, we obtain as in (2.21),

$$(2.23)\quad P(d_m(q_0))\big|_{I(q_0)} = P(R')\big|_{I(q_0)} .$$

This yields, on account of the definition (2.8),

$$(2.24)\quad S_k(R', I(q_0))\big|_{I(q_0)} = S_k(d_m(q_0), I(q_0))\big|_{I(q_0)} .$$

The assertion follows now in view of the trivial inclusion

$$S_k(d_m(q_o), I(q_0))\big|_{I(q_0)} \subset S_k(R, I(q_0))\big|_{I(q_0)}$$

by (2.22) and (2.24).

Since one has by (2.5), (2.8) and (2.18)

$$(2.25)\quad \dim S_k(R, \Omega) \leqq n!\,|R|$$

where $|\cdot|$ denotes the cardinality of a finite set, one would like to subdivide the H-cubes in possibly few layers of a given refinement in order to improve locally the approximation behaviour. This suggests considering in the following a certain subclass of refinements R. To this end, we call any H-cube q having the form

$$q = aH(Q) + x$$

where $a > 0$, $x \in R^s$, a "bottom cube". So, the bottom cubes in $R(H)$ are exactly those cubes which share an s-face, the "bottom face", with R^s. A refinement $R(H)$ is then called "economic" if and only if $\hat{R}(H) \backslash R(H)$ contains only bottom cubes, i.e., when constructing $R(H)$ from successive elementary subdivisions only bottom cubes are subdivided at each stage.

Finally, we shall also deal with the following restriction upon an m-type refinement R which forces R to be sufficiently gradual. To this end, let us call for any collection V of H-cubes $q \in V$ an "inner" H-cube (in V) if and only if all its neighbors (in V) have equal or smaller size. R is now said to be "properly nested" if and only if two neighboring H-cubes in R differ at most in one generation and R contains with any non-inner H-cube q also all its "brothers", $q' \in d_m(\hat{q})$.

3. LOCAL ERROR ESTIMATES

The objective of this section is to analyze the "local" approximation behaviour of elements in $S_k(R,\Omega)$ where R is an economic and properly nested refinement. To this end, let us introduce some more notation. For any collection V of H-cubes and $q \in V$, we set

$$U(q|V) = \{q' \in V: \ q' \text{ is a neighbor of } q \text{ in } V\}$$

and

$$\bar{U}(q|V) = \cup\{q': \ q' \in U(q|V)\} .$$

Furthermore, we denote by $S(f,V)$ a best approximation in $S_k(V,\Omega)$ to a given function f, i.e.,

$$\begin{aligned} \|f - S(f,V)\|_p(\Omega) &= \text{dist}_p(f,S_k(V,\Omega)) \\ &= \inf\{\|f - S\|_p(\Omega): \ S \in S_k(V,\Omega)\} \end{aligned}$$

for some fixed p, $1 \leq p \leq \infty$, $\|\cdot\|_p(\Omega)$ being the usual L_p-norm with respect to the domain Ω (with the familiar interpretation for $p = \infty$). In particular, we want to establish upper bounds for the local errors $\|f - A(f,V)\|_p(D)$, $D \subseteq \Omega$, where $A(f,V)$ is a certain global approximation in $S_k(V,\Omega)$ to f.

Recalling the definition of a layer given in section 2, the next result which still refers to uniform configurations of knot sets is implicitly contained in the proof of Theorem 5.1 in [8].

LEMMA 3.1. *Let* $\bar{L}_{q_0} = \cup\{q: q \in L_{q_0}\}$ *where* L_{q_0} *is a layer in some refinement* $R(H)$. *The following estimates hold for any* $f \in L_p(\Omega)$ *and any* $q \in L_{q_0}$.

$$(3.1) \quad \|f-S(f,L_{q_0})\|_p((q)_s \cap \Omega) \leq \begin{cases} C\|f\|_p((q)_s \cap \Omega); \\ \qquad (q)_s \setminus I(\bar{L}_{q_0}) \neq \emptyset, \\ C\,\mathrm{dist}_p(f,\Pi_k)(\bar{U}(q|L_{q_0}))_s \cap \Omega; \\ \qquad (q)_s \subseteq I(\bar{L}_{q_0}) \end{cases}$$

where the constant C *does not depend on* f *and* q.

In fact, essentially making use of (2.11), it is shown in [8] that (3.1) still holds when $S(f,L_{q_0})$ is replaced by the linear scheme

$$Q(f,L_{q_0})(x) = \sum_{P \in P(L_{q_0})} \lambda_P(f)M(x|P)$$

where λ_P are certain bounded linear functionals which are chosen so that $Q(\cdot,L_{q_0})$ reproduces Π_k on $I(\bar{L}_{q_0})$ (cf also [7]).

Our general estimates for the refinements R require the following specification of the position of some $q \in R$ relative to the remaining H-cubes. $q \in R$ is called "protected" if and only if

$$(3.2) \quad (q)_s \subseteq I(\cup\{\hat{q}': q' \in L_q\}).$$

It follows from (2.15)-(2.17) that a protected H-cube q is "surrounded" in L_q by H-cubes having the same size as q. However, a simple drawing for the case $s = k = 1$ shows that this is not sufficient for q being protected.

Now, given $R = R(H)$, let us denote for any $q \in \hat{R}$ by $q_0(q)$ the minimal ancestor of q in $\hat{R}$ which is a protected H-cube with respect to some refinement $R_0 \subseteq \hat{R}$. Setting

$$(3.3) \quad G(q|R) = U(q_0(q)|L_{q_0(q)}), \ \bar{G}(q|R) = \bar{U}(q_0(q)|L_{q_0(q)}0)$$

we have

$$(3.4)\quad \bar{G}(q|R) = \bar{U}(q|L_q)$$

if q is protected. In general, $G(q|R)$ consists of 3^n congruent H-cubes.

For a given economic m-type refinement we denote by $V_i = V_i(R)$, $i = 0,\dots,\ell(R)$, the collection of all H-cubes in R which differ from those in H_0 by exactly i generations. $\ell = \ell(R)$, the index of R, indicates then the latest generation occurring in R. The subset of all bottom cubes in V_i forms then the "bottom layer" $L_i = L_i(R)$, $i = 0,\dots,\ell(R)$. Note that for any $q_i \in V_i$

$$(3.5)\quad V_i = \cup\{L_q:\ q \in d_m(\hat{q}_i)\}\ .$$

The following lemma is easily derived from (2.13) or its consequences, (2.14)-(2.17).

LEMMA 3.2. <u>Let $R = R(H)$ be an economic and properly nested m-type refinement of H_0.</u>

i) <u>Let $q \in R$ be arbitrary. The elements of $G(q|R)$ differ from q in at most one generation.</u>

ii) <u>If $q \in L_i$ such that $(q)_s \cap (\bar{V}_{i+1})_s \neq \emptyset$, where $\bar{V}_i = \cup\{q':\ q' \in V_i\}$, then q is protected.</u>

iii) <u>If $q \in V_{i+1}$ is not protected, then $\hat{q}$ $(\in L_i)$ is protected.</u>

We are now in a position to analyze the following simple approximation scheme for an economic refinement R. Using the above notation, we set for $f \in L_p(\Omega)$

$$A_0(f) = S(f,H_0)$$

$$(3.6)\quad A_{j+1}(f) = S(f - A_j(f), V_{j+1}) + A_j(f),\ j = 0,\dots,\ell\text{-}1$$

$$A(f,R) = A_\ell(f)\ .$$

Recall that when dealing with economic refinements, the knot sets of (locally) smallest size are in general "overlapped" by knot sets of much larger size. However, it will turn out that when R is in addition properly nested the (locally) smallest size of the knot sets essentially determines the local approximation behaviour. In fact, the scheme (3.6), i.e., updating "coarser grid" approximations locally by finer grid approximations, provides best possible local error estimates.

THEOREM 3.1. Let $m = m(H)$ be finite and assume that the m-type refinement $R = R(H)$ (cf (2.18)) is economic and properly nested. The following estimates hold for any bottom cube $q \in R$.

$$(3.7) \quad \|f - A(f,R)\|_p((q)_s \cap \Omega) \leqq c \operatorname{dist}_p(f,\Pi_k)(\bar{G}(q|R))_s \cap \Omega$$

where the constant c does not depend on q, f and R.

Proof. (2.11) insures us that the assertion is true for $R = H_0$. Suppose now that (3.7) holds for all (m-type) refinements R which have at most index ℓ-1. Let R' be any properly nested, economic m-type refinement of degree ℓ with the corresponding collections $V_i = V_i(R)$ and the bottom layers L_i, $i = 0,\dots,\ell$, as above. We set $\bar{V}_i = \cup\{q' : q' \in V_i\}$. Let $q \in R'$ be a bottom cube. If $(q)_s \cap \bar{V}_\ell = \emptyset$ there is nothing to prove since then $S(f - A_{\ell-1}(f), V_\ell)(x) = 0$ for $x \notin (\bar{V}_\ell)_s$ (cf (3.6)). On the other hand, if $(q)_s \cap (\bar{V}_\ell) \neq \emptyset$, but $q \notin L_\ell$, then $q \in L_{\ell-1}$, since R' is properly nested. Hence, we get, in this case,

$$(3.8) \quad \begin{aligned} \|f - A(f,R')\|_p((q)_s \cap \Omega) &\leqq \|f - A_{\ell-1}(f)\|_p((q)_s \cap \Omega) \\ &+ \|S(f - A_{\ell-1}(f), V_\ell)\|_p((q)_s \cap (\bar{V}_\ell)_s \cap \Omega) . \end{aligned}$$

Now let R be the coarser refinement which is obtained by removing the partitions $d_m(\hat{q}')$, $q' \in V_\ell$, i.e., R has index ℓ-1. Clearly, R is still properly nested and, recalling the definition (3.3), our assumption assures

$$(3.9) \quad \|f - A_{\ell-1}(f)\|_p((q)_s \cap \Omega) \leqq C_1 \operatorname{dist}_p(f,\Pi_k)(\bar{G}(q|R))_s \cap \Omega .$$

Note that q is protected on account of Lemma 3.2, ii. Thus, combining (3.2) and (3.5), Lemma 3.1 becomes applicable and the second summand on the right hand side of (3.8) can be estimated as follows

$$\begin{aligned} &\|S(f - A_{\ell-1}(f), V_\ell)\|_p((q)_s \cap (\bar{L}_\ell)_s \cap \Omega) \\ (3.10) \qquad &\leqq c_2 \|f - A_{\ell-1}(f)\|_p((q)_s \cap \Omega) \\ &\leqq c_3 \operatorname{dist}_p(f, \Pi_k)_{(\bar{G}(q|R))_s \cap \Omega} . \end{aligned}$$

Since (3.3) provides, in this case, $\bar{G}(q|R) = \bar{G}(q|R')$, we have confirmed (3.7) so far for a constant $c_1 + c_3$. The same estimate is obtained quite analogously when $q \in L_\ell$ is not protected refering, in this case, to Lemma 3.2, iii. So let us assume, finally, that $q \in L_\ell$ is protected. (3.5) assures us that the following relation is valid

$$\|f - A_\ell(f)\|_p((q)_s \cap \Omega) = \operatorname{dist}_p(f - A_{\ell-1}(f), S_k(V_\ell, \Omega))_{(q)_s \cap \Omega}$$

Note that our hypothesis allows us now to apply Lemma 2.3 so that the right hand side of the above equality may be estimated by $\operatorname{dist}_p(f, S_k(V_\ell, \Omega))_{(q)_s}$. Observing (3.3), Lemma 3.1 provides again

$$(3.11) \quad \|f - A_\ell(f)\|_p((q)_s \cap \Omega) \leqq c_1 \operatorname{dist}_p(f, \Pi_k)_{(\bar{G}(q|R'))_s \cap \Omega}$$

where c_1 is the same constant as in (3.9). Combining this with the fact that R' is properly nested provides that c_1 and also $c_1 + c_3 = c_4$ do not depend on ℓ. Hence, the assertion follows from (3.9), (3.10) and (3.11).

REMARKS. (i) Recall that (3.7) covers the case of optimal smoothness (cf Prop. 2.1), i.e., $S_k(R(H), \Omega) \subset C^{k-1}(R^s)$. (ii) The diameter of a bottom cube in R may be viewed as the local "mesh size" of the configuration $P(R)$ (cf (2.7)). Theorem 3.1 shows that similarly, as in the univariate case (cf [2]), the local approximation behaviour on such an "elementary piece" $(q)_s \subset \Omega$ may be estimated by polynomial approximation with respect to a certain larger domain $\bar{G}(q|R)$. This in turn provides immediately, e.g., by means of the results in [4], local estimates in terms of (k+1)-th order moduli of continuity or corresponding K-functionals. Passing from $(q)_s$ to $\bar{G}(q|R)$ shows to what extent the global smoothness spoils the "local nature" of the spaces $S_k(R, \Omega)$. (iii) Let us now briefly comment on the hypothesis of Theorem 3.1.

Clearly, the collection of all bottom faces arising from an m-type refinement $\mathcal{R}(h)$ form an m-type partition $\Theta(\mathcal{R})$ of Ω consisting of s-cubes. It is not hard to show that if $\mathcal{R}=\mathcal{R}(H)$ is economic, the estimate

$$(3.12)\qquad |\mathcal{R}| \leq c|\Theta(\mathcal{R})|$$

holds uniformly in $\mathcal{R}$, so that in view of (2.25) the dimensionality of $S_k(\mathcal{R}(H),\Omega)$ may be related as usual to the cardinality of an s-dimensional partition rather than to the cardinality of an (s+k) dimensional one. (iv) Note that the structure of $S_k(\mathcal{R},\Omega)$ still enjoys much of the simplicity of spaces of (discontinuous) piecewise polynomials on cube partitions. The simple way of generating refinements without caring about the global smoothness and the local error estimate (3.7) suggest that this setting is very appropriate for smooth adaptive approximation. In fact, (3.7) and (3.12) will be used in a forthcoming paper to construct smooth adaptive schemes based on the spaces $S_k(\mathcal{R},\Omega)$ which are essentially equivalent to adaptive piecewise polynomial approximation (cf [3]). In particular, such algorithms for the smooth case will turn out to produce automatically properly nested refinements $\mathcal{R}(H)$. This confirms that the restriction to properly nested partitions in Theorem 3.1 is actually not essential.

REFERENCES

1. Allgower, E. and K. Georg, Triangulations by reflections with applications to approximation, Proc. Conf. Oberwolfach, Numerische Methoden der Approximations theorie, ed. by L. Collatz, G. Meinardus, H. Werner, Birkhäuser 1978.

2. de Boor, C., Splines as linear combinations of B-splines, Approximation Theory II, ed. by G. G. Lorentz, C. K. Chui, L. L. Schumaker, Academic Press (1976), 1-67.

3. de Boor, C. and J. R. Rice, An adaptive algorithm for multivariate approximation giving optimal convergence rates, J. Approx. Theory, 25 (1979), 337-359.

4. Dahmen, W., R. DeVore and K. Scherer, Multi-dimensional spline approximation, to appear in SIAM J. Numer. Anal.

5. Dahmen, W., On multivariate B-splines, to appear in SIAM J. Numer. Anal.

6. Dahmen, W., Polynomials as linear combinations of multivariate B-splines, Math. Z., 169 (1979), 93-98.

7. Dahmen, W., Konstrucktion mehrdimensionaler B-splines und ihre Anwendungen auf Approximationsprobleme, to appear in Proc. Conf. Oberwolfach, Numeriscne Methoden der Approximationstheorie , Birkhäuser, Basel.

8. Dahmen, W., Approximation by linear combinations of multivariate B-splines, in preparation.

9. Kuhn, H. W., Some combinatorial lemmas in topology, IBM J. Research and Develop., 45 (1960), 518-524.

10. Micchelli, C. A., A constructive approach to Kergin interpolation in R^k, University of Wisconsin at Madison, Math. Res. Center Report No. 1895, 1978, to appear in Rocky Mountain Journal of Mathematics as "A constructive approach to Kergin interpolation in R^k: Multivariate B-splines and Lagrange interpolation."

11. Micchelli, C. A., On a numerically efficient method for computing multivariate B-splines, Proc. Conf. Multivariate Approximation Theory, Oberwolfach, ed. by W. Schempp, K. Zeller, Birkhäuser (1979), 211-248.

Wolfgang Dahmen
Institut für Angewandte Mathematik
der Universität Bonn
Wegelerstraße 6
5300 Bonn, F.R. GERMANY

IBM Thomas J. Watson Research Center
PO Box 218
Yorktown Heights, NY 10598

APPROXIMATION BY SMALL RANK TENSOR PRODUCTS OF SPLINES

S. Demko

Approximation of smooth bivariate functions by classes of spline functions depending on a relatively small number of parameters is considered. Upper bounds for the error are proven and a computational method is proposed.

We consider the problem of approximating functions of two variables by tensor products of spline functions having small rank. For simplicity we assume that the univariate meshes involved are uniform and have an equal number of mesh points in each coordinate direction. The basic notation and definitions are as follows:

(1) $S_N: = S_{k,N}: = \{f \in C(I^2) : f(x,y) = \sum_i f_i(x)g_i(y)$ where f_i and g_i are polynomials of degree k-1 or less on each interval $(j/N, (j+1)/N)$, $0 \le j \le N-1$ and $f_i, g_i \in C^{k-2}(I)\}$ where $I: = [0,1]$.

For $f \in C(I^2)$,

$$||f||_\infty : = \sup\{|f(x,y)| : 0 \le x,y \le 1\}.$$

$$a_R(f): = \inf\{||f - \sum_{i=1}^{R} f_i \otimes g_i||_\infty : f_i, g_i \in C(I)\}$$

$$a_{R,N}(f): = \inf\{||f - \sum_{i=1}^{R} f_i \otimes g_i||_\infty : f_i, g_i \text{ are as in (1)}\}$$

$$\text{dist}(f,S_N): = a_{N,N}(f).$$

$f \otimes g$ is the function of two variables defined by $(f \otimes g)(x,y) = f(x)g(y)$ where f and g are univariate functions. The quantity $a_{R,N}(f)$ measures how well the function f can be approximated by a member of S_N whose rank is no greater than R. We shall be thinking of k as fixed throughout this paper and, accordingly, have suppressed the dependence of S_N and $a_{R,N}$ on k. It is probably best to think of k as being not too large.

ISBN 0-12-213650-0

A general member of S_N depends on $(N+k-1)^2$ (= dim S_N) parameters and for any function $f \in C^k(I^2)$ we have $\text{dist}(f,S_N) = O(N^{-k})$ as $N \to \infty$. In fact if $f \in C^m(I^2)$ for any $m \geq k$, then this same orderwise estimate holds and is, in general, sharp. However, it turns out that very smooth functions can be approximated to order N^{-k} by subfamilies of S_N depending on many fewer parameters than dim S_N. This fact can be useful when the approximating function is going to be transmitted and transmission costs or time is at a premium or when the approximating function is going to be stored for a long period of time and storage space is at a premium. For a spline function that is going to be repeatedly evaluated, it is probably best to use the B-spline basis for representing the function, in which case one need all dim S_N parameters.

It is possible to find <u>subspaces</u> of S_N of small dimension which approximate smooth functions well. This follows from:

PROPOSITION 1. <u>There is a constant depending on only k such that for every $K \subseteq C(I^2)$ there is an R dimensional subspace Z of S_N ($R \leq \dim S_N$) such that for every $f \in K$</u>

$$\inf_{z \in Z} ||f-z||_\infty \leq \text{const.}(d_R(K) + \text{dist.}(f,S_N))$$

<u>where $d_R(K)$ is the R-width of K in $C(I^2)$.</u>

PROOF. Let C be an upper bound for the projection constants of $\{S_N\}_N$ in $C(I^2)$. Let Y be an R-dimensional subspace of $C(I^2)$ which is ε-optimal for K in the sense of R-widths, P a projection onto S_N of norm no greater than C, then, the restriction of P to Y has at most R dimensional range in S_N. For $f \in K$ and $g \in Y$ we have

$$\begin{aligned} ||f-Pg|| &\leq ||f-Pf||+||P(f-g)|| \leq (1+||P||)\text{dist}(f,S_N) \\ &\qquad + ||P||\ ||f-g|| \\ &\leq (1+||P||)\text{dist}(f,S_N)+||P||(1+\varepsilon)d_R(K) \end{aligned}$$

if g is chosen well.

REMARKS: 1. If K is very thin, for example, if $d_R(K) \approx e^{-R}$ then there exist subspaces of S_N of dimension $\approx k \ln N$ which approximate K to the same order as all of S_N does. Unfortunately, Proposition 1 does not appear to provide us with any good constructive way of finding such subspaces.

2. Proposition 1 really has nothing to do with tensor products and can easily be reformulated in the context of general normed spaces.

It seems possible to obtain estimates for $a_{R,N}(f)$ from Proposition 1 by assuming that K and P have some tensor product structure. We shall by pass this approach and estimate $a_{R,N}(f)$ in terms of $\mathrm{dist}(f,S_N)$ and $a_R(f)$. Our approach also indicates the type of computation that should be done in order to construct a good small rank approximation to a given function.

Let $\{x_i : 1 \le i \le N+k-1\}$ be distinct points in I so that the projection $P : C(I^2) \to S_N$ determined by $Pf = s$ iff $f(x_i,x_j) = s(x_i,x_j)$ for all i,j has reasonably small norm. Again, we could have been more general here by taking collections of linear functionals $\{\lambda_i\}$, $\{\phi_i\}$ and considering the projection determined by $\{\lambda_i \otimes \phi_j\}$.

PROPOSITION 2. _There is a constant depending on only k such that for all $f \in C(I^2)$_

$$a_{R,N}(f) \le \mathrm{const.}\{\mathrm{dist}(f,S_N) + \inf\{||(f(x_i,x_j))-B||_{1,\infty} : \mathrm{rank}\ B \le R\}\}.$$

Here, $||M||_{1,\infty} := \max_{i,j} |m_{ij}|$ is the norm of the matrix M considered as an operator from ℓ_1^{N+k-1} into ℓ_∞^{N+k-1}.

PROOF. Let P be the interpolating projection defined above. Then, for all $s \in S_N$, $||s||_\infty \leqq ||P||\,||s(x_i,x_j)||_{1,\infty}$. So, for any $g \in S_N$ of rank R or less

$$||f-g||_\infty \leqq ||f-Pf||_\infty + ||Pf-Pg||_\infty \leqq (1+||P||)\,\mathrm{dist}(f,S_N) + ||P||\,||(f(x_i,x_j))-(g(x_i,x_j))||_{1,\infty}.$$

To complete the proof we need only observe that $g \in S_N$ has rank R or less if and only if $(g(x_i,x_j))$ has rank R or less.

The estimation of $a_{R,N}(f)$ is completed in

PROPOSITION 3. <u>There is a constant depending on only k such that for all $f \in C(I^2)$</u>

$$a_{R,N}(f) \le \text{const.}\{\text{dist}(f,S_N) + a_R(f)\}.$$

PROOF. f can be viewed as the kernel of an integral operator from the dual space of C(I), $[C(I)]^*$, into C(I). Let us denote this operator by f also so that $f(\mu)(x) = \int_0^1 f(x,y)d\mu(y)$ for $\mu \in [C(I)]^*$. We factor the matrix $(f(x_i,x_j))$ as follows. Let $S(z_1,\cdots,z_{N+k-1}) = \sum_i z_i \delta_{x_i}$ define a map ℓ_1^{N+k-1} into $[C(I)]^*$ (δ_{x_i} is Dirac measure) and let $Tg = (g(x_1),\cdots,g(x_{N+k-1}))$ map C(I) into ℓ_∞^{N+k-1}. Then, $(f(x_i,x_j)) = TfS$ and by well-known results on approximation numbers, cf. [4], we get that

$$\inf\{||(f(x_i,x_j))-B||_{1,\infty}:\text{rank } B\le R\}\le ||T||a_R(f)||S||=a_R(f).$$

EXAMPLE. Suppose f(x,y) maps $[C(I)]^*$ into the Sobolev space $W_2^m(I)$, then $a_R(f) = O(R^{-m})$ since the identity map from $W_2^m(I)$ into $L_\infty(I)$ has approximation numbers of this order. (cf. [3]). The assumption here is weaker than the assumption that $f \in C^m(I^2)$. If $m \ge k$, then taking $R \approx N^{k/m}$ yields $a_{R,N}(f) = O(N^{-k})$. This order of approximation is obtained with a family depending on only $O(N^{1+m/k})$ parameters.

While Proposition 3 provides upper bounds on $a_{R,N}(f)$, it is Proposition 2 that tells us what we need to compute good rank R spline approximations. Specifically, we should have a method for computing good rank R approximations to the matrix of data $(f(x_i,y_j))$. Fortunately, there is available a good numerical algorithm for computing such things. It is the Golub-Reinsch Singular Value Decomposition (SVD) algorithm, [2], which for a given rectangular matrix $A = (a_{ij})$ computes the <u>best</u> rank R approximation to A in the Euclidean operator norm $||A||_2 = \max\{|\lambda| : \lambda$ is an eigenvalue of $A^*A\}$. More specifically if A is a $p\times q$ matrix $p \ge q$ then the SVD computes non-negative numbers $\sigma_1 \ge \sigma_2 \ge \cdots \ge \sigma_q$ and orthonormal sets of column vectors $u_1,\cdots,u_q \in \mathbb{R}^p$ $v_1,\cdots v_q \in \mathbb{R}^q$ such that $A = \sum_i \sigma_i u_i v_i^T$. The best rank R approximation to A in the $||\cdot||_2$ norm is then $\sum_{i=1}^{R} \sigma_i u_i v_i^T$ and $||A-\sum_{i=1}^{R} \sigma_i u_i v_i^T||_2 = \sum_{i>R} \sigma_i$.

To construct spline functions from the vectors $\{u_i v_i\}$ obtained by applying the SVD to $(f(x_i,x_j))$ we construct for each vector u_ℓ a univariate spline (of degree k-1, smoothness k-2, on the mesh $\{i/N\}$) f_ℓ so that $f_\ell(x_i) = i^{th}$ coordinate of u_ℓ. We construct similar splines g_ℓ so that $g_\ell(x_i) = i^{th}$ coordinate of v_ℓ. Then, the function $h(x,y) = \sum_{\ell=1}^{R} f_\ell(x) g_\ell(y)$ is in S_N and $||(h(x_i,x_j))-(f(x_i x_j))||_{1,\infty} \le ||(h(x_i,x_j))-(f(x_i,x_j))||_2$. The bounds of Proposition 3 cannot be used to conclude that h approximates f to accuracy $\mathrm{dist}(f,S_N) + a_R(f)$. However, the argument used can be modified to obtain

PROPOSITION 4. Let $f \in C(I^2)$, then with $\{x_i\}$ as above.

(a) $\inf\{||(f(x_i,x_j))-B||_2 : \text{rank } B \le R\} \le (N+k-1) a_R(f)$

(b) the spline h determined by the above procedure satisfies $||h-f||_\infty \le \mathrm{const}(d(f,S_N) + Na_R(f))$.

PROOF. (b) is a consequence of (a). To prove (a) consider $(f(x_i,x_j))$ as a mapping from ℓ_2^{N+k-1} into itself. Let S and T be as in Proposition 3 so $(f(x_i,x_j)) = TfS$. But in the present context, the best we can say is that $||S||\ ||T|| \le N+k-1$.

REMARKS. 1. The use of the SVD here was motivated by Andrews' work on image transmission, [1]. We know of no algorithm that will compute the best rank R approximations to a given matrix in the $||\cdot||_{1,\infty}$ norm.

2. We don't know if the estimate (a) in Proposition 4 is sharp for matrices arising from smooth functions f. We can prove that if for all $i,j\,|f(x_i,x_j)-f(x_{i-1},x_j)| \le (N+k-1)^{-1}$ and $|f(x_i,x_{j+1})-f(x_i,x_j)| \le (N+k-1)^{-1}$ and $\max|f(x_i,x_j)| = 1$, then $||(f(x_i,x_j))||_{1,\infty} \le (N+k-1)^{-1}||(f(x_i,x_j))||_2$. However, the connection with Proposition 4 is unclear.

3. Some of the ideas presented here carry over to higher dimensions and other norms; we don't know of any tool like the SVD for more than two dimensions however.

REFERENCES

1. Andrews, H. C., Two dimensional transforms, in Picture Processing and Digital Filtering, T. S. Huang, ed., pp. 21-68, Springer-Verlag, New York, 1975.
2. Golub, G. and C. Reinsch, Singular value decomposition and least squares solutions, Numer. Math. 14(1970), 403-420.
3. Höllig, K., Approximationszahlen von Sobolev-Einbettungen, Math. Ann. 242(1979), 273-281.
4. Pietsch, A., s-numbers of operators in Banach spaces, Studia Math., 51(1974), 201-223.

S. Demko
School of Mathematics
Georgia Institute of Technology
Atlanta, Georgia 30332

VARIABLE KNOT, VARIABLE DEGREE SPLINE APPROXIMATION TO x^β

R. DeVore and K. Scherer

We determine the asymptotic rate of approximation of the function x^β by splines with free knots and variable degree in terms of the total number of free parameters.

0. INTRODUCTION

There are several examples of functions, which can be approximated more efficiently by splines with variable knots than by splines with fixed knots. These functions generally have singularities and the knots are bunched near the singularities [4], [2]. Another possible way to improve the approximation of a function is by varying the degree of the polynomials which make up the approximating spline. An initial study of problems of this type was made in [3]. Here, we want to consider the very special problem of approximating the function x^β, $\beta > 0$ by piecewise polynomials with variable knots and degree. Our main result is to determine the asymptotic behavior of the degree of approximation to x^β in terms of the number of free parameters used in the approximation.

If $f \in C[a,b]$, let $E_n(f) = E_n(f,[a,b])$ denote the error in approximating f by polynomials of order n (degree $n-1$) i.e.

$$(0.1) \quad E_n(f) = \inf_{P \in \mathbb{P}_n} ||f-P||_\infty [a,b]$$

where $\mathbb{P}_n$ is the space of polynomials of order n. Given $\underline{x} = (x_i)_0^m$, $0 = x_m < x_{m-1} < \cdots < x_0 = 1$ and integers $n_0, \ldots, n_{m-1}$, let $S_{\underline{n},\underline{x}}$ denote the space of piecewise polynomials with knots $\underline{x}$ and order $\underline{n} = (n_0, \ldots, n_{m-1})$. If $S \in S_{\underline{n},\underline{x}}$ then S is a polynomial of order n_i on

ISBN 0-12-213650-0

$[x_{i+1}, x_i]$, $i=0,\ldots,m-1$. If N is any positive integer, set

$$S_N = \bigcup_{\substack{\underline{n},\underline{x} \\ \sum n_i = N}} S_{\underline{n},\underline{x}} .$$

We are interested in giving asymptotic estimates for

$$E_N(f) = \inf_{S \in S_N} \|f - S\|_\infty [0,1]$$

for certain functions f. This involves finding asymptotically optimal knots $\underline{x}$ and order $\underline{n}$ so that $\sum n_i = N$ and

$$E_{\underline{n},\underline{x}}(f) \equiv \inf_{S \in S_{\underline{n},\underline{x}}} \|f - S\|_\infty [0,1] \sim E_N(f) .$$

Our main result is to show that for $f(x) = x^\beta$, $\beta > 0$, there are constants C_1 and $C_2 > 0$ depending only on β for which

$$(0.2) \quad C_2 N^{-2\beta-1} \delta^{2\sqrt{N\beta}} \le E_N(x^\beta) \le C_1 \delta^{2\sqrt{N\beta}}$$

with $\delta = \sqrt{2} - 1$. Thus, $E_N(x^\beta) \sim e^{-\gamma_\beta \sqrt{N} + 0(\log N)}$, $\gamma_\beta = -2\sqrt{\beta} \log \delta$.

1. POLYNOMIAL APPROXIMATION OF x^β

We shall need some rather fine estimates for $E_n(x^\beta;[a,b])$, especially how this quantity depends on $[a,b]$. We follow the ideas described in the classical book of Achieser [1] but take care to find the dependence of the constants on $[a,b]$.

LEMMA 1. *If* $0 < \beta < 1$, *there are constants* $C_1, C_2 > 0$ *depending only on* β, *such that for each* $n > 0$ *and each interval* $[a,b]$

$$(1.1) \quad C_1(a+b)^\beta n^{-2\beta-1} e^{-n\gamma} \le E_n(x^\beta,[a,b]) \le C_2(a+b)^\beta n^{-\beta-1} e^{-n\gamma}$$

with γ *defined by* $\cosh \gamma = \frac{b+a}{b-a}$.

Proof. By translation, we have for $\phi(t) = \left(t + \frac{b+a}{b-a}\right)^\beta$

$$E_n(x^\beta,[a,b]) = \left(\frac{b-a}{2}\right)^\beta E_n(\phi,[-1,1]) .$$

This gives via a comparison between L_2 and L_∞ norms that

$$\left(\frac{b-a}{2}\right)^\beta \left(\tfrac{1}{2} \sum_n^\infty |A_k|^2\right)^{\frac{1}{2}} \le E_n(x^\beta,[a,b]) \le \left(\frac{b-a}{2}\right)^\beta \sum_n^\infty |A_k| \tag{1.2}$$

with the A_k the Fourier-Chebyskev coefficients of ϕ. Making the substitution $t = \cos\theta$ and $\frac{b+a}{b-a} = \cosh\gamma$, we find

$$\begin{aligned} A_k &= \frac{2}{\pi}\int_{-1}^{1} \phi(t)\, T_k(t) \frac{dt}{\sqrt{1-t^2}} \\ &= \frac{2}{\pi}\int_0^\pi [\cos\theta + \cosh\gamma]^\beta \cos k\theta \, d\theta \\ &= \frac{2(-1)^k}{\pi}\int_{-\pi}^{0} [\cosh\gamma - \cos\theta]^\beta \cos k\theta \, d\theta \end{aligned} \tag{1.3}$$

This leads us to evaluate A_k by means of contour integration. Let $L(z) = \log|z| + i\,\mathrm{Arg}(z)$ be the branch of $\log(z)$ with branch cut the negative real axis $f(z) = e^{\beta L(\cosh\gamma - \cos z)}$. Then f is analytic inside the curve Γ described by:

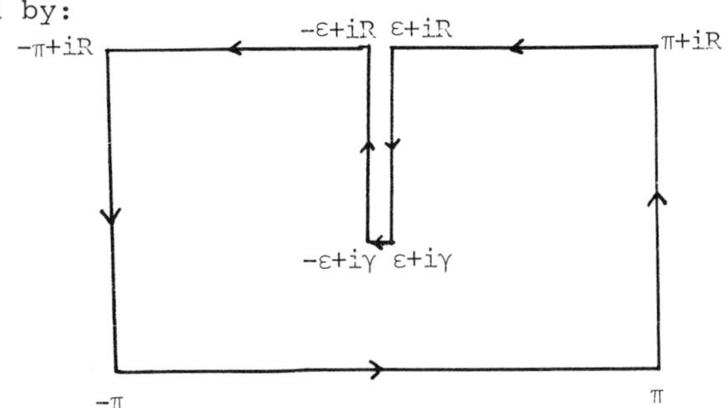

Hence,

$$\int_\Gamma f(z)e^{ikz}\, dz = 0 . \tag{1.4}$$

Since $f(z)\cos kz$ has period 2π,

$$(1.5)\quad \int_{\pi}^{\pi+iR} + \int_{-\pi+iR}^{-\pi} f(z)e^{ikz}\,dz = 0 .$$

Also, for $k \geq 1$,

$$\sup_{-\pi \leq x \leq \pi} |f(x+iR)e^{ik(x+iR)}| \to 0 \qquad (R \to \infty)$$

so that

$$(1.6)\quad \int_{\pi+iR}^{\varepsilon+iR} + \int_{-\varepsilon+iR}^{-\pi+iR} f(z)e^{ikz}\,dz \to 0 \qquad \varepsilon \to 0,\ R \to \infty .$$

Using (1.5) and (1.6) back in (1.4) and taking a limit as $\varepsilon \to 0$, $R \to \infty$, shows that

$$\begin{aligned}(1.7)\quad A_k &= \frac{(-1)^k}{\pi}\int_{-\pi}^{\pi} f(z)e^{ikz}\,dz \\ &= \frac{(-1)^k}{\pi}\int_{\gamma}^{\infty} e^{\pi i\beta}(\cosh y - \cosh\gamma)^{\beta} e^{-ky}\,i\,dy \\ &- \frac{(-1)^k}{\pi}\int_{\gamma}^{\infty} e^{-\pi i\beta}(\cosh y - \cosh\gamma)^{\beta} e^{-ky}\,i\,dy \\ &= \frac{2(-1)^{k+1}}{\pi}\sin\pi\beta\int_{\gamma}^{\infty}(\cosh y - \cosh\gamma)^{\beta} e^{-ky}\,dy \\ &= \frac{2(-1)^{k+1}}{\pi}\sin\pi\beta\, e^{-k\gamma}\int_{0}^{\infty}(\cosh(y+\gamma) - \cosh\gamma)^{\beta} e^{-ky}\,dy\end{aligned}$$

For $y \geq 1$, $\sinh\frac{y}{2}\sinh(\gamma+\frac{y}{2}) \leq \frac{1}{4}e^{\gamma+y}$ and for $y \leq 1$, $\sinh\frac{y}{2} \leq \frac{1}{2}y\,e^{y}$. Therefore, when $y > 0$

$$\cosh(y+\gamma) - \cosh\gamma = 2\sinh\frac{y}{2}\sinh(\gamma+\frac{y}{2}) \leq y\,e^{\frac{y}{2}}e^{\gamma+\frac{y}{2}} .$$

Using this in (1.7) gives the estimate for A_k

$$(1.8)\quad |A_k| \le \frac{2}{\pi} e^{-k\gamma} e^{\gamma\beta} \int_0^\infty y^\beta e^{-(k-\beta)y}\, dy$$

$$\le e^{\gamma\beta} e^{-k\gamma} (k-\beta)^{-\beta-1} \Gamma(\beta)$$

this gives the upper estimate (using (1.2))

$$E_n(x^\beta,[a,b]) \le C(b-a)^\beta e^{\gamma\beta} \sum_{k=n}^{\infty} (k-\beta)^{-\beta-1} e^{-k\gamma}$$

$$\le C(b-a)^\beta \left(\frac{b+a}{b-a}\right)^\beta (n-\beta)^{-\beta-1} e^{-n\gamma} \le C(a+b)^\beta n^{-\beta-1} e^{-n\gamma}$$

where we have used the definition of γ. This is the right-hand side of (1.1).

To prove the left-hand side of (1.1), we use the estimate

$$\cosh(y+\gamma) - \cosh\gamma = 2\sinh\frac{y}{2}\sinh(\gamma+\frac{y}{2})$$

$$= 2\sinh\frac{y}{2}[\sinh\gamma\cosh\frac{y}{2} + \cosh\gamma\sinh\frac{y}{2}]$$

$$\ge 2[\sinh\frac{y}{2}]^2\cosh\gamma \ge y^2\cosh\gamma$$

Hence, from (1.7),

$$|A_k| \ge \frac{2\sin\pi\beta}{\pi} e^{-k\gamma}(\cosh\gamma)^\beta \int_0^\infty y^{2\beta} e^{-ky}\, dy$$

$$\ge Ce^{-k\beta} k^{-2\beta-1}(\cosh\gamma)^\beta$$

where C depends only on β. Using this in (1.2) gives

$$E_n(x^\beta,[a,b]) \ge (a+b)^\beta \left(\sum_n^\infty e^{-2k\gamma} k^{-4\beta-2}\right)^{\frac{1}{2}} \ge C(a+b)^\beta e^{-n\gamma} n^{-2\beta-1}$$

where again we used the definition of γ. This is the lower estimate in (1.1), so Theorem 1 is proved.

2. MAIN RESULTS

The estimates of the last section can be used to determine the asymptotic behavior of $E_N(x^\beta)$. For this purpose we will need the following interesting lemma.

LEMMA 2. *If* $\delta = \sqrt{2} - 1$, *then for all* $x > 0$

$$\left(\frac{1-\delta^x}{1+\delta^x}\right)^x \geq \frac{1-\delta}{1+\delta} = \delta .$$

Proof. Consider the function $f(x) = \left(\frac{1-e^{-x}}{1+e^{-x}}\right)^x$ and with $y = e^x$ the related function $g(y) = \left(\frac{1-1/y}{1+1/y}\right)^{\ln y} = \left(\frac{y-1}{y+1}\right)^{\ln y}$. The function f has a minimum on $(0,\infty)$ at x_0 if and only if g has a minimum on $(1,\infty)$ at $y_0 = e^{x_0}$. Let $G(y) = \ln(g(y)) = \ln y \ln\left(\frac{y-1}{y+1}\right)$. Then, G and g achieve their minima at the same points. To find the minima of G , we note that $G(y) < 0$ on $(1,\infty)$ and $\lim_{y\to 1} G(y) = \lim_{y\to\infty} G(y) = 0$. Thus G achieves its minima at points y for which

$$(2.1)\qquad G'(y) = \left(\frac{2}{y^2-1}\right)\ln y + \frac{1}{y}\ln\left(\frac{y-1}{y+1}\right) = 0 .$$

There is symmetry in the function G which can be seen by using the transformation

$$(2.2)\qquad z = \frac{y+1}{y-1} \quad \text{for} \quad y > 1 .$$

Then $y = \frac{z+1}{z-1}$ and $G'\left(\frac{y+1}{y-1}\right) = -\frac{(y-1)^2}{2} G'(y)$, so that $G'(y) = 0$ if and only if $G'\left(\frac{y+1}{y-1}\right) = 0$.

It is easy to check that the fixed point $y_0 = \sqrt{2} + 1$ of (2.2) satisfies $G'(y_0) = 0$. In view of the symmetry noted above we can show that y_0 is the unique minimum of G if we can show that $G'(y) \geq 0$ for $y \geq y_0$, or in view of (2.1), it is enough to show that

$$H(y) = 2y \ln y - (y^2-1)\ln\left(\frac{y+1}{y-1}\right) \geq 0 \qquad y \geq y_0 .$$

This is true when $y = y_0$ and so it is enough to check $H'(y)$. We find

$$(2.3) \quad H'(y) = 4 + 2 \ln y - 2y \ln \frac{y+1}{y-1} \geq 4 + 2 \ln y - \frac{2y}{y-1} ,$$

where we have used the fact that $\ln\left(\frac{y+1}{y-1}\right) = \ln(1+\frac{2}{y-1}) \leq \frac{2}{y-1}$, $y > 1$. Now the right-hand side of (2.3) is > 0 when $y = y_0$ and is increasing for $y > y_0$, so that $H'(y) > 0$ and in turn $H(y) \geq 0$, $y \geq y_0$, as desired.

We have shown that the function f achieves a unique minimum when $x = \ln(\sqrt{2}+1)$. Thus, the function $\left(\frac{1-\delta^x}{1+\delta^x}\right)^x = \left(\frac{1-e^{x \ln \delta}}{1+e^{x \ln \delta}}\right)^x$ achieves its unique minimum when $x \ln \delta = -\ln(\sqrt{2}+1) = \ln(\sqrt{2}-1) = \ln \delta$, that is when $x = 1$. This proves the lemma.

We can now state and prove our main result.

THEOREM 1. <u>If</u> $\delta = \sqrt{2}-1$, <u>then there exist constants</u> C_1 <u>and</u> $C_2 > 0$ <u>depending only on</u> β <u>for which</u>

$$(2.4) \quad C_1 N^{-2\beta-1} \delta^{2\sqrt{N\beta}} \leq E_N(x^\beta) \leq C_2 \delta^{2\sqrt{N\beta}}, \quad 0 < \beta < 1, N \geq 1 .$$

<u>Proof</u>. We prove first the upper estimate. For any interval $I = [a,b]$, we have from Lemma 1 that

$$(2.5) \quad E_n(x^\beta, I) \leq Cb^\beta e^{-n\gamma} \leq Cb^\beta \left(\frac{b-a}{a+b+2\sqrt{ab}}\right)^n \leq Cb^\beta (\phi(\lambda))^n$$

where $\phi(\lambda) = \frac{1-\sqrt{\lambda}}{1+\sqrt{\lambda}}$ and $\lambda = \frac{a}{b}$.

With $\delta = \sqrt{2}-1$ as in Lemma 2, we have $\phi(\delta^2) = \delta$. For any integer $m > 0$, consider $\underline{x}$ with $x_i = \delta^{2i}$, $i = 0,1,\ldots,m$, $x_{m+1} = 0$ and $\underline{n}$ with $n_i = 1$ if $2(m-i)\beta < 2$ and $[2(m-i)\beta]$ otherwise. From (2.5), we have

$$(2.6)\quad E_{n_i}(x^\beta,[x_{i+1},x_i]) \le C\delta^{2i\beta}(\phi(\delta^2))^{n_i} = C\delta^{2i\beta+n_i}$$

$$\le C\delta^{-1}\,\delta^{2m\beta} \le C\delta^{2m\beta}\ ,\quad i < m\ .$$

For $i = m$, we have from Lemma 1,

$$E_1(x^\beta,[0,x_m]) \le C\delta^{2m\beta}\ .$$

Thus

$$(2.7)\quad E_{\underline{n},\underline{x}}(x^\beta) \le C\delta^{2m\beta}$$

Now, we want to check how large $\sum n_i$ is. Let $s = [\frac{1}{\beta}]$. If $2(m-i)\beta < 2$ then $(m-i) < s$. Hence,

$$\sum_{i=1}^{m+1} n_i \le n_{m+1} + \sum_{\beta(m-i)<1} n_i + \sum_{\beta(m-i)\ge 1} n_i$$

$$\le 1 + s + 2\beta\sum_{s}^{m} j \le 1 + s + 2\beta\left[\frac{m(m+1)}{2} - \frac{s(s+1)}{2}\right]$$

$$\le (s+1)(1-\beta s) + \beta m(m+1) \le (s+1)(1-\frac{s}{s+1}) + \beta m(m+1)$$

$$\le 1 + \beta m(m+1)\ .$$

Thus if $N \ge \beta m(m+1) + 1$ then because of (2.6) and (2.7)

$$E_N(x^\beta) \le C\delta^{2m\beta}\ .$$

Consider any $N \ge \beta^{-1}$. Choose m as the largest integer with $\beta m(m+1) \le N - 1$. Then $m \ge \sqrt{\frac{N}{\beta}} - 2$, and so

$$E_N(x^\beta) \le C\delta^{2m\beta} \le C\delta^{2(\sqrt{\frac{N}{\beta}}-2)\beta} \le C\delta^{2\sqrt{N\beta}}\ .$$

This inequality is automatically satisfied when $N \le \beta^{-1}$, so the upper estimate in (2.4) is proved.

To prove the lower estimate in (2.4), we will use the estimates (Lemma 1)

$$(2.8)\quad E_n(x^\beta,[a,b]) \ge Cn^{-2\beta-1}(a+b)^\beta\left(\frac{b-a}{b+a+2\sqrt{ab}}\right)^n \ge C_0 n^{-2\beta-1}\, b^\beta\, \phi(\lambda)\ ,$$

when $a > 0$

$$(2.9)\quad E_n(x^\beta,[0,b]) \ge C_0 n^{-2\beta-1}\, b^\beta$$

where ϕ and λ are as before and C_0 depends at most on β.

Suppose now that $\underline{x},\underline{n}$ are such that $N = \sum_0^k n_i$ and

$$(2.10)\quad E_{\underline{n},\underline{x}}(x^\beta) \le C_0 N^{-2\beta-1}\,\delta^{2\beta m}$$

with $\delta = \sqrt{2}-1$ and C_0 the constant in (2.8) and (2.9). We will estimate N from below in terms of m. Write $x_i = \delta^{2s_i}$, $i = 0,1,\ldots,k$. From (2.8) and (2.10) we find

$$(2.11)\quad n_i^{-2\beta-1}\,\delta^{2s_i\beta}\,\phi(\delta^{2(s_{i+1}-s_i)})^{n_i} \le N^{-2\beta-1}\,\delta^{2\beta m}\ ,$$

$i = 0,\ldots,k-1$.

Now each $n_i \le N$ and

$$\phi\left(\delta^{2(s_{i+1}-s_i)}\right)^{n_i} = \left(\frac{1-\delta^{(s_{i+1}-s_i)}}{1+\delta^{(s_{i+1}-s_i)}}\right)^{n_i\left(\frac{s_{i+1}-s_i}{s_{i+1}-s_i}\right)}$$

$$\ge \delta^{\frac{n_i}{s_{i+1}-s_i}} \qquad i = 0,\ldots,k-1$$

because of Lemma 2. Hence, (2.11) implies that

$$2s_i\beta + \frac{n_i}{(s_{i+1}-s_i)} \ge 2\beta m\ , \qquad i = 0,\ldots,k-1\ .$$

That is,

$$(2.12)\quad n_i \ge 2\beta(m-s_i)_+(s_{i+1}-s_i)\ , \qquad i = 0,\ldots,k-1$$

with the usual notation $x_+ = \max(x,0)$. Because of (2.9), we also have $s_k \geq m$. Therefore, we can estimate $\sum_{i=1}^{k-1} n_i$ by an upper Riemann sum for $2\beta \int_0^{s_k} (m-t)_+ dt = \beta m^2$. This gives

$$N = \sum_{i=0}^{k-1} n_i \geq \beta m^2 ,$$

That is,

$$m \leq \sqrt{\frac{N}{\beta}} .$$

We have just shown that for any $\underline{x}, \underline{n}$ with $\sum n_i = N$, we have

$$(2.13) \quad E_{\underline{n},\underline{x}}(x^\beta) \leq C_0 N^{-2\beta-1} \delta^{2\beta m} \quad \text{implies} \quad m \leq \sqrt{\frac{N}{\beta}} .$$

Thus,

$$E_N(x^\beta) \geq C_0 N^{-2\beta-1} \delta^{2\beta} \delta^{2\sqrt{N\beta}} \geq CN^{-2\beta-1} \delta^{2\sqrt{N\beta}}$$

for otherwise $m = \sqrt{\frac{N}{\beta}} + 1$ would give an $\underline{n}$ and $\underline{x}$ which contradicts (2.13). This completes the proof of the lower estimate in (2.4) and the theorem.

REFERENCES

1. Achieser, N. I., Theory of Approximation, Ungar, New York, 1956, 307 pp.

2. Burchard, H. and D. Hale, Piecewise polynomial approximation on optimal meshes, J. Approximation Theory 14 (1975), 128-147.

3. Dahmen, W. and K. Scherer, Best approximation by piecewise polynomials with variable knots and degrees, preprint.

4. Rice, J., On the degree of convergence of nonlinear spline approximation, Proc. Symp. on Approx. Theory, Madison, I. J. Schoenberg (Ed.), Academic Press., 1969, 349-365.

ACKNOWLEDGEMENT

The authors thank Professor Carl de Boor for helpful conversations about this work, especially his suggestion to use the transformation (2.2) in the proof of Lemma 2.

Ronald A. DeVore*
Department of Mathematics, Computer Science & Statistics
University of South Carolina
Columbia, South Carolina 29208

Karl Scherer
Institut für Angewandte Mathematik der Universität Bonn
53 Bonn, W. Germany

*This author was supported in part by the National Science Foundation Grant MCS 77 22982.

ON SHARP NECESSARY CONDITIONS FOR RADIAL FOURIER MULTIPLIERS

George Gasper and Walter Trebels

It is shown how a modification of the classical Hausdorff-Young inequality for Fourier transforms gives necessary criteria for radial Fourier multipliers in terms of smoothness conditions. These criteria improve results in [14], are best possible within the framework of weak bounded variation function spaces, and are comparable with some sufficient conditions.

It is well-known that interpolation techniques are very useful in multiplier theory; in particular the classical Hausdorff-Young inequality can be used to derive Fourier multiplier criteria. To become more precise let us first introduce our notation.

Let S denote the Schwartzian test function space of all infinitely differentiable functions on $\mathbb{R}^n$, $n\geq 1$, rapidly decreasing at infinity, and let S' denote its dual, the space of tempered distributions. Let $L^p(\mathbb{R}^n)$ be the standard Lebesgue space of p-th power integrable functions with norm

$$\|f\|_p = (\int_{\mathbb{R}^n} |f(x)|^p dx)^{1/p},\ 1\leq p<\infty,\ \|f\|_\infty = \operatorname{ess\,sup}_{\mathbb{R}^n}|f(x)| .$$

The Fourier transform of $f\in S$ is defined by

$$f^\wedge(\xi) = (2\pi)^{-n/2} \int_{\mathbb{R}^n} f(x)\, e^{-i\xi\cdot x}\, dx ,\qquad \xi\in\mathbb{R}^n .$$

Then the classical Hausdorff-Young inequality reads

$$\|f^\wedge\|_{p'} \leq \|f\|_p ,\quad 1\leq p\leq 2,\quad \frac{1}{p}+\frac{1}{p'} = 1 ;$$

for some applications of it in multiplier theory see, e.g. [5; p. 267], [9]. Let $\mathcal{F}^{-1}$ denote the inverse Fourier transform. A tempered distribution m is called a Fourier multiplier of type (p,r), notation $m\in M_p^r$, if

ISBN 0-12-213650-0

$$\|\mathcal{F}^{-1}[mf^\wedge]\|_r \le C\ \|f\|_p \quad , \qquad f\in S \ ;$$

the smallest constant C for which this inequality holds for all $f\in S$ defines the multiplier norm $\|m\|_{M_p^r}$ on M_p^r .

If S_{rad} denotes the subspace of S consisting of radial functions, then S_{rad} is dense in L^p_{rad} , the set of radial L^p-functions, whenever $1\le p<\infty$. The Fourier transform of $f\in S_{rad}$ may be computed [12; p. 155] via the modified Hankel transform of order $\alpha = (n-2)/2$ (here, in a slight abuse of notation, we set $f^\wedge(\rho) = f^\wedge(\xi)$, $\rho = |\xi|$, and $f(t) = f(x)$, $t = |x|$)

$$f^\wedge(\rho) = \mathcal{H}_{(n-2)/2}[f](\rho)$$

where, for $\alpha \ge -1/2$,

$$\mathcal{H}_\alpha[f](\rho) = \int_0^\infty f(t)(\rho t)^{-\alpha} J_\alpha(\rho t)\ t^{2\alpha+1}\ dt \ .$$

It is well-known that the Fourier transform of an arbitrary $f\in L^p_{rad}$ satisfies smoothness conditions up to a certain order (see, e.g., [10],[11],[12; p. 176],[13],[14]); hence, also a radial Fourier multiplier must satisfy certain smoothness conditions. But the smooth function $e^{i|\xi|^2} \notin M_p^p$, $p\neq 2$ (see [9]), shows that necessary criteria based upon smoothness alone cannot be good. An examination of the classical sufficient multiplier criteria of Marcinkiewicz, Mikhlin, and Hörmander makes it evident that one has to look for smoothness <u>and</u> growth conditions. Thus it is natural to head for necessary conditions which are comparable, e.g., with Hörmander's criterion [9] stated for radial Fourier multipliers as follows:

<u>If</u> $m(t)$ <u>is essentially bounded on</u> R^+ <u>and</u>

$$\|m\|_\infty + \sup_j \Big(\int_{2^{j-1}}^{2^j} |t^{[\frac{n}{2}]+1}\ m^{([\frac{n}{2}]+1)}(t)|^2\ \frac{dt}{t}\Big)^{1/2} < \infty \ ,$$

<u>then</u> $m(|\xi|) \in M_p^p$, $1<p<\infty$.

A result in this direction is contained in [14]; on account of an application of Hölder's inequality the necessary condition in [14] is not sharp. In this paper we replace the Hölder inequality argument by an interpolation argument which leads us to a modification of the classical Hausdorff-Young inequality.

To this end note that

$$\frac{d}{dt}\,(t^{-\alpha}J_\alpha(t)) = -t^{-\alpha}J_{\alpha+1}(t)$$

which immediately implies for $f\in S_{rad}$ that

$$(1)\qquad \frac{d}{d\rho}\,\mathcal{H}_\alpha[f](\rho) = -\rho\,\mathcal{H}_{\alpha+1}[f](\rho) \quad ;$$

in particular,

$$f^{\wedge\prime}(\rho) = -\rho\int_0^\infty (\rho t)^{-n/2}\,J_{n/2}(\rho t)f(t)\,t^{n+1}\,dt \quad .$$

Thus, since $J_\alpha(t) = O(t^{-1/2})$ as $t\to\infty$ and $J_\alpha(t) = O(t^\alpha)$ as $t\to 0$,

$$(2)\qquad \sup_{\rho>0}|\rho^{(n-1)/2}f^{\wedge\prime}(\rho)| \le \int_0^\infty |t^{(n+1)/2}f(t)|dt\ , \qquad f\in S_{rad} \quad .$$

On the other hand it follows by the Parseval formula [12, p. 16] with $\underline{\xi} = (\xi_1,\ldots,\xi_n,\xi_{n+1},\xi_{n+2})$ that

$$\int_0^\infty |f^{\wedge\prime}(\rho)|^2\,\rho^{n-1}\,d\rho = \int_0^\infty |\mathcal{H}_{n/2}[f](\rho)|^2\,\rho^{n+1}\,d\rho$$

$$= C\int_{\mathbb{R}^{n+2}} |f^{\wedge}(\underline{\xi})|^2\,d\underline{\xi} = C\int_{\mathbb{R}^{n+2}} |f(\underline{x})|^2\,d\underline{x}$$

$$= C\int_0^\infty |f(t)|^2\,t^{n+1}\,dt$$

and therefore

$$(3)\qquad \Big(\int_0^\infty |\rho^{(n-1)/2}f^{\wedge\prime}(\rho)|^2 d\rho\Big)^{1/2} = C\,\Big(\int_0^\infty |t^{(n+1)/2}f(t)|^2 dt\Big)^{1/2} \quad .$$

Applying the Riesz-Thorin interpolation theorem to (2) and (3) we obtain

$$\int_0^\infty |\rho^{(n-1)/2}f^{\wedge\prime}(\rho)|^{p'}d\rho\Big)^{1/p'} \le C\,\Big(\int_0^\infty |t^{(n+1)/2}f(t)|^p dt\Big)^{1/p}$$

with C independent of $f\in S$; in particular the following variant of the Hausdorff-Young inequality holds

$$(4)\qquad \Big(\int_0^\infty |\rho^{(n-1)/2}f^{\wedge\prime}(\rho)|^{p'}d\rho\Big)^{1/p'} \le C\|f\|_p,\ 1\le p = \frac{2(n-1)}{n+1} \le 2 \quad ,$$

for all $f \in L^p_{rad}$ since S_{rad} is dense. It is now easy to derive a necessary condition for radial Fourier multipliers from (4). Just consider a $\varphi \in S_{rad}$ with

$$\varphi^{\wedge}(\xi) = \begin{cases} 1, & 2^{-1} \le |\xi| \le 1 \quad , \\ 0, & |\xi| \le 2^{-2} \quad \text{or} \quad |\xi| \ge 2 \quad , \end{cases}$$

set $\varphi_j(x) = 2^{-jn}\,\varphi(2^{-j}x)$, and observe that, by [1], $\|\varphi_j\|_p \le C(2^j)^{n/p'}$. Then, with $m(\rho) = m(\xi)$, $\rho = |\xi|$, it follows from (4) that

$$\Big(\int_{2^{j-1}}^{2^j} |\rho^{(n-1)/2} m'(\rho)|^{p'} d\rho\Big)^{1/p'}$$

$$\le \Big(\int_0^\infty |\rho^{(n-1)/2} (m(\rho)\varphi_j^{\wedge}(\rho))'|^{p'} d\rho\Big)^{1/p'}$$

$$\le C\big\|\mathcal{F}^{-1}[m\varphi_j^{\wedge}]\big\|_p \le C\,\|m\|_{M_p^p}\,\|\varphi_j\|_p \le C\,2^{jn/p'}\,\|m\|_{M_p^p}$$

and hence

$$\sup_j \Big(\int_{2^{j-1}}^{2^j} |\rho m'(\rho)|^{p'} \frac{d\rho}{\rho}\Big)^{1/p'} \le C\,\|m\|_{M_p^p} \quad , \quad p = \frac{2(n-1)}{n+1} \quad .$$

Thus we have proved the case $k=1$ of the following

THEOREM 1. *Let m be a radial Fourier multiplier of type (p,p), $1 \le p \le 2$. Then*

$$(5) \quad \|m\|_\infty + \sup_j \Big(\int_{2^{j-1}}^{2^j} |\rho^k m^{(k)}(\rho)|^{p'} \frac{d\rho}{\rho}\Big)^{1/p'} \le C\,\|m\|_{M_p^p}$$

provided that k is a non-negative integer and $k = (n-1)(\frac{1}{p} - \frac{1}{2})$. (If $p=1$ this has to be interpreted as

$$\|m\|_\infty + \|\rho^k m^{(k)}(\rho)\|_\infty \le C\,\|m\|_{M_1^1} \;) \;.$$

Let us mention immediately that this theorem also gives a necessary condition for values of p for which $(n-1)(\frac{1}{p} - \frac{1}{2})$ is *not* an integer since $\|m\|_{M_r^r} \le \|m\|_{M_p^p}$, $1 \le p \le r \le 2$.

Proof. Since trivially $\|m\|_\infty \leq \|m\|_{M_p^p}$, we have to bound the integral in (5). Repeated applications of (1) show that

$$\mathcal{H}_{\frac{n}{2}+k-1}[f](\rho) = \sum_{j=1}^{k} c_{j,k}\,\rho^{j-2k}\,\mathcal{H}_{(n-2)/2}[f]^{(j)}(\rho) \quad .$$

Then, by a direct estimate as in the proof of (2),

$$\sup_{\rho>0} \Big|\sum_{j=1}^{k} c_{j,k}\,\rho^{j-k+(n-1)/2}\, f^{\wedge(j)}(\rho)\Big| \leq C \int_0^\infty |t^{k+(n-1)/2} f(t)|\,dt$$

and, by the Parseval formula as in the proof of (3),

$$\int_0^\infty \Big|\sum_{j=1}^{k} c_{j,k}\,\rho^{j-k+(n-1)/2}\, f^{\wedge(j)}(\rho)\Big|^2 d\rho = \int_0^\infty |\mathcal{H}_{k+(n-2)/2}[f](\rho)|^2\,\rho^{2k+n-1}\,d\rho = C\int_0^\infty |t^{k+(n-1)/2} f(t)|^2\,dt \quad .$$

Hence, by the Riesz-Thorin interpolation theorem, we have the following extension of (4)

$$(6)\quad \Big(\int_0^\infty \Big|\sum_{j=1}^{k} c_{j,k}\,\rho^{j-k+(n-1)/2}\, f^{\wedge(j)}(\rho)\Big|^{p_k'} d\rho\Big)^{1/p_k'} \leq C\,\|f\|_{p_k} \,,$$

$$1 \leq p_k = \frac{2(n-1)}{2k+n-1} \leq 2$$

for all $f\in L^{p_k}_{rad}$. Now let $m(|\xi|)\in M^{p_k}_{p_k}$. Then a repetition of the argument used in the case $k=1$ leads to

$$\Big(\int_{2^{j-1}}^{2^j} \Big|\sum_{i=1}^{k} c_{i,k}\,\rho^{i-k+(n-1)/2}\, m^{(i)}(\rho)\Big|^{p_k'} d\rho\Big)^{1/p_k'} \leq C\, 2^{jn/p_k'}\,\|m\|_{M^{p_k}_{p_k}} \quad .$$

Since $(n-1)/2-k = (n-1)/p_k'$, it follows from Minkowski's inequality that

$$\left(\int_{2^{j-1}}^{2^j} |\rho^k m^{(k)}(\rho)|^{p_k'} \frac{d\rho}{\rho}\right)^{1/p_k'}$$

$$\leq C\{\|m\|_{M_{p_k}^{p_k}} + \sum_{i=1}^{k-1} \left(\int_{2^{j-1}}^{2^j} |\rho^i m^{(i)}(\rho)|^{p_k'} \frac{d\rho}{\rho}\right)^{1/p_k'}\}$$

$$\leq C\{\|m\|_{M_{p_k}^{p_k}} + \sum_{i=1}^{k-1} \sup_{\rho>0} |\rho^{(i)} m^{(i)}(\rho)| \leq C \|m\|_{M_{p_k}^{p_k}}$$

where we used the fact, proved in [14], that

$$\sup_{\rho>0} |\rho^i m^{(i)}(\rho)| \leq C \|m\|_{M_p^p}, \quad 0 \leq i \leq k-1 < n(\tfrac{1}{p}-\tfrac{1}{2})-\tfrac{1}{2} .$$

Hence the proof is complete.

COROLLARY 2. *If* $m(|\xi|)\in M_p^r$, $1 \leq p < r < 2n/(n+1)$, $\frac{1}{r} = \frac{1}{p} + \frac{1}{s} - 1$, *then*

$$\|\rho^{n/s'} m(\rho)\|_\infty + \sup_j \left(\int_{2^{j-1}}^{2^j} |\rho^{k+n/s'} m^{(k)}(\rho)|^{r'} \frac{d\rho}{\rho}\right)^{1/r'}$$

$$\leq C \|m\|_{M_p^r}$$

provided that k *is a non-negative integer and* $k \doteq (n-1)(\frac{1}{r}-\frac{1}{2})$.

Proof. By Hölder's inequality and the previously mentioned estimates for $J_\alpha(t)$ we have that

$$\sup_{\rho>0} |\rho^{n/r'} f^\wedge(\rho)| \leq C \|f\|_r , \quad 1 \leq r < \frac{2n}{n+1} .$$

Thus, if $m(|\xi|)\in M_p^r$ and φ_j is as above,

$$\sup_{2^{j-1}\leq\rho\leq 2^j} |\rho^{n/r'} m(\rho)| \leq C \|\mathfrak{F}^{-1}[m\hat{\varphi}_j]\|_r$$

$$\leq C \|m\|_{M_p^r} \|\varphi_j\|_p \leq C\, 2^{jn/p'} \|m\|_{M_p^r}$$

with C independent of j, and so

$$\sup_{\rho>0} |\rho^{n/s'} m(\rho)| \leq C \|m\|_{M_p^r} .$$

Proof. Since trivially $\|m\|_\infty \leq \|m\|_{M_p^p}$, we have to bound the integral in (5). Repeated applications of (1) show that

$$\mathcal{H}_{\frac{n}{2}+k-1}[f](\rho) = \sum_{j=1}^{k} c_{j,k}\,\rho^{j-2k}\mathcal{H}_{(n-2)/2}[f]^{(j)}(\rho) \quad .$$

Then, by a direct estimate as in the proof of (2),

$$\sup_{\rho>0} \left|\sum_{j=1}^{k} c_{j,k}\,\rho^{j-k+(n-1)/2}\,f^{\wedge(j)}(\rho)\right|$$

$$\leq C \int_0^\infty |t^{k+(n-1)/2}\,f(t)|\,dt$$

and, by the Parseval formula as in the proof of (3),

$$\int_0^\infty \left|\sum_{j=1}^{k} c_{j,k}\,\rho^{j-k+(n-1)/2}\,f^{\wedge(j)}(\rho)\right|^2 d\rho$$

$$= \int_0^\infty |\mathcal{H}_{k+(n-2)/2}[f](\rho)|^2\,\rho^{2k+n-1}\,d\rho$$

$$= C \int_0^\infty |t^{k+(n-1)/2}\,f(t)|^2\,dt \quad .$$

Hence, by the Riesz-Thorin interpolation theorem, we have the following extension of (4)

$$(6) \quad \left(\int_0^\infty \left|\sum_{j=1}^{k} c_{j,k}\,\rho^{j-k+(n-1)/2}\,f^{\wedge(j)}(\rho)\right|^{p_k'} d\rho\right)^{1/p_k'} \leq C\,\|f\|_{p_k} \,,$$

$$1 \leq p_k = \frac{2(n-1)}{2k+n-1} \leq 2$$

for all $f \in L_{rad}^{p_k}$. Now let $m(|\xi|) \in M_{p_k}^{p_k}$. Then a repetition of the argument used in the case $k=1$ leads to

$$\left(\int_{2^{j-1}}^{2^j} \left|\sum_{i=1}^{k} c_{i,k}\,\rho^{i-k+(n-1)/2}\,m^{(i)}(\rho)\right|^{p_k'} d\rho\right)^{1/p_k'}$$

$$\leq C\,2^{jn/p_k'}\,\|m\|_{M_{p_k}^{p_k}} \quad .$$

Since $(n-1)/2-k = (n-1)/p_k'$, it follows from Minkowski's inequality that

$$\left(\int_{2^{j-1}}^{2^j} |\rho^k m^{(k)}(\rho)|^{p_k'} \frac{d\rho}{\rho}\right)^{1/p_k'}$$

$$\leq C\{\|m\|_{M_{p_k}^{p_k}} + \sum_{i=1}^{k-1} \left(\int_{2^{j-1}}^{2^j} |\rho^i m^{(i)}(\rho)|^{p_k'} \frac{d\rho}{\rho}\right)^{1/p_k'}\}$$

$$\leq C\{\|m\|_{M_{p_k}^{p_k}} + \sum_{i=1}^{k-1} \sup_{\rho>0} |\rho^{(i)} m^{(i)}(\rho)| \leq C \|m\|_{M_{p_k}^{p_k}}$$

where we used the fact, proved in [14], that

$$\sup_{\rho>0} |\rho^i m^{(i)}(\rho)| \leq C \|m\|_{M_p^p}, \quad 0 \leq i \leq k-1 < n(\tfrac{1}{p}-\tfrac{1}{2})-\tfrac{1}{2} .$$

Hence the proof is complete.

COROLLARY 2. *If* $m(|\xi|) \in M_p^r$, $1 \leq p < r < 2n/(n+1)$, $\frac{1}{r} = \frac{1}{p} + \frac{1}{s} - 1$, *then*

$$\|\rho^{n/s'} m(\rho)\|_\infty + \sup_j \left(\int_{2^{j-1}}^{2^j} |\rho^{k+n/s'} m^{(k)}(\rho)|^{r'} \frac{d\rho}{\rho}\right)^{1/r'}$$

$$\leq C \|m\|_{M_p^r}$$

provided that k *is a non-negative integer and* $k \doteq (n-1)(\frac{1}{r}-\frac{1}{2})$.

Proof. By Hölder's inequality and the previously mentioned estimates for $J_\alpha(t)$ we have that

$$\sup_{\rho>0} |\rho^{n/r'} f^\wedge(\rho)| \leq C \|f\|_r , \qquad 1 \leq r < \frac{2n}{n+1} .$$

Thus, if $m(|\xi|) \in M_p^r$ and φ_j is as above,

$$\sup_{2^{j-1} \leq \rho \leq 2^j} |\rho^{n/r'} m(\rho)| \leq C \|\mathfrak{F}^{-1}[m\varphi_j^\wedge]\|_r$$

$$\leq C \|m\|_{M_p^r} \|\varphi_j\|_p \leq C\, 2^{jn/p'} \|m\|_{M_p^r}$$

with C independent of j , and so

$$\sup_{\rho>0} |\rho^{n/s'} m(\rho)| \leq C \|m\|_{M_p^r} .$$

Analogously, the remaining integral is estimated by using the corresponding estimate in the previous proof. Thus the Corollary is established.

REMARKS. 1) The above Theorem (and Corollary) can be extended to fractional values of k within the framework of weak bounded variation function spaces (WBV-spaces; see [7]) as follows: Define for $0<\delta<1$ and a locally integrable function g the fractional integral

$$I_\omega^\delta(g)(t) = \begin{cases} \frac{1}{\Gamma(\delta)} \int_t^\omega (s-t)^{\delta-1} g(s)\, ds, & 0<t<\omega \\ 0, & t\geq\omega \end{cases}$$

and the fractional derivatives (with $[\gamma]$ denoting the integer part of γ)

$$g^{(\delta)}(t) = \lim_{\omega\to\infty} -\frac{d}{dt} I_\omega^{1-\delta}(g)(t) ,$$

$$g^{(\gamma)}(t) = \left(\frac{d}{dt}\right)^{[\gamma]} g^{(\gamma-[\gamma])}(t) , \quad t,\gamma>0 ,$$

whenever the right sides exist (see [3],[7]). Then for $\gamma>0$, $1\leq q\leq\infty$, consider

$$\begin{aligned} WBV_{q,\gamma} = \{g\in L^\infty\cap C(0,\infty): \; & I_\omega^{1-\delta}(g)\in AC_{loc},\ \omega>0, \quad \text{if } \delta>0 , \\ & g^{(\delta)},\ldots,g^{(\gamma-1)}\in AC_{loc} \quad \text{if } \gamma\geq 1 , \\ & \text{and } \|g\|_{q,\gamma;W} < \infty\} \end{aligned}$$

where $\delta = \gamma-[\gamma]$,

$$\|g\|_{q,\gamma;W} = \|g\|_\infty + \sup_{j\in\mathbb{Z}} \left(\int_{2^{j-1}}^{2^j} |t^\gamma g^{(\gamma)}(t)|^q \frac{dt}{t}\right)^{1/q} , \quad 1\leq q<\infty,$$

and $\|g\|_{\infty,\gamma;W} = \|g\|_\infty + \|t^\gamma g^{(\gamma)}(t)\|_\infty$.

In a paper (to appear) on necessary conditions for Hankel multipliers it will be proved that if $m\in M_p^p$ is radial, notation $m\in M_p^p(\text{rad})$, and we set $m(\rho) = m(\xi)$, $\rho = |\xi|$, then $m\in WBV_{p',\gamma}$ for $0\leq\gamma\leq(n-1)|\frac{1}{p}-\frac{1}{2}|$ and $\|m\|_{p',\gamma;W} \leq C\,\|m\|_{M_p^p}$, i.e.,

(7) $$M_p^p(\text{rad}) \subset WBV_{p',\gamma} , \quad 0 \leq \gamma \leq (n-1)\left|\frac{1}{p}-\frac{1}{2}\right| , \quad 1<p<\infty .$$

Since $t^{-\alpha}J_\alpha(t)$ and all of its derivatives are locally absolutely continuous away from the origin and $M_p^p = M_{p'}^{p'}$, the above Theorem (i.e. relation (5)) is just (7) in the case when $k = \gamma = (n-1)|\frac{1}{p}-\frac{1}{2}|$ is an integer.

2) In [14] it is proved that if $1 \le p < 2n/(n+1)$ then

$$(7') \quad M_p^p(\text{rad}) \subset WBV_{\infty,\kappa} , \quad 0 \le \kappa < n(\frac{1}{p}-\frac{1}{2})-\frac{1}{2} .$$

Since, by [7; Theorem 4], $WBV_{p',\gamma} \subset WBV_{\infty,\kappa}$, $1/p' < \gamma-\kappa$, (7) clearly improves (7').

3) To show that (7) is sharp consider the example $|\xi|^{-\mu}J_\mu(|\xi|)$. For $\mu = -\frac{1}{2}$ it is bounded and hence belongs to M_2^2; for $\mu = \frac{n}{2}-1$ it is the Fourier transform of the (bounded) surface measure of the unit sphere $\{x \in \mathbb{R}^n : |x|=1\}$, see [13]. Interpolating between these two end points H.Dappa (personal communication) has shown (see also [4]) via the Stein interpolation theorem for analytic families of operators [12; p. 205] that

$$(8) \quad K_\mu(|\xi|) \equiv |\xi|^{-\mu} J_\mu(|\xi|) \in M_p^p , \quad \mu \ge (n-1)|\frac{1}{p}-\frac{1}{2}|-\frac{1}{2} .$$

Clearly $K_\mu(t)$, $t>0$, is a C^∞-function. Furthermore all derivatives (including the fractional ones) behave like $O(t^{-\mu-1/2})$ for large t and so $K_\mu \notin WBV_{p',\gamma}$ when $\gamma > \mu+\frac{1}{2}$, which shows that the differentiation order γ in (7) cannot be increased. At the same time (7) tells us that (8) is best possible: $K_\mu(|\xi|) \in M_p^p$ if and only if $\mu \ge (n-1)|\frac{1}{p}-\frac{1}{2}|-\frac{1}{2}$.

By Hölder's inequality it easily follows that $WBV_{q_1,\gamma} \subset WBV_{q_2,\gamma}$ if $q_1 > q_2$; therefore the question arises if one can improve (7) by increasing the parameter $q=p'$. It can be seen that this is not possible by using the Riesz means $r_\mu(|\xi|) = (1-|\xi|)_+^\mu \equiv (\max\{1-|\xi|,0\})^\mu$: Since one can easily show that $r_\mu^{(\gamma)}(t) = C(1-t)_+^{\mu-\gamma}$, $r_\mu \in WBV_{q,\gamma}$ with $\gamma = (n-1)|\frac{1}{p}-\frac{1}{2}|$ and $q > p' > 2$ would require (on account of the L^q_{loc} integrability of $r_\mu^{(\gamma)}$ at $t=1$) that $\mu > \gamma-\frac{1}{q} = n|\frac{1}{p}-\frac{1}{2}|-\frac{1}{2} + (\frac{1}{p'}-\frac{1}{q})$ which contradicts the fact [13] that $r_\mu \in M_p^p$ when $\mu > n|\frac{1}{p}-\frac{1}{2}|-\frac{1}{2}$, $\mu \ge \frac{n-1}{2(n+1)}$.

Thus $q \le p'$ and (7) is sharp within the framework of weak bounded variation function spaces. As a further application of (7) we obtain the well-known result that $r_\mu \notin M_p^p$ if $\mu \le n|\frac{1}{p}-\frac{1}{2}|-\frac{1}{2}$.

4) It is shown in [8] that

$$\text{(9)} \quad WBV_{2,\gamma} \subset M_p^p(\text{rad}) \ , \quad 1<p<\infty \ , \quad \gamma>n/2 \ .$$

By applying the complex interpolation method to (9) and $L^\infty = M_2^2$ one may show (cf. [2],[7]) that

$$WBV_{q,\gamma} \subset M_p^p(\text{rad}) \quad , \quad \gamma>n|\frac{1}{p}-\frac{1}{2}| \ , \quad q\gamma>n \ .$$

The examples $r_\mu(|\xi|)$ and $e^{i|\xi|^\nu}(1+|\xi|)^{-\nu\gamma}$ [6] suggest the conjecture that for $1<p<2$

$$\text{(10)} \quad WBV_{q,\gamma} \subset M_p^p(\text{rad}) \ , \quad q>1 \ , \quad \gamma \ge \max\{\frac{1}{q}, \gamma_c(p), \gamma_c(p)+\frac{1}{q}-\frac{1}{2}\}$$

where the critical index $\gamma_c(p)$ is defined by $\gamma_c(p) = n|\frac{1}{p}-\frac{1}{2}|$. It can be shown with the aid of the last two examples that (10) cannot be improved by lowering q or γ , i.e. (10) (if true) would be best possible within the framework of the WBV-spaces. A comparison between (7) and (10) shows that the conjectured best possible sufficient condition and the best possible necessary one on M_p^p, $1<p<2$, are governed precisely by the Sobolev type embedding theorem for WBV-spaces [7; Theorem 4].

5) Analogous comments hold for M_p^r(rad)-multipliers.

REFERENCES

1. Butzer, P.L., R.J. Nessel, and W. Trebels, On radial M_p^q-Fourier multipliers. In: Math. Structures - Computational Math. - Math. Modelling, pp. 187-193, Sofia 1975.

2. Connett, W.C. and A.L. Schwartz, The theory of ultraspherical multipliers. Memoirs, Amer. Math. Soc. 9 (1977), no. 183.

3. Cossar, J., A theorem on Cesàro summability, J. London Math. Soc. 16 (1941), 56-68.

4. Dappa, H., Spezielle Fourier-Multiplikatoren, Diplomarbeit, TH Darmstadt 1979.

5. Edwards, R.E., Fourier Series, Vol. 2, New York 1967.

6. Fefferman, C. and E.M. Stein, H^p spaces of several variables, Acta Math. 129 (1972), 137-193.

7. Gasper, G. and W. Trebels, A characterization of localized Bessel potential spaces and applications to Jacobi and Hankel multipliers. Studia Math. 65 (1979), 17-52.

8. Gasper, G. and W. Trebels, Multiplier criteria of Hörmander type for Fourier series and applications to Jacobi series and Hankel transforms. Math. Ann. 242 (1979), 225-240.

9. Hörmander, L., Estimates for translation invariant operators in L^p-spaces. Acta Math. 104 (1960), 93-139.

10. Schoenberg, I.J., Metric spaces and completely monotone functions. Ann. of Math. (2) 39 (1938), 811-841.

11. Schwartz, A.L., The smoothness of Hankel transforms, J. Math. Anal. Appl. 28 (1969), 500-507.

12. Stein, E.M. and G. Weiss, Introduction to Fourier analysis on Euclidean spaces. Princeton 1971.

13. Tomas, P.A., On radial Fourier multipliers, Thesis, Cornell Univ. 1974.

14. Trebels, W., Some necessary conditions for radial Fourier multipliers. Proc. Amer. Math. Soc. 58 (1976), 97-103.

G. Gasper*
Department of Mathematics
Northwestern University
Evanston, Illinois 60201
U.S.A.

W. Trebels
Fachbereich Mathematik
Technische Hochschule
D-6100 Darmstadt
Federal Republic of Germany

* Supported in part by NSF Grant MCS 76-06635 A01.

APPROXIMATING BIVARIATE FUNCTIONS AND MATRICES BY NOMOGRAPHIC FUNCTIONS

M. v. Golitschek

We study certain discrete problems in nomography and present an efficient algorithm for computing best nomographic approximations.

There has been some recent interest ([1]-[7]) in the following approximation problem: Given a continuous function f on the unit square $Q = [0,1]^2$ and a continuous function g of one real variable, find the distance λ;

$$(1) \qquad \lambda := \inf_{u,v \in C[0,1]} \max_{(x,y) \in Q} |f(x,y) - g(u(x) + v(y))|.$$

Functions of form $g(u(x) + v(y))$ are called nomographic.

EXAMPLE 1. In fact, all papers mentioned above are concerned with the case of $g(t) = t$, i.e., the approximation by functions of form $u(x) + v(y)$. Little is done in the general case.

EXAMPLE 2. If $g(t) = e^t$, then (1) is equivalent to the problem of approximating a given function f by products $U(x) \cdot V(y)$ of positive functions U and V.

EXAMPLE 3. If $g(t) = \ln t$, the one-sided problem

$$\inf_{u,v}\{\delta \,|\, 0 \leq f(x,y) - \ln(u(x) + v(y)) \leq \delta \text{ for all } (x,y) \in Q\}$$

is equivalent to

$$\inf_{u,v}\left\{\delta \,\middle|\, 1 \leq \frac{F(x,y)}{u(x) + v(y)} \leq \delta \text{ for all } (x,y) \in Q\right\},$$

where F is a given positive function on Q.

ISBN 0-12-213650-0

By discretization, $0 \leq x_1 < x_2 < \ldots < x_m \leq 1$, $0 \leq y_1 < y_2 < \ldots < y_n \leq 1$, $a_{ij} := f(x_i, y_j)$, $u_i := u(x_i)$, $v_j := v(y_j)$, (1) leads to the following matrix problem formulated in a slightly more general setting.

PROBLEM A. Let $A = (a_{ij})$ be a real (m,n)-matrix, E a subset of $\{(i,j) | i=1,\ldots,m , j=1,\ldots,n\}$, D an open interval, and g a continuous real-valued function in D. Find the distance λ^*,

$$(2) \quad \lambda^* := \inf_{u_i+v_j \in D} \max_{(i,j)\in E} |a_{ij} - g(u_i + v_j)|.$$

The purpose of this short note is to characterize λ^* and to present an effective algorithm for the computation of λ^* under the following conditions for the function g:

(3a) g is strictly increasing and continuous in D,

(3b) the inclusion

$$\left\{z \in R \,\middle|\, \min_{(i,j)\in E} a_{ij} \leq z \leq \max_{(i,j)\in E} a_{ij}\right\} \subseteq \{g(t) | t \in D\}$$

is satisfied.

Remark 1. Condition (3a), the monotonicity of g, is essential for the results of this paper. Condition (3b) could be weakened, but is sufficient to guarantee that our algorithm below can be applied without difficulties.

The closed simple paths in E play an essential role in our paper, since they are so-called H-sets for the classes $g(u(x) + v(y))$ of functions if g has property (3a) (see Lemma 1 below).

Definition: A sequence p of 2k distinct knots in E,

$$(4) \quad p = (i_1,j_1),(i_1,j_2),(i_2,j_2),\ldots,(i_k,j_k),(i_k,j_{k+1})$$

with $j_{k+1} = j_1$, $i_r \neq i_s$, $j_r \neq j_s$ for $r, s = 1,\ldots,k$, $r \neq s$, is called a closed simple path in E. The collection of all closed simple paths in E is denoted by P.

LEMMA 1. *Suppose that* $g \in C(D)$ *has property* (3a) *and that* p *is a closed simple path in* E *of form* (4). *If there exist numbers* M, $M \geqq 0$, *and* u_{i_r}, v_{j_r}, $(r = 1,2,\ldots,k)$ *such that the inequalities*

$$(5) \quad a_{i_r,j_r} - g\left(u_{i_r} + v_{j_r}\right) \geqq M, \quad a_{i_r,j_{r+1}} - g\left(u_{i_r} + v_{j_{r+1}}\right) \leqq -M,$$

$r = 1,2,\ldots,k$, *hold with at least one strict inequality in* (5), *then*

$$(6) \quad \lambda_p := \inf_{u_i+v_j \in D} \max_{(i,j)\in p} |a_{ij} - g(u_i + v_j)| > M$$

where the first equality in (6) *serves to define* λ_p.

Proof. It follows by (3a) and (5) that we can find a positive number δ and numbers u'_{i_r}, v'_{j_r} such that $u'_{i_1} = u_{i_1}$ and

$$(7) \quad \begin{aligned} a_{i_r,j_r} - g\left(u'_{i_r} + v'_{j_r}\right) &\geqq M + 2\delta, \\ a_{i_r,j_{r+1}} - g\left(u'_{i_r} + v'_{j_{r+1}}\right) &\leqq -M - 2\delta, \end{aligned}$$

$r = 1,2,\ldots,k$, holds. Suppose that $\lambda_p \leqq M$. Then there exist numbers u''_{i_r} and v''_{j_r} such that

$$(8) \quad \begin{aligned} a_{i_r,j_r} - g\left(u''_{i_r} + v''_{j_r}\right) &\leqq M + \delta, \\ a_{i_r,j_{r+1}} - g\left(u''_{i_r} + v''_{j_{r+1}}\right) &\geqq -M - \delta, \end{aligned}$$

$r = 1,2,\ldots,k$, is satisfied. Obviously, we may assume that $u''_{i_1} = u'_{i_1}$. Then, by (7) and (8), we obtain the inequalities

$$(9) \quad \begin{aligned} g\left(u''_{i_r} + v''_{j_r}\right) &> g\left(u'_{i_r} + v'_{j_r}\right), \\ g\left(u''_{i_r} + v''_{j_{r+1}}\right) &< g\left(u'_{i_r} + v'_{j_{r+1}}\right), \end{aligned}$$

$r = 1,2,\ldots,k$, and thus, by (3a) and $u''_{i_1} = u'_{i_1}$, the sequence of inequalities

$$v''_{j_1} > v'_{j_1};\ v''_{j_2} < v'_{j_2};\ u''_{i_2} > u'_{i_2};\ v''_{j_3} < v'_{j_3}; \ldots ; \ldots$$

$$u''_{i_k} > u'_{i_k};\ v''_{j_{k+1}} < v'_{j_{k+1}}.$$

Since $j_{k+1} = j_1$, the first and last inequality contradict each other. That conclusdes the proof of Lemma 1.

It will be shown in our theorem below that the set P of closed simple paths even contains an extremal path with $\lambda_p = \lambda^*$. For the computation of the distance λ^* we suggest the following algorithm:

ALGORITHM. Let without loss of generality $n \leq m$. Choose a number $M_0 \geq \lambda^*$, for instance $M_0 := (\max a_{ij} - \min a_{ij})/2$. Set $M_1 := 0$ and $M := (M_1 + M_0)/2$.

STEP 0. Set $u_{i0} := v_{i0} := 0$ for $i \in I_0 := \{1,2,\ldots,m\}$ and $j \in J_0 := \{1,2,\ldots,n\}$.

STEP 1. Perform alternately steps 1a and 1b for $k=1,2,\ldots,n$.

<u>1a</u>. Compute the set $I_k = I_k(M)$,

$$I_k := \left\{ i \in I_0 \,\middle|\, \max_{j \in J_0} \{a_{ij} - g(u_{i,k-1} + v_{j,k-1})\} > M \right\},$$

and set $u_{ik} := u_{i,k-1}$ for $i \in I_0 \setminus I_k$. For $i \in I_k$ compute the unique number u_{ik} for which

$$\max_{j \in J_0} \{a_{ij} - g(u_{ik} + v_{j,k-1})\} = M \tag{10}$$

is satisfied.

<u>1b</u>. Compute the set $J_k = J_k(M)$,

$$J_k := \left\{ j \in J_0 \,\middle|\, \min_{i \in I_0} \{a_{ij} - g(u_{ik} + v_{j,k-1})\} < -M \right\},$$

and set $v_{jk} := v_{j,k-1}$ for $j \in J_0 \setminus J_k$. For $j \in J_k$ compute the unique number v_{jk} for which

$$\min_{i \in I_0} \{a_{ij} - g(u_{ik} + v_{jk})\} = -M \tag{11}$$

is satisfied.

If the set J_k is empty, then we realize that

$$\max_{(i,j)\in E} |a_{ij} - g(u_{ik} + v_{jk})| \leq M$$

and hence $\lambda^* \leq M$. Then we set $M_0: = M$, choose the new number $M: = (M_1 + M_0)/2$ and go to Step 0.

STEP 2. If the set J_n is nonempty we know by Lemma 2 below that $\lambda^* > M$. Hence, we set $M_1: = M$, choose the new number $M: = (M_1 + M_0)/2$ and go to Step 0.

Remark 2. The numbers M_0 are always upper and the numbers M_1 are always lower bounds of distance λ^*. The maximum in (10) is attained for an index $j \in J_{k-1}$ and the minimum in (11) for an index $i \in I_k$. Hence it suffices to calculate the maximum in 1a only for all $j \in J_{k-1}$ and the minimum in 1b only for all $i \in I_k$.

The above algorithm is essentially based on the following observation:

LEMMA 2. *If the set $J_n = J_n(M)$ of our above algorithm is nonempty, then distance λ^* is greater than M.*

Proof. Let $j \in J_n$. Then we can find a sequence

$$j_n: = j, i_{n-1}, j_{n-1}, \ldots, j_2, i_1, j_1, i_0, j_0 \tag{12}$$

of indices such that $j_k \varepsilon J_k$, $i_k \varepsilon I_k$ and the minimum in (11) is attained by (i_{k-1}, j_k) and the maximum in (10) is attained by (i_{k-1}, j_{k-1}), $k=n,n-1,\ldots,1$, i.e.

$$\begin{aligned} a_{i_{k-1},j_k} - g(u_{i_{k-1},k} + v_{j_k,k}) &= -M \\ a_{i_{k-1},j_{k-1}} - g(u_{i_{k-1},k} + v_{j_{k-1},k-1}) &= M. \end{aligned} \tag{13}$$

We realize that

$$\begin{aligned} 0 < u_{i_0,1} < u_{i_1,2} < \cdots < u_{i_k,k+1} < \cdots < u_{i_{n-1},n} \\ 0 = v_{j_0,0} > v_{j_1,1} > \cdots > v_{j_k,k} > \cdots > v_{j_n,n}. \end{aligned} \tag{14}$$

At least one of the following two cases occurs:

Case 1. There exist indices r and s, $0 \leqq r < s \leqq n-1$, such that $j_{s+1} = j_r$ and the sequence

$$p := (i_r,j_r),(i_r,j_{r+1}),(i_{r+1},j_{r+1}),\ldots,(i_s,j_s),(i_s,j_{s+1}) \tag{15a}$$

is a closed simple path of form (4).

Case 2. There exist indices r and s, $0 \leqq r < s \leqq n-2$, such that $i_{s+1} = i_r$ and the sequence

$$\begin{aligned} p := {} & (i_r,j_{s+1}),(i_r,j_{r+1}),(i_{r+1},j_{r+1}),\ldots, \\ & (i_s,j_s),(i_s,j_{s+1}) \end{aligned} \tag{15b}$$

is a closed simple path of form (4).

If Case 1 occurs, we set $u_{i_k} := u_{i_k,k+1}$, $v_{j_k} := v_{j_k,k}$ for $k = r,r+1,\ldots,s$. Then it follows by (13) that

$$\begin{aligned} a_{i_k,j_k} - g\left(u_{i_k} + v_{j_k}\right) &= M, \quad k = r,\ldots,s \\ a_{i_k,j_{k+1}} - g\left(u_{i_k} + v_{j_{k+1}}\right) &= -M, \quad k = r,\ldots,s-1 \end{aligned} \tag{16}$$

Since $j_{s+1} = j_r$ and $v_{j_{s+1},s+1} < v_{j_r,r}$ it follows from (13) and the monotonicity of g that

$$a_{i_s,j_{s+1}} - g\left(u_{i_s} + v_{j_{s+1}}\right) < -M. \tag{17}$$

Lemma 1, (16), and (17) lead to the inequality $\lambda_p > M$ and hence to $\lambda^* \geqq \lambda_p > M$.

If Case 2 occurs, we set $u_{i_k} := u_{i_k,k+1}$, $k = r,\ldots,s$, and $v_{j_k} := v_{j_k,k}$, $k = r+1,\ldots,s+1$. Then strict inequality $\lambda_p > M$ follows by the same argument as in Case 1 since $i_{s+1} = i_r$ and $u_{i_{s+1},s+2} > u_{i_r,r+1}$ and hence

$$\begin{aligned} a_{i_r,j_{s+1}} - g\left(u_{i_r} + v_{j_{s+1}}\right) > {} \\ a_{i_{s+1},j_{s+1}} - g\left(u_{i_{s+1},s+2} + v_{j_{s+1},s+1}\right) = M. \end{aligned}$$

That concludes the proof of Lemma 2.

THEOREM. *After the N-th iteration of the above algorithm the estimate* $M = M(N)$ *of* λ^* *satisfies the inequality*

$$(18) \quad |\lambda^* - M| \leqq M_0 2^{-N-1}.$$

The distance λ^* *can be characterized by*

$$(19) \quad \lambda^* = \max_{p \in P} \lambda_p.$$

Proof. Inequality (18) is an immediate consequence of the definition of M after each iteration. Set $1^* := \max_{p \in P} \lambda_p$. Then, $\lambda^* > 1^*$ since $p \subset E$ for any $p \in P$. Suppose that $\lambda^* > 1^*$. Then apply the above algorithm with a number M, $1^* < M < \lambda^*$. Since $J_n(M)$ must be nonempty, we find a sequence of form (12) and a closed simple path p of form (15a) or (15b) with $\lambda_p > M$. But that contradicts our assumption $1^* < M$. That concludes the proof of our theorem.

Remark 3. The above algorithm can be used to find an extremal closed simple path. We only have to apply our algorithm with a number M satisfying

$$\lambda^* > M > \max\{\lambda_p | p \in P \text{ with } \lambda_p < \lambda^*\}.$$

Then each closed simple path p constructed by (12), (15a) or (15b) is extremal.

Remark 4. For the special case $g(u + v) := u + v$, the author ([6],[7]) has proved that one iteration of the above algorithm suffices to calculate λ^*.

Remark 5. Under conditions (3) the one-sided approximation (PROBLEM B),

$$(20) \quad \inf_{u_i + v_j \in D} \{\lambda | 0 \leqq a_{ij} - g(u_i + v_j) \leqq \lambda \text{ for all } (i,j) \in E\},$$

can also be solved by our algorithm if we define in Step 1b

$$J_k := \left\{ j \in J_0 \middle| \min_{i \in I_0} \{a_{ij} - g(u_{ik} + v_{j,k-1})\} < 0 \right\}$$

and replace (11) by

$$(11)' \quad \min_{i \in I_0} \{a_{ij} - g(u_{ik} + v_{jk})\} = 0.$$

Then Lemma 2 and the above theorem are also valid for the one-sided problem (20). Lemma 1 also holds if we replace in (5) the upper bound (-M) by 0 and in (6) the length λ_p of p by

$$\lambda_p' := \inf_{u_i + v_j \in D} \{\lambda \,|\, 0 \leq a_{ij} - g(u_i + v_j) \leq \lambda \text{ for all } (i,j) \in p\}.$$

<u>Remark 6.</u> Let $A = (a_{ij})$ be a nonnegative (m,n)-matrix. For $g(t) = e^t$, Problems A and B are equivalent to the problem of approximating A by the subclass of all positive (m,n)-matrices of rank 1.

REFERENCES

1. Aumann, G., Über Approximative Nomographie, I and II, Bayer. Akad. Wiss. Math. - Nat.Kl.S.B. (1958), 137-155, Ibid. (1959), 103-109.

2. Cheney, E. W. and W. A. Light, On the approximation of a bivariate function by the sum of univariate functions, Report CNA-140, Center for Numerical Analysis, University of Texas at Austin, August 1978, to appear, J. Approximation Theory.

3. Cheney, E. W. and M. v. Golitschek, On the algorithm of Diliberto and Straus for approximating bivariate functions by univeriate ones, Numer. Funct. Anal. and Optimiz. 1 (1979), 341-363.

4. Diliberto, S. P. and E. G. Straus, On the approximation of a function of several variables by the sum of functions of fewer variables, Pacific J. Math. 1 (1951), 195-210.

5. Fulkerson, D. R. and P. Wolfe, An algorithm for scaling matrices, SIAM Review 4 (1962), 142-146.

6. Golitschek, M. v., Approximation of functions of two variables by the sum of two functions of one variable, Proceedings of Conference on Numerical Methods in Approximation Theory (conducted by L. Collatz, G. Meinardus, and H. Werner), Oberwolfach, March 1979.

7. Golitschek, M. v., An algorithm for scaling matrices and computing the minimum cycle mean in a digraph. Preprint Nr. 51, Mathem. Institute de Universität Würzburg, June 1979.

8. Golomb, M., Approximation by functions of fewer variables, On Numerical Approximation (R. Langer, ed.), University of Wisconsin Press, Madison, Wisconsin, (1959), 275-327.

M. v. Golitschek
Institut für Angewandte Mathematik und Statistik
der Universität Würzburg
8700 Würzburg
GERMANY (Fed. Rep.)

L_∞-BOUNDS OF L_2-PROJECTIONS ON SPLINES

B. Güsmann

Nitsche and de Boor proved the existence of L_∞-norms of L_2-projections in terms of a global mesh ratio. Modifying their proofs, we give sufficient conditions for the existence of a uniform bound for a set of meshes. The elements of such a set need not be quasiuniform.

1. INTRODUCTION

Results on L_∞-bounds of L_2-projections on splines can be found in papers of Nitsche [11], Douglas, Dupont, Wahlbin [6] and de Boor [4]. Let us define a mesh π of $I = [0,1]$ by $\pi:\ 0 = x_0 < x_1 < \ldots < x_n = 1$ and furthermore $I_i = [x_i, x_{i+1}]$, $S(r,k,\pi) = \{v \in C^k\colon \forall v \text{ is a polynomial of degree} \leq r \text{ in } I_i, \forall i\}$. The L_2-projection is defined by

$$P_2^\pi\colon \begin{cases} L_\infty(I) \to S(r,k,\pi) \\ u \to u_\pi\,, \end{cases}$$

$(u_\pi, v) = (u,v)$, $\forall v \in S(r,k,\pi)$ and the L_∞-norms are

$$\|u\|_{L_\infty(I)} = \sup_{x\in I} |u(x)|$$

$$\|P_2\|_{L_\infty} = \sup_{\substack{u\in L_\infty \\ u\neq 0}} \frac{\|P_2^\pi u\|_{L_\infty(I)}}{\|u\|_{L_\infty(I)}}\,.$$

The results are as follows:

(1.1) Let P_2^π be the projection on $S(r,0,\pi)$, then $\|P_2^\pi\|_{L_\infty} \leq C$ independent of π [4].

(1.2) Let P_2^π be the projection on $S(2,1,\pi)$, then $\|P_2^\pi\|_{L_\infty} \leq C$ independent of π [2].

ISBN 0-12-213650-0

(1.3) If $C_g := \max_{i,j} \frac{h_i}{h_j}$, $h_i = x_{i+1} - x_i$, is uniformly bounded for a set of meshes Π, then the projection on $S(r,k,\pi)$ fulfills $\|P_2^\pi\|_{L_\infty} \leqq C$ independent of $\pi \in \Pi$ [11], [6], [4].

Following the proofs of [11] and [4], it is stated in section 3 that a result like (1.3) can be obtained with weaker assumptions. Examples of non-quasiuniform sets of meshes satisfying our assumptions are given.

For applications of the results of spline interpolation see [2], [3]. Applications to Galerkin projections are pointed out in section 4.

2. PRELIMINARIES

Π is called locally quasiuniform if

a) $C_\ell = \max_{|i-j|=1} \frac{h_i}{h_j}$ is bounded uniformly for all $\pi \in \Pi$.

In the following we shall need the property

b) For a given $\varepsilon > 0$ there is an integer m such that

$$\frac{h_i}{|x_i - x_j|} \leqq \varepsilon \text{ if } |i-j| \geqq m, \ \forall \pi \in \Pi.$$

For the extension of Nitsche's result [11] we need weight-functions

$$(2.1) \quad p_\alpha(x) := (\rho^2 + (x - \bar{x})^2)^\alpha, \ \rho, \ \alpha \in \mathbb{R}, \ \bar{x} \in [0,1]$$

and the estimate

$$(2.2) \quad |p_\alpha^{(\nu)}(x)| \leqq C(\alpha,\nu) \cdot p_{\alpha-\nu/2}(x).$$

$C(\alpha,\nu)$ depends not on ρ and $\bar{x}$. Let j be defined by $x_j \leqq \bar{x} < x_{j+1}$. For $\bar{x} = 1$ we put $j := n - 1$. $(j = j(\pi))$.

If, for a $K \in \mathbb{R}$, we define $\rho := K \cdot h_j$, then we have $p_\alpha = p_{\alpha,\pi}$. For a set of meshes Π and a constant K we define

$$\kappa := \kappa(K) := \sup_{\pi \in \Pi} \max_i h_i \|p_{-1/2,\pi}\|_{L_\infty(I_i)}.$$

The most important properties of $p_{\alpha,\pi}$ are noted in Lemma 1 and Lemma 2.

LEMMA 1. *Let* $\varepsilon > 0$ *be given and let* Π *satisfy* a) *and* b) *for this* ε. *Then there is* $K > 0$, *such that with* $\rho = K\cdot h_j$, $\kappa \leq \varepsilon$.

Proof [10]. We define

$$\bar{p}_{\alpha,\pi,i} := \max_{x\in I_i} p_{\alpha,\pi}(x)$$

$$\underline{p}_{\alpha,\pi,i} := \min_{x\in I_i} p_{\alpha,\pi}(x) .$$

LEMMA 2. *Let* $\bar{I}_i$ *be the union of* I_i *with* t *neighboring intervals and let* Π *satisfy* a) *and* b) *for* $\varepsilon = \frac{1}{2\cdot C(\alpha,1)}$. (see (2.2) for $C(\alpha,\nu)$). *Then there is* $K > 0$ *and a* $C(t)$ *such that with* $\rho = K\cdot h_j$,

$$\frac{\bar{p}_{\alpha,\pi,m}}{\underline{p}_{\alpha,\pi,n}} \leq C(t)\ ,\forall m,n:\quad I_m, I_n \subset \bar{I}_i,\ \forall\pi \in \Pi,\ \forall i\ .$$

Proof. The proof follows by (2.2), Lemma 1 and the continuity of $p_{\alpha,\pi}$.

For $u \in L_2(I_i)$ we define

$$\|u\|_{\alpha,I_i} := \left(\int_{I_i} p_\alpha u^2 dx\right)^{1/2} .$$

If $u \in L_2(I_i)$ but $u \notin L_2(I)$, then

$$\|u\|^2_{\alpha,I} = \sum_i \|u\|^2_{\alpha,I_i}\ ,\quad H^r_\pi = \{u:\quad u \in H^r(I_i),\ \forall i\}.$$

By the approximation results in [1] and by Lemma 2, it follows that for every $u \in H^{k+1} \cap H^{r+1}_\pi$ there is a $v \in S(r,k,\pi)$ such that

$$\text{(2.3)}\qquad \|v-u\|_{\alpha,I_i} \leq C\ h_i^{r+1}\|u^{(r+1)}\|_{\alpha,\bar{I}_i}.$$

The number of subintervals in $\bar{I}_i$ depends only on r and $C = C(C_\ell, r)$ is independent of π. For $v \in S(r,k,\pi)$ the inverse inequality for weighted norms reads

$$(2.4)\quad \| v^{(\mu)} \|_{\alpha, I_i} \leqq C\, h_i^{\nu-\mu} \| v^{(\nu)} \|_{\alpha, I_i},$$

$0 \leqq \nu < \mu \leqq r$, $C = C(r)$ independent of π.

3. SUFFICIENT CONDITIONS FOR UNIFORM BOUNDS

THEOREM 1. <u>If Π satisfies</u> a) <u>and</u> b) <u>for all</u> $\varepsilon > 0$, <u>then for</u> $u \in L_\infty(I)$

$$\| P_2^\pi u \|_{\alpha, I} \leqq 2 \| u \|_{\alpha, I}, \quad \pi \in \Pi \quad (p_\alpha = p_{\alpha,\pi},\ \rho = K \cdot h_j \text{ for suitable } K).$$

<u>Proof</u>. For all $v \in S(r,k,\pi)$ we have [11]:

$$(3.1)\quad \| u_\pi \|^2_{\alpha, I} \leqq 3 \| u \|^2_{\alpha, I} + 8 \| p_\alpha u_\pi - v \|^2_{-\alpha, I}.$$

With v from (2.3):

$$\| p_\alpha u_\pi - v \|^2_{-\alpha, I_i} \leqq C(r, C_\ell) h_i^{2(r+1)} \| (p_\alpha u_\pi)^{(r+1)} \|^2_{-\alpha, \bar{I}_i}.$$

By Leibniz's rule and (2.2) we get

$$(3.2)\quad \begin{aligned} h_i^{2(r+1)} \| (p_\alpha u_\pi)^{(r+1)} \|^2_{-\alpha, I_i} &\leqq C(r) \sum_{\nu=0}^{r} h_i^{2(r+1)} \cdot \| p^{(r+1-\nu)} u_\pi^{(\nu)} \|^2_{-\alpha, I_i} \\ &\leqq C(\alpha, r) \sum_{\nu=0}^{r} h_i^{2(r+1)} \| p_{-(r+1-\nu)} \|_{L_\infty(I_i)} \| u_\pi^{(\nu)} \|^2_{\alpha, I_i}. \end{aligned}$$

In the last sum we estimate

$$h_i^{2(r+1)} \| p_{-(r+1-\nu)} \|_{L_\infty(I_i)} \leqq h_i^{2\nu} \kappa^{2(r+1-\nu)}$$

and by (2.4)

$$\| u_\pi^{(\nu)} \|^2_{\alpha, I_i} \leqq C(r) h_i^{-2\nu} \| u_\pi \|^2_{\alpha, I_i}.$$

Finally, (3.2) is continued with the assumption $\kappa < 1$:

$$h_i^{2(r+1)} \| (p_\alpha u_\pi)^{(r+1)} \|_{-\alpha,I_i}^2 \leqq C(\alpha,r)\kappa^2 \| u_\pi \|_{\alpha,I_i}^2 .$$

The complete estimate of $\| p_\alpha u_\pi - v \|_{-\alpha,I}^2$ is:

$$\begin{aligned} \| p_\alpha u_\pi - v \|_{-\alpha,I}^2 &= \sum_i \| p_\alpha u_\pi - v \|_{-\alpha,I_i}^2 \\ &\leqq C(r,C_\ell) \sum_i h_i^{2(r+1)} \| (p_\alpha u_\pi)^{(r+1)} \|_{-\alpha,\bar{I}_i}^2 \\ &\leqq C(r,C_\ell) \sum_i \sum_{I_j \subset \bar{I}_i} h_j^{2(r+1)} \| (p_\alpha u_\pi)^{(r+1)} \|_{-\alpha,I_j}^2 \\ &\leqq C(\alpha,r,C_\ell)\kappa^2 \sum_i \| u_\pi \|_{\alpha,\bar{I}_i}^2 \\ &\leqq C(\alpha,r,C_\ell)\kappa^2 \| u_\pi \|_{\alpha,I}^2 . \end{aligned}$$

With $C_1 := C(\alpha,r,C_\ell)$ and (3.1), we get

$$(3.3) \quad \| u_\pi \|_{\alpha,I}^2 \leqq 3 \| u \|_{\alpha,I}^2 + 8C_1\kappa^2 \| u_\pi \|_{\alpha,I}^2 .$$

According to Lemma 1, we choose K such that

$$(3.4) \quad \kappa^2 \leqq \frac{1}{32 \cdot C_1} .$$

This means that we need property b) with

$$(3.5) \quad \varepsilon^2 = \frac{1}{32 \cdot C_1} .$$

From (3.3) and (3.4) follows

$$\| u_\pi \|_{\alpha,I} \leqq 2 \| u \|_{\alpha,I} .$$

Remark: Instead of assuming that Π has property b) for all $\varepsilon > 0$, it follows from Lemma 2 and (3.5) that it is sufficient for the proof of Theorem 1 if

(3.6) Π has property b) for $\varepsilon = \min\left(\frac{1}{2\cdot C(\alpha,1)}, \sqrt{\frac{1}{32\cdot C_1}}\right)$,

$C(\alpha,1)$ according to (2.2).

Note that up to this part of the paper the only assumption on $\bar{x}$ is $\bar{x} \in [0,1]$ and there is no assumption on α.

THEOREM 2. *Let* Π *satisfy* a) *and* (3.6) *for* $\alpha = -1$, *then*

$$\|P_2^{\pi}u\|_{L_\infty(I)} \leqq C\|u\|_{L_\infty(I)}, \quad \forall\pi \in \Pi$$

with C *independent of* $\pi \in \Pi$.

Proof. We have fixed α to $\alpha = -1$. Let $\bar{x}$ be such that $u_\pi(\bar{x}) = \|u_\pi\|_{L_\infty(I)}$. (Without loss of generality, we can assume that $u_\pi(\bar{x})$ is positive.) Let j be the index with $\bar{x} \in I_j$. With the inverse estimate

$$\|u_\pi'\|_{L_\infty(I_j)} \leqq C_2h_j^{-1}\|u_\pi\|_{L_\infty(I_j)}$$

we get for $x \in I_j$

$$u_\pi(x) \geqq \|u_\pi\|_{L_\infty(I)}(1 - C_2h_j^{-1}|\bar{x}-x|). \tag{3.7}$$

Let $Q \subset I_j$ be such that

$$|\bar{x}-x| \leqq \tfrac{1}{2}C_2^{-1}h_j, \quad \forall x \in Q. \tag{3.8}$$

This implies $|\bar{x}-x| \leqq \rho$.
From (3.7) and (3.8), we get

$$\|u_\pi\|^2_{-1,I} = \int_I p_{-1}u_\pi^2dx \geqq \frac{1}{4}\cdot\|u_\pi\|^2_{L_\infty(I)}\cdot\frac{1}{2}\cdot\frac{1}{\rho^2}\cdot\frac{1}{2}C_2^{-1}h_j.$$

Together with

$$\|u\|^2_{-1,I} \leqq 4\|u\|^2_{L_\infty(I)}\cdot\frac{1}{\rho} \qquad [11]$$

and Theorem 1, we get

$$\|u_\pi\|^2_{L_\infty(I)} \leqq \frac{\rho}{h_j} \cdot 128 \cdot C_2 \cdot \|u\|^2_{L_\infty(I)} \cdot$$

As $\rho = K \cdot h_j$ we have proved Theorem 2.

EXAMPLE 1. For $q \in (0,1)$, define $f = [\frac{q}{1-q} + 1]$ (brackets mean greatest integer less than or equal to ...), Π_q is a set of meshes π_n of $[0,1]$, π_n: $0 = x_0 < x_1 < \ldots < x_{f+(n-1)} = 1$. Here, the points $x_0, \ldots, x_f = q^{n-1}$ are equidistant with distance $h = \frac{q^{n-1}}{f}$, while $x_{f+(i-1)} = q^{n-i}$, $i = 1, \ldots, n$.

THEOREM 3. <u>There is a</u> $\delta \in [0,1]$ <u>such that, for</u> $q \in (1-\delta,1)$, Π_q <u>fulfills</u> a) <u>and</u> (3.6).

<u>Proof</u>. Because $q^{n-2} - q^{n-1} \geqq h \geqq q^{n-1} - q^n$, a) is fulfilled with $C_\ell = \frac{1}{q}$. Next we assume $q \geqq \frac{1}{2}$, that is $C_\ell \leqq 2$, which means that ε in (3.6) can be given independent of q.

Now we put $\delta = \frac{1}{4}\varepsilon$ and $q \in (1-\delta,1)$. There is an m such that

$$\frac{h_i}{|x_i - x_j|} \leqq \varepsilon \text{ if } |i-j| \geqq m, \ \forall \pi \in \Pi_q .$$

A case study gives the conditions

$$\text{(3.9)} \quad \frac{1 - q + \varepsilon q}{\varepsilon} \leqq \left(\frac{1}{q}\right)^{m-1},$$

$$\text{(3.10)} \quad \frac{1}{\varepsilon} \leqq m,$$

$$\text{(3.11)} \quad q - \frac{1}{4} \geqq q^{m-1}.$$

EXAMPLE 2. Let $\beta \geqq 1$. Π_β is the set of meshes

π_n: $0 = x_0 < \ldots < x_n = 1$ where $x_i = (\frac{i}{n})^\beta$, $i = 0, \ldots, n$.

Π_β fulfills a) and b) for all $\varepsilon > 0$ [10].

For a mesh π: $0 = x_0 < \ldots < x_n = 1$ we define the knots $t_{-r} = t_{-r+1} = \ldots = t_0 = 0$, $t_m = t_{m+1} = \ldots = t_{m+r} = 1$. In the interior of [0,1], we put for every x_i r-k knots, t_j. Then it is $t_i \leqq t_{i+1} \leqq \ldots \leqq t_{i+r}$, $t_i < t_{i+r}$ in the interior.

THEOREM 4. *There is a $\lambda \in [0,1]$ such that $\|P_2^\pi\|_{L_\infty}$ is uniformly bounded for a set of meshes Π if*

c) $$\max_j \sum_i \lambda^{|i-j|} \sqrt{\frac{t_{j+r+1} - t_j}{t_{i+r+1} - t_i}} \leqq C_3 \,, \forall \pi \in \Pi,$$

C_3 independent of π.

Proof. If we would use the condition

$$\max_i \sum_j \lambda^{|i-j|} \sqrt{\frac{t_{j+r+1} - t_j}{t_{i+r+1} - t_i}} \leqq C_3 \,, \forall \pi \in \Pi,$$

instead of c), the proof would follow out of [4, (2), (3), Lemma 1].

The proof of Theorem 4 with condition c) follows with an additional argument about the inverse of a band matrix [5].

EXAMPLE 3. Let $q \in (\lambda^2,1)$ and Π_q as defined in Example 1, then there is a C_3 such that Π_q fulfills c) [7].

EXAMPLE 4. Let $\beta \geqq 1$ and Π_β as defined in Example 2, then there is C_3 such that Π_β fulfills c) [7].

4. GALERKIN PROJECTIONS

This section is concerned with the boundary value problem

(4.1) $-u'' + a_1u' + a_0u = f$, $u(0) = u(1) = 0$, $a_\nu \in C^\nu[0,1]$.

$u_G = P_G^\pi u$ is the Galerkin projection of the solution of u of (4.1) on

$$S_0(r+1,k+1,\pi) = \{v \in S(r+1,k+1,\pi)\colon \; v(0) = v(1) = 0\},$$

that is

$$(u_G', v') + (a_1 u_G' + a_0 u_G, v) = (f,v),\ \forall v \in S_0(r+1,k+1,\pi).$$

P_2^π is the L_2-projection on $S(r,k,\pi)$. We assume that the homogenous problem (4.1) has only the trivial solution and $u \in W^{k+2,\infty} \cap W_\pi^{r+2,\infty}$. There is a constant C_4 such that the following theorem is true.

THEOREM 5. *Let* Π *satisfy* $\max_i h_i \leq C_4$, $\forall \pi \in \Pi$, *and for* C_5: $\|P_2^\pi\|_{L_\infty} \leq C_5$, $\forall \pi \in \Pi$. *Then there is a constant* C_6 *such that*

$$(4.2)\quad \|P_G u - u\|_{L_\infty(I)} \leq C_6 \max_i \bar{h}_i^{\,r+2} \|u^{(r+2)}\|_{L_\infty(\bar{I}_i)},\ \forall \pi \in \Pi,$$

$\bar{I}_i$ *is the union of* I_i *with* $t = t(r)$ *neighboring intervals,*

$$\|u^{(r+2)}\|_{L_\infty(\bar{I}_i)} = \max_{I_j \subset \bar{I}_i} \|u^{(r+2)}\|_{L_\infty(I_j)}.$$

The proof can be derived from [8].

Theorem 2, together with Theorem 5, is an extension of [10, Theorem 4.1], where the following additional assumption is used: There are constants $C < \infty$, $\alpha < 0$ such that

$$\forall \pi \in \Pi,\ \forall_j \sum_{\substack{i=0\\ i\neq j}}^{n-1} \left(\frac{h_j}{h_i}\right)^{-2\alpha+2m-1} \left(\frac{h_i}{|x_i - x_j|}\right)^{-2\alpha} \leq C.$$

The estimate (4.2) leads up to an adaptive Galerkin procedure. This is pointed out in [9]. With a suitable approximation $\tilde{u}$ of $u^{(r+2)}$ the estimate

$$\|\tilde{u} - u^{(r+2)}\|_{L_\infty(I)} = O(h^{(r+1)/(r+2)}),\ h = \max_i h_i$$

is proved independently of the global mesh ratio C_g, if C_5 is independent of C_g.

REFERENCES

1 de Boor, C., On Uniform approximation by splines, J. Approximation Theory, 1 (1968), 219-235

2 de Boor, C., On the convergence of odd-degree spline interpolation, J. Approximation Theory, 1 (1968), 452-463

3 de Boor, C. Bounding the error in spline interpolation, SIAM Rev., 16, No. 4 (1974), 531-544

4 de Boor, C., A bound on the L_∞-norm of L_2-approximation by splines in terms of a global mesh ratio, Math. Comp., 30, No. 136 (1976), 765-771

5 Demko, S., Inverses of band matrices and local convergence of spline projections, SIAM J. Numer. Anal., 14 (1977), 616-619

6 Douglas, J., Dupont, T., Wahlbin, L., Optimal L_∞ error estimates for Galerkin approximations to solutions of two-point boundary value problems, Math. Comp., 29, No. 130 (1975), 475-483

7 Güsmann, B., Galerkin-Projektionen auf Splines mit ungleichmäßigen Gittern, Dissertation, Universität des Saarlandes (1978)

8 Güsmann, B., Bounds of Galerkin projections on splines with highly nonuniform meshes, Bericht A 78/21, FB 10 Universität des Saarlandes (1978)

9 Güsmann, B., Ein adaptives Galerkin-Verfahren mit einer Fehlerabschätzung für auftretende Ableitungen, Bericht A 79/06, FB 10 Universität des Saarlandes (1979)

10 Natterer, F., Uniform convergence of Galerkin's method for splines on highly nonuniform meshes, Math. Comp., 31, No. 138 (1977), 457-468

11 Nitsche, J. A., L_∞-convergence of finite element approximation, 2. Conference on Finite Elements, Rennes, France (1975)

B. Güsmann
Abt. FE 324
Messerschmitt-Bölkow-Blohm GmbH
D-8000 München 80

DIAMETERS OF CLASSES OF SMOOTH FUNCTIONS

Klaus Höllig

In this paper we investigate the asymptotic behaviour of diameters of classes of smooth functions. We consider a discretization technique which reduces the problem to the estimation of the corresponding diameters in finite sequence spaces and apply this method to the estimation of n-width, approximation numbers and entropy.

O. INTRODUCTION

It may be easily shown that for any s-number in the sense of Pietsch [17]

$$s_n(W_p^r[0,1]^N, L_q[0,1]^N) \sim n^{-r/N}, \quad p=q$$

and one might think that the Sobolev embedding theorem will give the asymptotic orders for arbitrary p and q. Actually for different values of p and q the orders are much better than one would expect and they depend on the special s-number. The first result of this type was due to Birman and Solomjak [5] who computed completely the entropy of Sobolev embeddings

$$\varepsilon_n(W_p^r[0,1]^N, L_q[0,1]^N) \sim n^{-r/N}.$$

It is indeed surprising that for example the L_∞-ε-nets for the W_1^r and W_∞^r unit balls have asymptotically the same cardinality. For this reason Birman and Solomjak conjectured in their paper that the asymptotic behaviour of n-width should be better than the standard orders of the usual approximation processes. But despite of some special cases [8,11,15] the complete problem remained unsolved till 1977 when Kasin [12] discovered

$$d_n(W_2^r[0,1]^N, L_\infty[0,1]^N) \sim n^{-r/N}$$

and with this result determined the orders of n-width for all Sobolev embeddings. Finally the author [9,10] and independently Maiorov [16] under some stronger assumption on r

ISBN 0-12-213650-0

showed

$$\delta_n(W_p^r[0,1]^N, L_p, [0,1]^N) \sim n^{-r/N + 1/p - 1/2}, \quad r > N$$

and computed the precise asymptotic behaviour of linear n-width, that is of approximation numbers.

We present here a unified approach to all these results and show that they are closely related to the corresponding diameter problems in finite sequence spaces l_p^m. Diameter problems have been discretizated in various ways [4,8,11,13, 14,15]. The essential idea of our approach dates back to Maiorov [14]. The modification here makes extensive use of B-splines and their remarkable basis properties which allow to pass in a very natural way from function spaces to finite dimensional spaces. Together with existing Jackson and Bernstein inequalities for the corresponding spline subspaces we can easily reduce the diameter problem to such one for finite sequence spaces. In particular the lower estimates become very simple. We then demonstrate the advantage of our method on the estimates of n-width, approximation numbers and entropy. The treatment of entropy in this way is new and provides an easy proof of Birman and Solomjak's results. We have formulated our results for $\Omega = [0,1]^N$ only, but applying extension theorems for Sobolev and Besov spaces [1,19] they can be extended to sufficiently smooth domains.

1. DEFINITIONS, BASIC PROPERTIES OF SPLINE FUNCTIONS

By X_p^r, $r \in \mathbb{R}_+$, $1 \le p \le \infty$ we denote either one of the spaces $W_p^r[0,1]^N$ or $B_{pq}^r[0,1]^N$ with the usual norm

$$||f|W_p^r|| = ||f|L_p|| + \sup_{|\alpha|=r} ||D^\alpha f|L_p||, \quad r \in \mathbb{N}.$$

For the definitions of Sobolev spaces for non-integral r we refer to [1].

$$||f|B_{pq}^r|| = ||f|L_p|| + \{ \int_0^\infty (t^{-r}\omega_k(f,t)_p)^q \frac{dt}{t} \}^{1/q}, \quad k>r.$$

For a continuous linear operator T: X → Y one may define various s-numbers [17]. Here we restrict ourselves to three asymptotic charakteristics of the operator T which we believe to be of main interest. Their definitions are n-width

$$d_n(T) = \inf_{\substack{\dim S \leq n \\ S \subset Y}} \sup_{||x|X|| \leq 1} \inf_{y \in S} ||Tx-y|Y||$$

approximation numbers

$$\delta_n(T) = \inf_{\text{rg } P \leq n} ||T-P|X \to Y||,$$

where the "inf" is taken over all linear operators from X to Y with rank $\leq$ n.

entropy numbers [6]

$$\varepsilon_n(T) = \inf\{\varepsilon \mid \exists y_1,\ldots,y_{2^n} \in Y \text{ with } T(B(X)) \subset \bigcup_{\nu=1}^{2^n} y_\nu + \varepsilon B(Y)\},$$

where B(X) denotes the closed unit ball of the B-space X.

In what follows a_n stands for either of the three quantities d_n, δ_n, ε_n. Only the first two are s-numbers but also the entropy numbers share some of the properties of s-numbers, in particular those stated below.

LEMMA 1 [6,17]. *For every linear operator* T *the numbers* a_n *form a monoton decreasing sequence.*

(1.1) $||T|| = a_0(T) \geq a_1(T) \geq \ldots$

If $T: X \to Y$ *admits the factorization* $X \xrightarrow{T_0} X' \xrightarrow{T_1} Y' \xrightarrow{T_2} Y$, *then*

(1.2) $a_n(T) \leq ||T_0||\ a_n(T_1)\ ||T_2||.$

The numbers a_n *are additive, that is*

(1.3) $a_{n_1+n_2}(T_1+T_2) \leq a_{n_1}(T_1) + a_{n_2}(T_2)$

for two linear operators $T_i: X \to Y$.

We need some basic facts about spline functions which we will list below. By C we denote absolute positive constants which may depend on p,q,r,N,l but not on n,m,k. We write $a \sim b$ if the estimate $C_1 a \leq b \leq C_2 a$ holds. Divide $[0,1]^N$ into cubes of length 2^{-k}. The corresponding space of smooth splines of degree l $S_k = S_k(N,l) \subset C^{l-1}[0,1]^N$ has the dimension $d_k = d_k(N,l)$ with

(1.4) $d_k \sim 2^{kN}.$

S_k is spanned by tensor products of one dimensional B-splines $M_\nu = M_\nu(k,N,l)$ [3] normalized in such a way that

$$\sum_{\nu=1}^{d_k} M_\nu(x) = 1, \qquad x \in [0,1]^N.$$

We identify $s=\sum c_\nu M_\nu \in S_k$ with the sequence of coefficients c_ν and denote by $I_k=I_k(N,l)$: $S_k \to \mathbb{R}^d$ the corresponding isomorphism. The B-spline basis is well conditioned as the following Lemma shows.

LEMMA 2. *Denote by l_p^m $\mathbb{R}^m$ with the p-norm. The norm of a spline and its coefficients with respect to the B-spline basis are related by*

$$(1.5) \qquad ||I_k|L_p \to l_p^{d_k}|| \le C\, 2^{kN/p}$$

$$(1.6) \qquad ||I_k^{-1}|l_p^{d_k} \to L_p|| \le C\, 2^{-kN/p}.$$

This lemma is an easy generalization of the well known one dimensional case [3]. Combining (1.5) and (1.6) we obtain

$$(1.7) \qquad ||s|L_p|| \le C\, 2^{kN(1/q - 1/p)_+} ||s|L_q||, \qquad s \in S_k.$$

We have used that $l_q^m \hookrightarrow l_p^m$, $p \ge q$ and $L_q \hookrightarrow L_p$, $p \le q$.

2. ESTIMATES FOR a_n

Suppose $r - N/p + N/q > 0$ and denote by J: $X_p^r \to L_q$ the embedding operator. Under very general conditions we obtain lower and upper estimates for $a_n(X_p^r,L_q) := a_n(J|X_p^r \to L_q)$ by $a_n(l_p^m,l_q^m)$ the corresponding numbers for the unit matrix I^m: $l_p^m \to l_q^m$. In the following the smooth splines $s \in S_k$ are of sufficiently high degree l such that S_k is contained in X_p^r.

THEOREM 1 (lower estimate). *If a Bernstein inequality holds, i. e.*

$$(2.1) \qquad ||s|X_p^r|| \le C\, 2^{kr}\, ||s|L_p||, \qquad s \in S_k$$

then

$$(2.2) \qquad a_n(X_p^r,L_q) \ge C\, m^{-r/N + 1/p - 1/q}\, a_n(l_p^m,l_q^m), \qquad m > n.$$

Proof. We choose k in such a way that $k = \inf\{j|d_j \ge m\}$ and factorize the identity map I^m: $l_p^m \to l_q^m$ by

$$l_p^m \xrightarrow{i} l_p^{d_k} \xrightarrow{I_k^{-1}} S_k \cap L_p \xrightarrow{i} S_k \cap X_p^r \xrightarrow{J} L_q \xrightarrow{P_k} L_q \cap S_k \xrightarrow{I_k} l_p^{d_k} \xrightarrow{p} l_q^m.$$

Here i and p are the canonical inclusions and projections and P_k may be any family of bounded projections on S_k. By $S_k \cap X$ we denote S_k with the norm of the space X. We apply (1.2), (1.4), (1.5), (1.6) and Bernstein's inequality and obtain

$$\begin{aligned} a_n(l_p^m, l_q^m) &\le ||i||\ ||I_k^{-1}||\ C\ 2^{kr}\ a_n(X_p^r, L_q)\ ||P_k||\ ||I_k|| \times \\ &\quad ||p|| \\ &\le C_1\ 2^{-kN/p}\ 2^{kr}\ a_n(X_p^r, L_q)\ 2^{kN/q} \\ &\sim m^{r/N\ -1/p\ +\ 1/q}\ a_n(X_p^r, L_q) \end{aligned}$$

which proves the theorem.

We remark that for $X_p^r = W_p^r$ and $X_p^r = B_{pq}^r$ a Bernstein inequality is obviously satisfied. The proof is merely an application of Markov's inequality for polynomials.

THEOREM 2 (upper estimate). _If a linear Jackson inequality holds, i. e. if there exist linear operators_ P_k: $X_p^r \to S_k$ _such that_

(2.3) $||J-P_k|X_p^r \to L_p|| \le C\ 2^{-kr}$

then the following estimates hold

(2.4) $a_n(X_p^r, L_q) \le a_{d_k}(P_k) + C\ n^{-r/N\ +\ (1/p\ -\ 1/q)_+}$

(2.5) $a_n(X_p^r, L_q) \le C\ n^{-r/N\ +\ 1/p\ -\ 1/q} \times$
$$\sum_{\nu=-k}^{\infty} 2^{-\nu(r\ -\ N/p\ +\ N/q)}\ a_{n_\nu}(l_p^{d_{k+\nu}}, l_q^{d_{k+\nu}}),$$

where the integers k, n_ν _have to be chosen subject to the conditions_

$$d_k \sim n, \qquad \sum_{\nu=-k}^{\infty} n_\nu \le C_0\ n.$$

In the estimate (2.4) _the term_ $a_{d_k}(P_k)$ _vanishes for the s-numbers_ $a_n = d_n$ _and_ $a_n = \delta_n$ _since_ rg $P_k \le d_k$ _and is only significant for entropy numbers._

Proof. The embedding J: $X_p^r \to L_q$ admits the representation

$$J = \sum_{\nu=0}^{\infty} T_\nu,$$

where $T_0 = P_0$, $T_\nu = P_\nu - P_{\nu-1}$ and in view of (1.7) and (2.3)

$$||T_\nu|X_p^r \to L_q|| \le C\ 2^{-\nu r\ +\ \nu(N/p\ -\ N/q)_+}.$$

For the proof of (2.4) we choose $k = \sup\{j | d_j \le n\}$ and apply (1.1) and (1.3).

$$a_n(J|X_p^r \to L_q) \leq a_{d_k}(J) \leq a_{d_k}(\sum_{\nu=0}^{k} T_\nu) + a_0(\sum_{\nu=k+1}^{\infty} T_\nu)$$
$$= a_{d_k}(P_k) + ||\sum_{\nu=k+1}^{\infty} T_\nu|X_p^r \to L_q||$$
$$\leq a_{d_k}(P_k) + C\ 2^{-kr + k(N/p - N/q)_+} \sum_{\nu=1}^{\infty} 2^{-\nu r + \nu(N/p-N/q)_+}$$

Since $2^{kN} \sim d_k \sim n$ and $\sum_{\nu=1}^{\infty} \ldots < C$ this gives (2.4).
To prove (2.5) we choose again $d_k \sim n$ and apply (1.3).

$$a_n(X_p^r, L_q) \leq \sum_{\nu=-k}^{\infty} a_{n_\nu}(T_{k+\nu}|X_p^r \to L_q)$$

We factorize $T_\mu : X_p^r \to L_q$

$$X_p^r \xrightarrow{T_\mu} L_p \cap S_\mu \xrightarrow{I_\mu} l_p^{d_\mu} \xrightarrow{i} l_q^{d_\mu} \xrightarrow{I_\mu^{-1}} L_q \cap S_\mu$$

and apply (1.2).

$$a_{n_\nu}(T_\mu) \leq ||T_\mu||\ ||I_\mu||\ a_{n_\nu}(l_p^{d_\mu}, l_q^{d_\mu})\ ||I_\mu^{-1}||$$

Because of lemma 2, (2.3) and $2^{kN} \sim d_k \sim n$ we finally obtain

$$a_{n_\nu}(T_{k+\nu}) \leq C\ 2^{-(k+\nu)(r-N/p)}\ a_{n_\nu}(l_p^{d_{k+\nu}}, l_q^{d_{k+\nu}})\ 2^{-(k+\nu)N/q}$$
$$\leq C\ n^{-r/N + 1/p - 1/q}\ 2^{-\nu(r-N/p+N/q)}\ a_{n_\nu}(l_p^{d_{k+\nu}}, l_q^{d_{k+\nu}})$$

which completes the proof of the theorem.

We remark that for $X_p^r = W_p^r$ and $X_p^r = B_{pq}^r$ Jackson's inequality is a consequence of the standard estimate for the quasiinterpolant $P_k : L_p \to S_k$ [7]

$$||f - P_k f|L_p|| \leq C\,\omega_r(f, 2^{-k})_p .$$

For the s-numbers $a_n = d_n, \delta_n$ we may simplify the estimate (2.5) as follows.

COROLLARY 1. <u>For any s-number</u> a_n <u>the estimate</u>

(2.5') $$a_n(X_p^r, L_q) \leq C\ n^{-r/N + 1/p - 1/q} \times \sum_{\nu=1}^{\infty} 2^{-\nu(r - N/p + N/q)}\ a_{n_\nu}(l_p^{d_{k+\nu}}, l_q^{d_{k+\nu}})$$

<u>holds if we choose the integers</u> k, n_ν <u>subject to the conditions</u>

$$d_k \sim n, \quad \sum_{k=1}^{\infty} n_\nu \leq C_0\, n .$$

Proof. We define $n_\nu = d_{k+\nu}$, $\nu \leq 0$ and since

$$\sum_{\nu=-k}^{0} n_\nu \sim \sum_{\nu=-k}^{0} 2^{(k+\nu)N} \sim 2^{kN} \sim d_k \sim n$$

the condition $\sum_{k=1}^{\infty} n_\nu \leq C_0\, n$ implies $\sum_{\nu=-k}^{\infty} n_\nu \leq C_1\, n$ and we can apply (2.5). Since for any s-number $a_{d_{k+\nu}}(l_p^{d_{k+\nu}}, l_q^{d_{k+\nu}}) = 0$ $\sum_{\nu=-k}^{\infty} 2^{-\nu(r - N/p + N/q)} a_{n_\nu}(l_p^{d_{k+\nu}}, l_q^{d_{k+\nu}})$ reduces to $\sum_{\nu=1}^{\infty} \cdots$.

We proceed now to the applications.

3. n-WIDTH

For the n-width of the identity matrix $I^m: l_p^m \to l_q^m$ we have the asymptotic results [12,17,18]

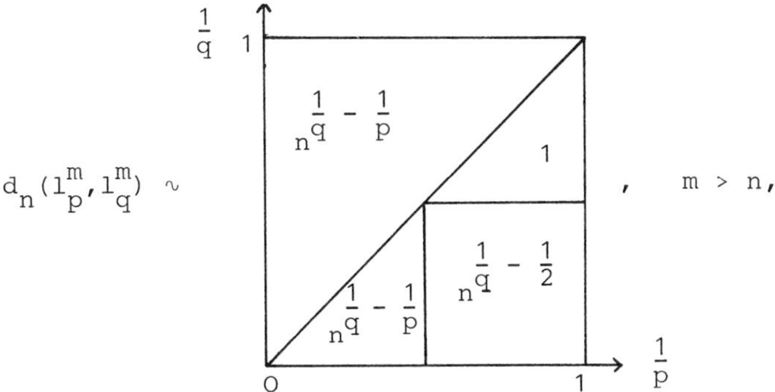

where the constants in the upper estimates depend on the quotient m/n.

The precise orders for the n-width of the Sobolev embeddings were discovered by Kašin [12]

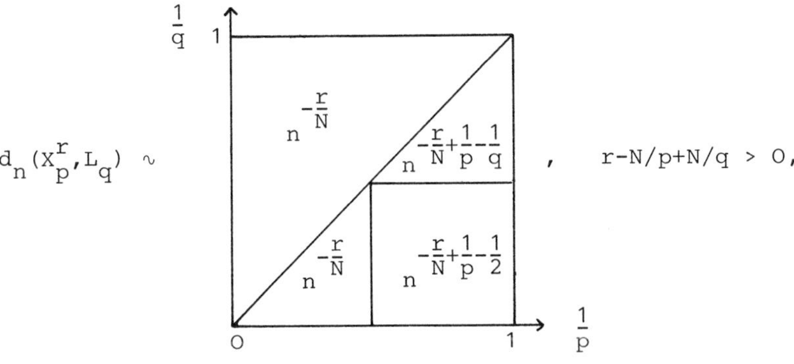

with the additional restrictions $r - N/2 > 0$ if $2 \leq p < q$ and $r - N/p > 0$ if $p < 2 < q$.

In order to illustrate the discretization technique of the preceeding section we now show how the latter diagram can be deduced from the former. Concerning the lower estimates this is done by (2.2) if we take $m = 2n$ in the first diagram. To prove the upper estimates we consider three cases:

1. $p \geq q$ or $q \leq 2$

Here we can apply (2.4).

2. $2 \leq p < q$

The embeddings $X_p^r \hookrightarrow X_2^r$, $L_\infty \hookrightarrow L_q$ imply $d_n(X_p^r, L_q) \leq d_n(X_2^r, L_\infty)$.

3. $p < 2 < q$

We may use the embeddings $X_p^r \hookrightarrow X_2^{r-N/p+N/2}$ and $L_\infty \hookrightarrow L_q$ to obtain

$$d_n(X_p^r, L_q) \leq C\, d_n(X_2^{r-N/p+N/2}, L_\infty)$$

Hence it suffices to prove for $r - N/2 > 0$

$$d_n(X_2^r, L_\infty) \leq C\, n^{-r/N}.$$

To this end we substitute Kašin's estimate [12]

$$d_n(l_2^m, l_\infty^m) \leq C\, n^{-1/2} \left(\ln \frac{m}{n}\right)^{3/2}, \qquad m > n$$

into (2.5') and make a suitable choice of the parameters n_ν. We omit the details here since a similar procedure will be discussed in chapter 5.

4. APPROXIMATION NUMBERS

This case has been discussed in detail in [9,10], see also [16]. Therefore we will only state the results.

Concerning finite sequence spaces we have [9,10,17,18]

$$\delta_n(l_p^m, l_q^m) \sim$$

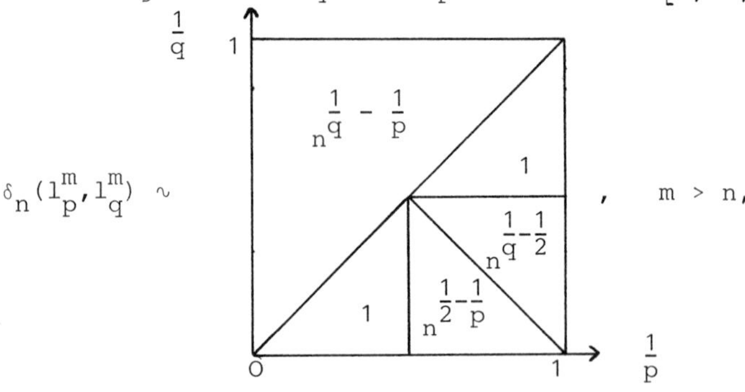

$, \qquad m > n,$

where the constants in the upper estimates depend on the quotient m/n.

The precise asymptotic orders for the Sobolev embeddings are [9,10,16]

$\delta_n(X_p^r, L_q) \sim$ 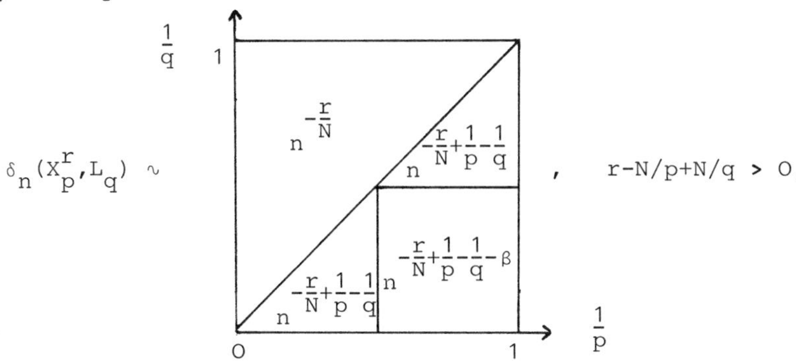 , $r - N/p + N/q > 0$,

with $\beta = \beta(p,q) = \mathrm{Min}\,\{1/2 - 1/q\ ,\ 1/p - 1/2\}$ and the additional restriction $r > N$ in the case $p<2<q$.

As in the case of n-width we can use the embedding theorem to reduce the problem. Indeed, to estimate $\delta_n(X_p^r, L_q)$ from above in the nontrivial cases it suffices to prove

$$(4.1) \quad \delta_n(X_p^r, L_{p'}) \leq C\, n^{-r/N + 1/p - 1/2}, \quad r > N.$$

This may be done by substituting the estimate of the author [9,10]

$$\delta_n(l_p^m, l_{p'}^m) \leq C\, n^{1/2 - 1/p} \left(\frac{m}{n}\right)^{2/p'} \left(\frac{\ln m}{\ln n}\right)^{\theta}, \quad m > n$$

into the estimate (2.5'). Maiorov [16] proved (4.1) by another method under a stronger assumption on r, $r > 3N/2$.

5. ENTROPY

We want to show how our discretization method can be applied to give a simple proof of Birman and Solomjak's important result [5]

$$(5.1) \quad \varepsilon_n(W_p^r, L_q) \sim n^{-r/N}, \quad r - N/p + N/q > 0$$

in the case $r > N$.

Since for $r - N/p + N/q > 0$ $X_\infty^r \subsetneq X_p^r \subsetneq L_q \subsetneq L_1$ we have $\varepsilon_n(X_\infty^r, L_1) \leq C\, \varepsilon_n(X_p^r, L_q)$. To estimate the left hand side from below we use (2.2) with $m = 2n$ and apply a result of Carl and Pietsch [6]

(5.2) $\varepsilon_n(l_\infty^m, l_1^m) \geq C\ m.$

Hence

$$\varepsilon_n(X_\infty^r, L_1) \geq C\ n^{-r/N-1}\ n = C\ n^{-r/N}$$

which gives the lower bound. Quite analogously for the upper estimate in the case $r > N$ it suffices to show

(5.3) $\varepsilon_n(X_1^r, L_\infty) \leq C\ n^{-r/N}.$

To this end we have to refine the corresponding estimate for $\varepsilon_n(l_1^m, l_\infty^m)$ in [6]

$$\varepsilon_n(l_1^m, l_\infty^m) \leq C\ n^{-1}\ \ln m.$$

LEMMA 3.

(5.4) $\varepsilon_n(l_1^m, l_\infty^m) \leq \begin{cases} C\ n^{-1}\ (\ln \frac{m}{n})^2, & 2n \leq m \\ C\ m^{-1}\ 2^{-n/m}, & 2n \geq m \end{cases}$

Proof. First observe that the second estimate is a consequence of the first. Indeed, if

$$\varepsilon_n(l_1^{2n}, l_\infty^{2n}) \leq C\ n^{-1},$$

that is 2^n l_∞^{2n}-balls of radius $C \cdot n^{-1}$ cover $B(l_1^{2n})$, then $2^n \cdot 2^{2n \cdot j}$ l_∞^{2n}-balls of radius $C \cdot n^{-1} \cdot 2^{-j}$ cover $B(l_1^{2n})$. In other words

$$\varepsilon_{n+2jn}(l_1^{2n}, l_\infty^{2n}) \leq C\ n^{-1}\ 2^{-j}$$

which implies

$$\varepsilon_{jm}(l_1^m, l_\infty^m) \leq C\ m^{-1}\ 2^{-j}.$$

To prove the first estimate we give an upper bound for the number $M(n,m)$ of l_∞^m-balls of radius n^{-1} which cover $B(l_1^m)$. Since

$$B(l_1^m) \subset \bigcup_{\substack{j \in \mathbb{Z}^m \\ \sum_{\nu=1}^m |j_\nu| \leq n}} \left(\frac{j_1}{n}, \ldots, \frac{j_m}{n}\right) + \frac{1}{n}\ B(l_\infty^m)$$

clearly

$$\begin{aligned} M(n,m) &\leq \#\{j \in \mathbb{Z}^m \mid \sum_{\nu=1}^m |j_\nu| \leq n\} \\ &\leq \#\{j \in \mathbb{N}_0^m \mid \sum_{\nu=1}^m j_\nu \leq 2n\}. \end{aligned}$$

The last inequality is true because to each pair $(-j_\nu,+j_\nu)$ we may associate the pair $(2j_\nu,2j_\nu-1)$. Therefore

$$M(n,m) \leq \sum_{k=0}^{2n} \#\{j \in \mathbb{N}_0^m \mid \sum_{\nu=1}^{m} j_\nu = k\} = \sum_{k=0}^{2n} \binom{k+m-1}{m-1}$$
$$\leq 2n \binom{2n+m-1}{m-1}.$$

By induction on m we show now

(1) $\quad 2n \binom{2n+m-1}{m-1} \leq (m/n)^{Cn}, \qquad 2n < m.$

A direct computation shows that (1) is true for $m = 2n+1$.

$$2n \binom{4n}{2n} = 2n\ 2^{2n} \frac{1\cdot 3\cdots(4n-1)}{(2n)!} \leq 2n\ 2^{2n}\ 2^{2n} \leq (m/n)^{5n}.$$

To establish (1) for arbitrary $m > 2n$ it suffices to show

$$\{2n\binom{2n+m}{m}\}/\{2n\binom{2n+m-1}{m-1}\} \leq \{(m+1)/n\}^{Cn}/\{m/n\}^{Cn}, \text{ i.e.}$$

$$\frac{2n+m}{m} \leq \left(\frac{m+1}{m}\right)^{Cn}.$$

Multiplying both sides with m^{Cn} shows that $C \geq 2$ will do. Since $(m/n)^{Cn} = 2^{Cn\ \ln(m/n)}$ (1) shows by the definition of entropy numbers

(2) $\quad \varepsilon_{C\,n\,\ln(m/n)}(l_1^m, l_\infty^m) \leq n^{-1}, \qquad 2n < m.$

One may easily prove the following inequality

$$1 \leq \frac{1}{\ln a} \left(\ln \frac{a}{C \ln a}\right)^2, \qquad a \geq C_0$$

where C_0 depends on the given constant C. Assuming for the moment $m \geq C_0\ n$ we may replace the right hand side of (2) by

$$\frac{C}{C\ n\ \ln(m/n)} \left(\ln \frac{m}{C\ n\ \ln(m/n)}\right)^2$$

and the substitution $C\ n\ \ln(m/n) \to n$ proves the lemma for $(m/n) \geq C_0$. But changing in (5.4) the constant C we get obviously rid of the restriction $m \geq C_0\ n$.

We are now ready to prove (5.3).

Proof of (5.3). We apply (2.5) in the case $(p,q) = (1,\infty)$.

(1) $\quad \varepsilon_n(X_1^r, L_\infty) \leq C\ n^{-r/N+1} \sum_{\nu=-k}^{\infty} 2^{-\nu(r-N)}\ \varepsilon_{n_\nu}(l_1^{d_{k+\nu}}, l_\infty^{d_{k+\nu}})$

We define $n_\nu = C_0\ 2^{kN-\alpha|\nu|}$ where we choose $\alpha>0$ less than $\mathrm{Min}(N,r-N)$ and C_0 such that $n_\nu \geq d_{k+\nu}$, $\nu\leq 0$ which is possible because of $d_{k+\nu} \sim 2^{kN-|\nu|N}$ for $\nu\leq 0$. Since $2^{kN}\sim n$ the restriction

$$\sum_{\nu=-k}^{\infty} n_\nu \leq C_1\ n$$

of (2.5) is obviously satisfied.

To estimate the right hand side of (1) we split the sum into two parts

$$\sum_{\nu=-k}^{\infty} \ldots = \sum_{\nu=-k}^{0} \ldots + \sum_{\nu=1}^{\infty} \ldots .$$

Since $n_\nu \geq d_{k+\nu}$ for $\nu \leq 0$ we use the second estimate of (5.4) to obtain for the first term

$$(2) \quad \sum_{\nu=-k}^{0} \ldots \leq \sum_{\nu=-k}^{0} 2^{-\nu(r-N)} \varepsilon_{(C_0 2^{kN} 2^{\alpha\nu})} (1_1^{d_{k+\nu}}, 1_\infty^{d_{k+\nu}})$$

$$\leq C_1 \sum_{\nu=-k}^{0} 2^{-\nu(r-N)} 2^{-(k+\nu)N} 2^{-C_2(2^{kN+\alpha\nu-kN-\nu N})}$$

$$\leq C_1 \, 2^{-kN} \sum_{\nu=0}^{k} 2^{\nu r} 2^{-C_2 \cdot 2^{\nu(N-\alpha)}}$$

$$\leq C_3 \, n^{-1} \sum_{\nu=0}^{\infty} 2^{\nu r} 2^{-C_2 \cdot 2^{\nu(N-\alpha)}} \leq C_4 \, n^{-1} .$$

To estimate the second term we apply the other part of (5.4)

$$(3) \quad \sum_{\nu=1}^{\infty} \ldots \leq \sum_{\nu=1}^{\infty} 2^{-\nu(r-N)} \varepsilon_{(C_0 2^{kN} 2^{-\alpha\nu})} (1_1^{d_{k+\nu}}, 1_\infty^{d_{k+\nu}})$$

$$\leq C_5 \sum_{\nu=1}^{\infty} 2^{-\nu(r-N)} 2^{-kN+\alpha\nu} (\nu(\alpha+N))^2$$

$$= C_5 \, 2^{-kN} \sum_{\nu=1}^{\infty} 2^{-\nu(r-N-\alpha)} (\nu(\alpha+N))^2 \leq C_6 \, n^{-1} .$$

Combining the estimates (1),(2) and (3) we end up with

$$\varepsilon_n(X_1^r, L_\infty) \leq C \, n^{-r/N + 1} (C_4 n^{-1} + C_6 n^{-1}) \leq C_7 \, n^{-r/N} .$$

REFERENCES

1. Adams, R.A., Sobolev spaces, Academic Press, New York, London 1975.

2. Bergh, J. and J.Löfström, Interpolation spaces, Springer, Berlin, Heidelberg, New York 1976

3. Boor, C.de, Splines as linear combinations of B-splines, In: Approximation theory II, G.G.Lorentz, C.K.Chui, L.L.Schumaker, eds., Academic Press 1976, 1-47.

4. Boor, C.de, R. DeVore and K. Höllig, Mixed norm n-width, to appear in Proc. AMS.

5. Birman, M.S. and M.Z. Solomjak, Piecewise polynomial approximation of functions of the classes W_p^α, Mat. Sbornik 73 (1967), 295-317 = Math. USSR-Sbornik 2 (1967) 295-317

6. Carl, B. and A. Pietsch, Entropy numbers of operators in Banach spaces, In: General topology and its relations to modern analysis and algebra, Proc. Fourth Prague Topological Sympos., Prague 1976, 21-33, Lecture Notes in Math. 609, Springer, Berlin 1977.

7. Dahmen W., R. DeVore and K. Scherer, Multi-dimensional spline approximation, to appear in Siam Journal on Numerical Analysis.

8. Gluskin, E.D., On a problem concerning diameters, Dokl. Akad.Nauk SSSR 219 (1974) = Soviet Math.Dokl. 15 (1974), 1592-1596.

9. Höllig, K., Approximationszahlen von Sobolev-Einbettungen, Mathematische Annalen 242 (1979), 273-281.

10. Höllig, K., Approximationszahlen von Sobolev-Einbettungen, Dissertation, Bonn 1979.

11. Ismagilov, R.S., Diameters of sets in normed linear spaces and the approximation of functions by trigonometric polynomials, Uspekhi Mat.Nauk 29:3 (1974), 161-178 = Russian Math.Surveys 29:3 (1974), 169-186.

12. Kašin, B.S., The width of certain finite dimensional sets and classes of smooth functions, Izv.Akad.Nauk SSSR 41 (1977), 334-351 = Math. USSR Izvestija 11 No.2 (1977).

13. Lubitz, C., s-Zahlen von Sobolev-Einbettungen, Diplomarbeit, Bonn 1978.

14. Maiorov, V.E., Discretization of the problem of diameters, Uspekhi Mat.Nauk 30 (1975), 179-180.

15. Maiorov, V.E., Theorems of representation and best approximation in the classes W_p^r and H_p^r, Dokl.Akad.Nauk SSSR 228 No.2 (1976), 708-712.

16. Maiorov, V.E., On linear diameters of Sobolev classes, Dokl.Akad.Nauk SSSR 243 No.5 (1978) = Soviet Math.Dokl. 19 No.6 (1978), 1491-1494.

17. Pietsch, A., s-numbers of operators in Banach spaces, Studia Mathematica 51 (1974), 201-223.

18. Pinkus, A., Matrices and n-width, Technion preprint No. MT 403 (1974).

19. Stein, E.M., *Singular integrals and differentiability properties of functions*, Princeton University Press 1970.

K. Höllig
Institut für Angewandte Mathematik / SFB
Wegelerstr. 6
53 Bonn
Federal Republic of Germany

ON ESTIMATES OF DIAMETERS

B. S. Kashin

This paper contains highlights of the report delivered by the author at the conference.

If K is a compact set in a Banach space X, then the quantity

$$(1) \quad d_n(K,X) = \inf_{L,\ \dim L \leq n} \Delta_X(K,L)$$

is called the n-th diameter of this set in the sense of Kolmogorov; in the above definition

$$(2) \quad \Delta_X(K,L) = \sup_{y \in K} \inf_{z \in L} \|y-z\|_X .$$

Starting in the fifties, dozens of papers have been dedicated to the problem of estimating diameters (c.f. [11], [5]). In recent years, estimates of diameters have found applications in the qualitative analysis of various computational algorithms as well as in functional analysis and in the investigation of distribution of the eigenvalues of various operators. In this context, the problem arises to estimate the diameters of certain compact sets of functions in the L^q-metric, $1 \leq q \leq \infty$ ($L^\infty = C$).

A number of papers by various authors are dedicated to the best-known problem of this kind, namely to the problem of estimating the quantities

$$(3) \quad d_n(W_p^r, L^q(0,1)),\ 1 \leq p,\ q \leq \infty,\ r > 0$$

where W_p^r stands for the class of smooth functions f(x) which possess r-1 absolutely continuous derivatives and $\|f\|_{L^p(0,1)} + \|f^{(r)}\|_{L^p(0,1)} \leq 1$; here, for fractional r, the

ISBN 0-12-213650-0

derivative $f^{(r)}$ is understood in the generalized sense. The quantity (3) is interesting only when W_p^r is compact in L^q, i.e. if $r > \frac{1}{p} - \frac{1}{q}$.

An investigation of the orders of magnitude of the quantity (3) as $r > 1$, $n \to \infty$ was completed by the author in 1976. The exact result has the following form: If $1 \leq p, q \leq \infty$, $rp > 1$, then

$$d_n(W_p^r, L^q) \sim \begin{cases} n^{-r} & \text{if } p \geq q \text{ or } 2 < p < q \\ n^{-r-\frac{1}{2}+\frac{1}{p}} & \text{if } p \leq 2 < q \\ n^{-r-\frac{1}{q}+\frac{1}{p}} & \text{if } 1 \leq p < q \leq 2 . \end{cases} \tag{4}$$

Relation (4) has been proved in the following way: In the case $q \leq \max(p,2)$, it was proved by the joint efforts of several authors (for details c.f. [11]) that

$$d_n(W_p^r, L^q) = \Delta_{L^q}\left(W_p^r, T^{[n/2]}\right), \tag{5}$$

where

$$T^n = \{P(t) : \; P(t) = a_0 + \sum_{k=1}^{n} (a_k \cos kt + b_k \sin kt)\}.$$

The estimate

$$\Delta_{L^q}(W_p^r, T^n) \sim \begin{cases} n^{-r+\frac{1}{p}-\frac{1}{q}} & \text{if } p < q \\ n^{-r} & \text{if } p \geq q \end{cases}$$

has been known since the thirties and thus the estimate for $d_n(W_p^r, L^q)$ from above was not difficult in this case. For estimates of the diameter (3) from below in the case $p \geq q$, topological considerations were applied by Tihomirov, and later by other authors, which gave not only the proof of these estimates from below but also the computation of the exact value of (3) in several cases. For $1 \leq p < q \leq 2$, the estimates from below may be obtained fairly simply. The starting point here was obtained in the beginning of the fifties by W. Rudin and S.B. Stechkin, who proved (4) for $p=1$, $q=2$. The complete case $1 \leq p < q \leq 2$ is contained in [2].

It turned out that the behavior of the diameters (3) in the case $q > \max(p,2)$, is entirely different from that in the first case. In 1974, R. S. Ismagilov proved that

$$d_n(W_1^2, C) \leq C_\varepsilon n^{-\frac{6}{5}+\varepsilon} \leq Cn^{-1} \leq \Delta_C(W_1^2, T^n);\ n > n_0.$$

This result implies (c.f. E. D. Gluskin [1]) that already in the case $p = 1$, $q = \infty$, $r = 2$ the relation (5) is no longer true and, consequently, the trigonometric system is not a good approximating subspace for the class W_1^2 in the uniform norm. The same deficiency is possessed by the spaces of splines and algebraic polynomials. It was shown by the author that, in general, for $q > \max(p,2)$, spaces, which are in a certain sense randomly chosen n-dimensional subspaces, are close to the best and, on that basis, the orders of the diameters (3) were computed.

Usually, for estimates of the diameter (3), a discretization of the problem was used, namely the reduction of it to the estimate of the diameters

$$(6)\qquad d_n(B_p^m, \ell_q^m),$$

where $B_p^m = \{x \in R^m\colon\ \|x\|_{\ell_p^m} \leq 1\}$ and, for $x = \{x_i\} \in R^m$

$$\|x\|_{\ell_q^m} = \begin{cases} \left(\sum_1^m |x_i|^q\right)^{1/q}, & 1 \leq q < \infty \\ \max|x_i|, & q = \infty. \end{cases}$$

That is, the asymptotic behavior of $d_n(W_1^r, L^2)$ has been determined in [10] by using the result $d_n(B_1^m, \ell_2^m) = (\frac{m-n}{n})^{1/2}$. A sufficiently sharp method of discretization is contained in [1] and in [8]. The main difficulty in the estimation of the diameters (3) consists in the sharp evaluation of (6). Note that with the aid of estimates of the diameters (6) it is possible to estimate the diameters of other classes of functions with that or other degrees of accuracy.

In case $q \geqq \max(p,2)$, routine subspaces could not supply satisfactory estimates for (6). In 1974, a probabilistic approach was proposed by the author; this approach may be briefly described as follows. A sufficiently rich family, $\Omega_{m,n}$, of n-dimensional subspaces L in R^m, is considered endowed with some measure μ (for example, the set of all n-dimensional subspaces in R^m with Haar measure or the set of subspaces $L \subset R^m$, $\dim L = n$, which are described, using the standard basis, by the matrix with elements ± 1, and $\mu L = 2^{-mn}$); the quantity

$$I_{p,q} = \int_{\Omega_{m,n}} \Delta_{\ell_q}(B_p^m, L)\, d\mu$$

is evaluated.

It is clear that

$$(7) \quad d_n(B_p^m, \ell_q^m) = \inf_{L,\, \dim L = n} \Delta_{\ell_q}(B_p^m, L) \leqq I_{p,q}.$$

It turns out that the difference between the left- and right-hand sides in (7) is "not large", and thus if we get a sharp estimate of the integral $I_{p,q}$, we get a good estimate for the diameter (6), as well. For example, if we take as $\Omega_{m,n}$ the set of all n-dimensional subspaces, then it may be proved that for arbitrary p and q, $1 \leqq p,\ q \leqq \infty$, as $n \to \infty$, the estimate

$$d_n(B_p^{2n}, \ell_q^{2n}) \sim I_{p,q} \sim \begin{cases} 1 & \text{if } 1 \leqq p < q \leqq 2 \\ n^{-\frac{1}{2}+\frac{1}{q}} & \text{if } p < 2 < q \\ n^{-\frac{1}{p}+\frac{1}{q}} & \text{if } p \geqq \min(2,q) \end{cases}$$

is valid.

Thus, random n-dimensional subspaces provide a good approximation for the ball B_p^{2n} in the ℓ_q^{2n} metric for all p and q, while the n-dimensional space of discrete trigonometric polynomials provides a good approximation for B_p^{2n} only if $q \leqq \max(2,p)$. Using this method, the author [5] proved the following:

THEOREM 1. *If* $m > n$, *then*

$$\tfrac{1}{2}n^{-1/2} < d_n(B_2^m, \ell_\infty^m) < Cn^{-1/2}(1+\ln(m/n))^{3/2}.$$

Estimates of the diameters $d_n(B_2^m, \ell_\infty^m)$ turn out to be the most important for applications; thus Theorem 1 was efficient enough to compute the orders of magnitude of the diameters (3) for all p and q.

A special feature of the behavior of the diameters (6) is their weak dependence upon m. The latter fact may be seen from Theorem 1 and is still more clear from the results in [4].

THEOREM 2. *If* $\gamma n < m < n^\lambda$, $\gamma > 1$, $\lambda > 1$, *then*

$$c_\gamma n^{-1/2} < d_n(B_1^m, \ell_\infty^m) < c_\lambda n^{-1/2}.$$

Another peculiarity is the fact that from the estimates of $d_n(B_p^m, \ell_\infty^m)$, sufficiently sharp estimates of $d_n(B_p^m, \ell_q^m)$ for $q < \infty$ follow in a trivial way. For example, the inequality $\|x\|_{\ell_q^m} \leqq m^{1/q}\|x\|_{\ell_\infty^m}$ and Theorem 2 imply that for $q > 2$

$$d_n(B_1^m, \ell_q^m) \leqq C_q \min(1, m^{1/q}\, n^{-1/2}).$$

On the other hand, the estimate of $d_n(B_1^m, \ell_q^m)$ from below has the same form. More precisely, the following statement holds:

THEOREM 3. *If* $m \geqq 2n$ *and* $q > 2$, *then*

$$\tfrac{1}{4}\min(1, m^{1/q}\, n^{-1/2}) \leqq d_n(B_1^m, \ell_q^m) \leqq C_q \min(1, m^{1/q}\, n^{-1/2}).$$

It follows that the general behavior of the diameters (6) up to logarithm factors is now clear, but exact estimates uniform in m and n are still not known in the most interesting case, $q = \infty$.

The following problems are of special interest to the author:

1. Determine the exact order (as $n \to \infty$) of the quantity $d_n\left(B_2^{n^2}, \ell_\infty^{n^2}\right)$.

2. Determine, for all m and n, the order of magnitude of $d_n(B_1^m, \ell_\infty^m)$.

It should be noted in connection with problem 2 that $d_n(B_1^m, \ell_\infty^m) \leqq Cn^{-1/2}(1+\ell n(m/n))^{1/2}$ for all m and n (see [5]) and that the exponent 1/2 in the factor $(1+\ell n(m/n))$ cannot be decreased since $d_n\left(B_1^{K^n}, \ell_\infty^{K^n}\right) \geqq 1/4$ for K sufficiently large (see [9]).

It is known that the following quantity is in a certain sense dual to the diameter (1):

$$(8) \quad d^n(K,X) = \inf_{L,\ \operatorname{codim} L=n} \ \sup_{y \in L \cap K} \|y\|_X$$

For example, the following relation holds:

$$d_n(B_p^m, \ell_q^m) = d^n(B_{q'}^m, \ell_{p'}^m), \quad \frac{1}{p} + \frac{1}{p'} = \frac{1}{q} + \frac{1}{q'} = 1.$$

Here, we present the following simple result which provides an estimate from below of the quantity (8) in the case $X = \ell_2^n$ and which may be applied to a large variety of sets K.

THEOREM 4 (c.f. [6]). *For any central-symmetric and convex set $K \subset B_2^m$, and any integer r, $1 \leqq r \leqq m$, we have*

$$\inf_{L,\ \dim L \geqq r} \ \sup_{y \in L \cap K} \|y\|_{\ell_2^m} \geqq \left[V(K)(V_r V_{m-r})^{-1}\right]^{1/2},$$

where $V(K)$ denotes volume of the set K, and $V_s = \pi^{s/2}\Gamma^{-1}(\frac{s}{2}+1)$ is volume of the ball B_2^s.

The following result provides a lower bound for diameters in $_2^n$ of sets which lie on the Euclidean sphere in R^n.

THEOREM 5 (c.f. [1]). *Given* $\varepsilon > 0$, *there is a number* $\delta(\varepsilon) > 0$ *such that for any* $G \subset \{x: \|x\|_{\ell_2^n} = 1\}$ $1 \leq n \leq \infty$ *and any array* $\{i_r\}_{r=1}^{n-j}$, $1 \leq i_1 < \dots < i_{n-j} \leq n$ *and* $j \leq n\delta(\varepsilon)$, *with the property* $G \cap \{x \in R^n: x_{i_1} = \dots = x_{i_{n-j}} = 0\} \neq \phi$, *we have*

$$1 - \varepsilon \leq d_j(G, \ell_2^n) \leq 1.$$

Theorem 5 is a consequence of the next result. Given a matrix $A = (a_{ij})$, $1 \leq i \leq i'$, $1 \leq j \leq j'$, let

$$\|A\| = \sup_{\sum_{i=1}^{i'} x_i^2 = 1 = \sum_{j=1}^{j'} y_j^2} \sum_{i=1}^{i'} \sum_{j=1}^{j'} a_{ij} x_i y_j .$$

THEOREM 6. *If* $A = (a_{ij})$, $1 \leq i \leq m$, $1 \leq j \leq n$ *and* $m > n$, *then there is a set* $\Omega = \{i_\nu\}_{\nu=1}^{n}$ *such that*

$$\|(a_{ij}),\ i \in \Omega,\ 1 \leq j \leq n\| \leq C \ell n^{-1/2}(m/n).$$

REFERENCES

1. Gluskin, E., On a problem concerning diameters, Sov. Math. Doklady, 15 (1974), 1592-1596.
2. Ismagilov, R. S., Diameters of sets in normed linear spaces and the approximation of functions by trigonometric polynomials, Russian Math. Surveys, 29 (1974), 169-186.
3. Kashin, B. S., On Kolomogorov diameters of octohedra, Doklady Akad. Nauk., 214 (1974), 1024-1026.
4. Kashin, B. S., On diameters of octohedra, Uspehi Mat. Nauk., 30:4 (1975), 251-252.
5. Kashin, B. S., The diameters of certain finite dimensional sets and classes of smooth functions, Isv. Akad. Nauk. SSSR, Ser. Mat., 41 (1977) #2, 334-351.
6. Kashin, B. S., On some properties of spaces of trigonometric polynomials with uniform norm, Trudy Mat. Inst. Akad. Nauk., 145 (1979).
7. Kashin, B. S., On one property of bilinear forms, Report, Akad. Nauk. Gruz. SSSR, 93 (1979).
8. Maiorov, V., Discretization of the problem of diameters, Uspehi Mat. Nauk., 30 (1975), 179-180.
9. Pietsch, A., Nuclear Locally Convex Spaces, 1972, Springer Vertag.
10. Stechkin, S. B., On the best approximation of given classes of functions by arbitrary polynomials, Uspehi Mat. Nauk., 9:1, 133-134.
11. Tikomirov, V. M., Some Problems on Approximation Theory, State U. Moscow, Moscow, 1976.

B.S. Kashin
Stecklov Institute
Moscow, USSR

BOUNDS FOR THE ERROR IN TRIGONOMETRIC HERMITE INTERPOLATION

P. E. Koch and T. Lyche

We study the error in trigonometric interpolation to nonperiodic smooth functions on arbitrary intervals of length less than 2π.

1. INTRODUCTION AND SUMMARY

In this paper we study trigonometric analogs of the formula

$$(1.1)\quad g(x) - f(x) = \prod_{j=1}^{m} (x-x_j) D^m g(\xi)/m!$$

for polynomial interpolation to a smooth function g at m points $x_1,x_2,\ldots,x_m$. We consider any interval $[a,b]$ with $b-a < 2\pi$ using the system

$$(1.2)\quad S_{2n+1} = \{ \sum_{k=-n}^{n} a_k e^{ikx} \mid a_k \in \mathbb{C} , x \in [a,b]\}.$$

If $f \in S_{2n+1}$ agrees with a (smooth, nonperiodic) function g at $x_1,\ldots,x_m$ then

$$(1.3)\quad g(x) - f(x) = c(x) \prod_{j=1}^{m} (2\sin\tfrac{1}{2}(x-x_j)) L_m g(\xi)/m!$$

where $m=2n+1$,

$$(1.4)\quad L_{2n+1} = D(D^2+1^2)(D^2+2^2) \ldots (D^2+n^2) \quad , \quad D = d/dx ,$$

and the function c is independent of g. The existence of (1.3) follows from results in [2] and [3]. Here we give in section 2 and 3 upper and lower bounds for c on $[a,b]$ In particular it follows that if $b-a < h$ and if h is small then $c(x) = 1 + O(h^2)$.

ISBN 0-12-213650-0

We also consider an even number of interpolation points using the system

(1.5) $\tilde{S}_{2n} = \{f(x) + a_n\sigma_{2n}(x) \mid f \in S_{2n-1}, a_n \in \mathbb{C}, x \in [a,b]\}$

where $\sigma_{2n} = \sin(n(x - \sum_{j=1}^{2n} x_j/2n))$.

σ_{2n} is chosen because

1. The interpolant exists and is unique.
2. If $x_k = -\pi + (k-1)\pi/n$ $k=1,2,\ldots,2n$ then we get the familiar system $\{1, \cos x, \sin x, \ldots, \cos(n-1)x, \sin(n-1)x, \cos nx\}$
3. $\tilde{S}_{2n}$ has a minimal amplitude property ([3], [4]).

A Newton form for the interpolants in $\tilde{S}_{2n}$ and S_{2n+1} was given in [3].

We show in section 4 that (1.3) with $m=2n$ is valid for interpolation in $\tilde{S}_{2n}$ provided we replace the operator L_m by

(1.6) $\tilde{L}_{2n}g(y) = (\cos\tfrac{1}{2}(x-y)D + n\sin\tfrac{1}{2}(x-y))L_{2n-1}g(y)$.

Both for S_m and $\tilde{S}_m$ we have

(1.7) $c(x) = \frac{m}{2} \int T(y)\,dy$.

$T = T_{0,m}$ is a trigonometric B-spline of order m. T can be defined recursively [2] or via certain divided differences [2], [3]. Let

(1.8) $S_{2n} = \sum_{k=1-n}^{n} \{a_k e^{i(k-\frac{1}{2})x} \mid a_k \in \mathbb{R}, x \in [a,b]\}$

Then T is realvalued, piecewise in S_m, and its $(m-1)$. derivative has jumps at $x_1, \ldots, x_m$, and x. Moreover T has local support and is nonnegative.

The results are also valid for interpolation in S_{2n} and

(1.9) $\tilde{S}_{2n+1} = \{f(x) + a_{n+1}\cos((n+\tfrac{1}{2})(x - \sum_{j=1}^{2n+1} x_j/(2n+1)))\}$

and we consider these spaces as well.

2. TRIGONOMETRIC B-SPLINES

In this paper m will be a fixed positive integer, $[a,b]$ an interval with $b-a < 2\pi$, and $y_o, \ldots, y_m$ are nondecreasing points in $[a,b]$, with $y_o < y_m$. We let $s_j(y) = \sin\frac{1}{2}(y-y_j)$, $\hat{s}_j(y) = \sin\frac{1}{2}(y_j-y)$, and $c_j(y) = \cos\frac{1}{2}(y-y_j)$ for any $y \in [a,b]$. Suppose $z_o, \ldots, z_m$ are distinct points in $[a,b]$, and that g is a given function on $[a,b]$. Define trigonometric divided differences by

(2.1a) $[z_o, z_1, \ldots, z_m]_t g$

$$= 2^{m-1} \frac{\det\begin{pmatrix} 1 & \cos x & \sin x & \ldots & \cos nx & \sin nx & g(x) \\ z_o & z_1 & z_2 & \ldots & z_{m-2} & z_{m-1} & z_m \end{pmatrix}}{\det\begin{pmatrix} \cos\frac{1}{2}x & \sin\frac{1}{2}x & \ldots & \cos(n+\frac{1}{2})x & \sin(n+\frac{1}{2})x \\ z_o & z_1 & \ldots & z_{m-1} & z_m \end{pmatrix}}$$

if $m=2n+1$ is odd, and if $m=2n$ then

(2.1b) $[z_o, z_1, \ldots, z_m]_t g$

$$= 2^{m} \frac{\det\begin{pmatrix} \cos\frac{1}{2}x & \sin\frac{1}{2}x & \ldots & \cos(n-\frac{1}{2})x & \sin(n-\frac{1}{2})x & g(x) \\ z_o & z_1 & \ldots & z_{m-2} & z_{m-1} & z_m \end{pmatrix}}{\det\begin{pmatrix} 1 & \cos x & \sin x & \ldots & \cos nx & \sin nx \\ z_o & z_1 & z_2 & \ldots & z_{m-1} & z_m \end{pmatrix}}$$

Here we have used the abbreviation

$$\det\begin{pmatrix} f_o & f_1 & \ldots & f_m \\ z_o & z_1 & \ldots & z_m \end{pmatrix}$$

for the determinant

$$\begin{vmatrix} f_o(z_o) & f_1(z_o) & \ldots & f_m(z_0) \\ \cdot & \cdot & & \cdot \\ \cdot & \cdot & & \cdot \\ f_o(z_m) & f_1(z_m) & \ldots & f_m(z_m) \end{vmatrix} .$$

If $y_o, \ldots, y_m$ is a nondecreasing rearrangement of $z_o, \ldots, z_m$ then $[y_o, \ldots, y_m]_t = [z_o, \ldots, z_m]_t$. The definition of $[y_o, \ldots, y_m]_t$ is extended by continuity to the case where some or all of the y_j's are equal.

Since any $f \in S_m$ has at most $m-1$ zeros counting multiplicities $[y_o, \ldots, y_m]_t g$ is well defined provided g is sufficiently smooth. We also note that $[y_o, \ldots, y_m]_t f = 0$ for any $f \in S_m$. The t-differences can be used to give a Newton form for the trigonometric Hermite interpolation polynomial and they can be computed recursively in a difference scheme [3].

Let for fixed $x \in [a,b]$ $\phi_m(y) = \sin\frac{1}{2}(y-x)_+^{m-1}$ where

$$\sin\tfrac{1}{2}(y-x)_+^{m-1} = \begin{cases} \sin\frac{1}{2}(y-x)^{m-1} & y \geq x \\ 0 & \text{otherwise} \end{cases} \qquad (0_+^0 = 0)$$

and define

$$T(x) = T_{o,m}(x) = \begin{cases} [y_o, \ldots, y_m]_t\, \phi_m & y_m > y_o \\ 0 & \text{otherwise} \end{cases} \qquad x \in [a,b]$$

For $1 \leq k \leq m$ and $0 \leq j \leq m-k$ we define $T_{j,k}$ similarly by taking a divided difference over $y_j, \ldots, y_{j+k}$. T is called a trigonometric B-spline of order m. It can be computed from the recurrence relation

$$(2.2) \quad T_{o,m} = (s_o T_{o,m-1} + \hat{s}_m T_{1,m-1})/s_o(y_m)$$

starting with

$$(2.3) \quad T_{o,1}(y) = \begin{cases} 1/s_o(y_1) & y_o \leq y < y_1 \\ 0 & \text{otherwise.} \end{cases}$$

The following lemma is used to bound the sup norm of T.

LEMMA 2.1. *Let* $\hat{T}_{o,m} = s_o(y_m) T_{o,m}$. *Then for* $1 \leq k \leq m$

$$(2.4) \quad \hat{T}_{o,m} = \sum_{j=o}^{m-k} c_{j,k} \hat{T}_{j,k}$$

where $c_{o,m} = 1$ *and with* $c_{-1,k+1} = c_{m-k,k+1} = 0$

$$(2.5) \quad c_{j,k} = (s_j c_{j,k+1} + \hat{s}_{j+k} c_{j-1,k+1})/s_j(y_{j+k}).$$

Moreover if $y_j \leq y \leq y_{j+k}$ *then*

$$(2.6) \quad 0 \leq c_{j,k}(y) \leq 1/\prod_{r=k}^{m-1} \cos(h_r/4)$$

where

(2.7) $h_k = \max\{y_{j+k} - y_j \mid 0 \le j \le m-k\}$.

Proof. (2.4) holds for $k=m$, and if it holds for $k=n+1$ then it follows from the recurrence (2.2) written in the form $\hat{T}_{j,n+1} = s_j T_{j,n} + \hat{s}_{j+n+1} T_{j+1,n}$ that (2.4) holds for $k=n$ with $c_{j,n}$ given by (2.5). But if $y_j \le y \le y_{j+k}$ then for $j=0,1, \ldots ,m-k$

$$0 \le (s_j(y) + \hat{s}_{j+k}(y))/\hat{s}_{j+k}(y_j) \le 1/\cos(h_k/4)$$

and the theorem follows.

The next lemma will be used to give upper and lower bounds for the L_1 norm of T.

LEMMA 2.2. Let $m \ge 2$. Then

(2.8) $m\int T_{o,m}(y)c_m(y)dy/2 = I_{o,m-1}$, $y_m > y_o$

(2.9) $m\int T_{o,m}(y)c_o(y)dy/2 = I_{1,m-1}$, $y_m > y_o$

(2.10) $I_{\frac{1}{2},m-1}/\cos((y_m-y_o)/4) \le I_{o,m} \le I_{\frac{1}{2},m-1}/\cos^2((y_m-y_o)/4)$

where

$$I_{j,k} = \begin{cases} k\int T_{j,k}(y)dy/2 & y_{j+k} > y_j \\ 1 & \text{otherwise} , \end{cases}$$

and $I_{\frac{1}{2},k} = (I_{o,k} + I_{1,k})/2$.

Proof. We use (2.2) and the differentiation formula

$$2DT_{o,m}/(m-1) = (c_o T_{o,m-1} - c_m T_{1,m-1})/s_o(y_m)$$

([2 p.274]). Solving for $T_{o,m-1}$ and $T_{1,m-1}$ we find

(2.11) $T_{o,m-1} = c_m T_{o,m} + \frac{2}{m-1}\hat{s}_m DT_{o,m}$

(2.12) $T_{1,m-1} = c_o T_{o,m} - \frac{2}{m-1}s_o DT_{o,m}$.

To prove (2.8) we write (2.11) in the form

$$T_{o,m-1} = \frac{m}{m-1}c_m T_{o,m} + \frac{2}{m-1}D(\hat{s}_m T_{o,m})$$

Integrating we find

$$(2.13)\quad \frac{m-1}{2}\int T_{o,m-1}(y)\,dy = \frac{m}{2}\int c_m(y)\,T_{o,m}(y)\,dy - s_o(y_m)\,T_{o,m}(y_o+)$$

Now if $y_o < y_{m-1}$ then $T_{o,m}(y_o+) = 0$, and (2.13) reduces to (2.8) . If $y_o = y_{m-1}$ then by definition $T_{o,m-1} = 0$ while using (2.2) and induction

$$T_{o,m}(y) = \begin{cases} \hat{s}_m(y))^{m-1}/(s_o(y_m))^m & y_o \leq y < y_m \\ 0 & \text{otherwise} \end{cases}$$

so that $\hat{T}_{o,m}(y_o+) = 1$. Since $I_{o,m-1} = 1$ (2.8) follows. Using (2.12) the proof of (2.9) is similar. Clearly (2.10) holds for $y_o = y_m$. Suppose $y_m > y_o$. Adding (2.8) and (2.9) we find

$$(2.14)\quad \frac{m}{2}\cos\frac{1}{4}(y_m-y_o)\int\cos\tfrac{1}{2}(y-\tfrac{1}{2}(y_o+y_m))\,T_{o,m}(y)\,dy = I_{\frac{1}{2},m-1}\ .$$

But then

$$\cos^2((y_m-y_o)/4)\,I_{o,m} \leq I_{\frac{1}{2},m-1} \leq \cos((y_m-y_o)/4)\,I_{o,m}$$

and this is equivalent to (2.10).

Let $||\cdot||_p$ denote the usual L_p norm on $[a,b]$. We can now prove the main result of this section.

PROPOSITION 2.1. *Suppose* $y_m > y_o$. *Then*

$$(2.15)\quad \underline{\pi}_{m-1}/m \leq s_o(y_m)\,||T_{o,m}||_\infty \leq \overline{\pi}_{m-1}$$

$$(2.16)\quad \underline{\pi}_m \leq \frac{m}{2}||T_{o,m}||_1 \leq \min(\overline{\pi}_m^2, \frac{m}{2}\hat{\pi}_m)$$

$$(2.17)\quad ||T_{o,m}||_p \leq ||T_{o,m}||_1^{1/p}\,||T_{o,m}||_\infty^{1-1/p} \qquad 1 < p < \infty$$

where

$$\overline{\pi}_k = 1/\prod_{j=1}^{k} \cos(h_j/4) \quad , \qquad \hat{\pi}_m = (y_m - y_o)\overline{\pi}_{m-1}/s_o(y_m) \quad ,$$

and

$$\underline{\pi}_k = 1/\prod_{j=1}^{k} \cos(\underline{h}_j/4) \quad , \qquad \underline{h}_j = \min\{y_{r+j}-y_r \mid 0 \le r \le m-j\} \quad .$$

Proof. Using (2.4) , (2.3) , and (2.6) we have

$$||\hat{T}_{o,m}||_\infty = ||\sum_{j=o}^{m-1} c_{j,1}\hat{T}_{j,1}||_\infty = \max_{o\le j\le m-1} \max_{y_j\le y\le y_{j+1}} c_{j,1}(y)$$
$$\le \overline{\pi}_{m-1} \ .$$

This proves the upper bound in (2.15) . Moreover $||T_{o,m}||_1 \le (y_m-y_o)||T_{o,m}||_\infty$ and the last upper bound in (2.16) follows from (2.15). To show the first upper bound in (2.16) we apply (2.10) repeatedly and obtain

$$I_{o,m} \le 2^{1-m} \sum_{k=o}^{m-1} \binom{m-1}{k} I_{k,1}/\prod_{k=2}^{m} \cos^2(h_k/4) \ .$$

But if $h = (y_{k+1}-y_k)/2 > 0$ then

$$1/\cos\tfrac{h}{2} \le I_{k,1} = h/\sinh \le 1/\cos^2 \tfrac{h}{2} \ .$$

Hence the first upper bound in (2.16) follows. The proof of the lower bound in (2.16) is similar. For the lower bound in (2.15) we have

$$\sin\tfrac{1}{2}(y_m-y_o) = \tfrac{1}{2}\cos\tfrac{1}{4}(y_m-y_o) \int_{y_o}^{y_m} \cos\tfrac{1}{2}(y-y_{\frac{1}{2}m})\,dy$$

where $y_{\frac{1}{2}m} = (y_o+y_m)/2$. Thus by (2.14)

$$||\hat{T}_{o,m}||_\infty \ge \tfrac{1}{2}\cos\tfrac{1}{4}(y_m-y_o)\int\cos\tfrac{1}{2}(y-y_{\frac{1}{2}m})T_{o,m}(y)\,dy = \tfrac{1}{m}I_{\frac{1}{2},m-1}.$$

But now the lower bound in (2.15) follows from the lower bound in (2.16) . Finally for (2.17) we apply Hölders inequality to the product $T^p = T\cdot T^{p-1}$.

3. TRIGONOMETRIC INTERPOLATION I

Let $g \in C^m$ where C^m is the class of real valued m times continuously differentiable functions on $[a,b]$. Given m nondecreasing points $x_1, x_2, \ldots, x_m$ in $[a,b]$ let

$$U_m(g) = \{ f \in C^m \mid f^{(r_i)}(x_i) = g^{(r_i)}(x_i) , \quad i=1,\ldots,m\}$$

where $r_i = \max\{j \mid x_{i-j}=x_i\}$. Let f be the unique Hermite interpolant to g in S_m at $x_1,\ldots,x_m$. I.e. $f \in S_m \cap U_m(g)$. In [3] it was shown that for m odd and any $x \in [a,b]$

$$(3.1) \quad g(x) - f(x) = \prod_{k=1}^{m} s_k(x)[x_1,\ldots,x_m,x]_t g .$$

We have a similar proof for m even. An alternative derivation of (3.1) valid for any positive integer m can be given by transforming the divided difference form of the error in polynomial interpolation to trigonometric form. For

$$f(x) = e^{-\frac{1}{2}i(m-1)x} f_p(e^{ix})$$

where f_p is the polynomial of degree less than m which agree with h given by

$$h(e^{ix}) = e^{\frac{1}{2}i(m-1)x} g(x)$$

at the points $e^{ix_1}, \ldots, e^{ix_m}$. Thus

$$g(x) - f(x) = e^{-\frac{1}{2}i(m-1)x}(h(e^{ix}) - f_p(e^{ix}))$$

$$= e^{-\frac{1}{2}i(m-1)x} \prod_{k=1}^{m} (e^{ix} - e^{ix_k})[e^{ix_1}, \ldots, e^{ix_m}, e^{ix}]h .$$

Using $e^{ix} - e^{ix_k} = 2ie^{\frac{1}{2}i(x+x_k)} s_k(x)$ we find

$$g(x) - f(x) = \prod_{k=1}^{m} s_k(x)(2i)^m \exp(\tfrac{1}{2}i(x+\sum_{k=1}^{m} x_k)) \cdot [e^{ix_1}, \ldots, e^{ix_m}, e^{ix}]h .$$

But according to [2] (see (2.7) and (2.11)) polynomial and trigonometric divided differences are related as follows

$$[e^{ix_1}, \dots, e^{ix_m}, e^{ix}]h = (2i)^{-m}\exp(-\tfrac{1}{2}i(x+\sum_{k=1}^{m} x_k))\cdot [x_1,\dots,x_m,x]_t g \quad .$$

Hence we obtain (3.1).

Fix x and let $y_o, \dots, y_m$ be a nondecreasing rearrangement of $x_1, \dots, x_m, x$. According to [2 p.276] we have

$$(3.2) \quad [x_1,\dots,x_m,x]_t g = \frac{2^{m-1}}{(m-1)!}\int T_{o,m}(y) L_m g(y)\,dy$$

where $T_{o,m}$ is the trigonometric B-spline of section 2, and L_m is given by (1.4) for $m=2n+1$ and

$$(3.3) \quad L_{2n} = (D^2+\tfrac{1}{2}^2)(D^2+(2-\tfrac{1}{2})^2) \dots (D^2+(n-\tfrac{1}{2})^2) \ .$$

Combining (3.1) and (3.2) we have an integral representation for the error

$$(3.4) \quad g(x) - f(x) = \prod_{k=1}^{m} (2s_k(x))\cdot\frac{m}{2}\int T_{o,m}(y) L_m g(y)\,dy/m! \ .$$

Since $T_{o,m}$ is nonnegative we obtain from proposition 2.1

THEOREM 3.1. _For any_ $x \in [a,b]$ _there is a_ $\xi \in [a,b]$ _such that_

$$(3.5) \quad g(x) - f(x) = c(x)\prod_{k=1}^{m} (2s_k(x)) L_m g(\xi)/m!$$

where $c(x) = m\int T_{o,m}(y)\,dy/2$ _is bounded as follows_

$$\underline{\pi}_m \le c(x) \le \min(\overline{\pi}_m^{2} , \tfrac{m}{2}\overline{\pi}_m) \ .$$

We note that if $h = b-a$ then $\overline{\pi}_m \le 1/\cos^m(h/4)$. Hence we obtain the bound

$$(3.6) \quad ||g-f||_\infty \le (4\tan(h/4))^m ||L_m g||_\infty/m! \qquad (h \le \pi)$$

If $h = b-a$ is small then $\underline{\pi}_m = 1 + O(h^2) = \overline{\pi}_m$. Thus $c(x) = 1 + O(h^2)$ and (3.5) can be written

$$g(x) - f(x) = (1 + O(h^2)) \prod_{k=1}^{m} (x-x_k) L_m g(\xi)/m!$$

which is similar to the error expression for polynomial interpolation. Suppose also g is periodic on $[a,b]$, i.e $D^j g(a) = D^j g(b)$ $j=0,1,\ldots,m-1$ with $\int g=0$ if m is even. Using an analog of the Euler MacLaurin type formula (16.38) in [1] for the interval $[a,b]$ it can be shown that

$$||L_m g - D^m g||_\infty \le \text{const}_m \cdot h^2 ||D^m g||_\infty \quad .$$

Hence trigonometric interpolation behaves very much like polynomial interpolation when h is small and $g \in C^m$ is periodic on $[a,b]$.

Consider next the case where $b-a$ is large $b-a \simeq 2\pi$. For simplicity of argument assume $m=2n$ is even and that the interpolation points are uniformly spaced

$$x_k = -\pi + (k-1)\pi/n \qquad k=1,2,\ldots,2n.$$

If $x \in (x_m, \pi)$ then $\underline{\pi}_m = 1/\prod_{k=1}^{m} \cos\frac{1}{4}(x-x_k)$ so that

$$|g(x) - f(x)| \ge \prod_{k=1}^{m} (4\sin\tfrac{1}{4}(x-x_k))|L_m g(\xi)|/m! \ .$$

Moreover $\hat{\pi}_m = (x-x_1)\overline{\pi}_{m-1}/s_1(x)$ and

$$\overline{\pi}_{m-1} = 1/\prod_{k=1}^{m-1} \cos\tfrac{1}{4}(x_m-x_k) \le 1/(\cos\tfrac{1}{4}(x_m-x_1)\prod_{k=2}^{m-1} \cos\tfrac{1}{4}(x-x_k)) \ .$$

Since $2s_m(x) \le 4\cos((x_m-x_1)/4)$ and $x - x_1 < 2\pi$ we obtain

$$\begin{aligned} |g(x) - f(x)| &\le \tfrac{m}{2}\hat{\pi}_m \cdot \prod_{k=1}^{m} (2s_k(x))|L_m g(\xi)|/m! \\ &\le 8\pi m \prod_{k=2}^{m-1} (4\sin\tfrac{1}{4}(x-x_k))|L_m g(\xi)|/m! \ . \end{aligned}$$

Note that g is not assumed to be periodic in these estimates Note also that if g is periodic with rapidly decaying Fourier coefficients then the term $L_m g(\xi)$ can be bounded in terms of sums of Fourier coefficients.

4. TRIGONOMETRIC INTERPOLATION II

In this section we consider interpolation in the $\tilde{S}_m$ spaces given by (1.5) and (1.9). In [3] the unique interpolant $f \in \tilde{S}_{2n} \cap U_{2n}(g)$ was constructed and it was shown that for any $x \in [a,b]$

$$(4.1) \quad g(x) - f(x) = \prod_{k=1}^{m} s_k(x)\lambda[x_1,\ldots,x_m,x]_t g$$

where

$$(4.2) \quad \lambda[x_1,\ldots,x_m,x]_t g = ([x_2,\ldots,x_m,x]_t g - c_1(x)[x_1,\ldots,x_m]_t g)/s_1(x) \quad .$$

λ is a divided difference with properties similar to the t-differences ([3]). However it is not a symmetric function of its arguments. I.e. in (4.2) $x_1,\ldots,x_m$ can be taken in any order, but x must be the last point.

It can also be shown that there is a unique $f \in \tilde{S}_{2n+1} \cap U_{2n+1}(g)$, and that (4.1) holds with $m=2n+1$. The construction of this f can be done as in [3] for the $\tilde{S}_{2n}$ case.

The following theorem gives a representation for λ.

THEOREM 4.1. *Let* $x_1, \ldots ,x_m$, x *be* $m+1$ *points in* $[a,b]$ *not all equal. Then*

$$(4.3) \quad \lambda[x_1,\ldots,x_m,x]_t g = \frac{2^{m-1}}{(m-1)!}\int T_{o,m}(y)\tilde{L}_m g(y)\,dy$$

where

$$\tilde{L}_m g(y) = (\cos\tfrac{1}{2}(x-y)D + \tfrac{m}{2}\sin\tfrac{1}{2}(x-y))L_{m-1}g(y) \quad .$$

Here $T_{o,m}$ *is the* B-spline *on* $x_1,\ldots,x_m,x$ *and* L_{m-1} *is given by* (1.4) *and* (3.3) .

Proof. Since λ is independent of the ordering of $x_1,\ldots,x_m$ we assume $x \neq x_1$. Assume also that no m points of $x_1,\ldots,x_m,x$ are equal. Using (4.2) , (3.2) , (2.11) , and (2.12) we have (note that (2.11) and (2.12) are valid when $y_o,\ldots,y_m$ are not ordered) with $y_{j-1}=x_j$,$j=1,\ldots,m$, $y_m=x$

$$(m-1)!\lambda[y_o,\dots,y_m]_t g/2^{m-1}$$

$$= \frac{m-1}{2}\int (T_{1,m-1} - c_o(y_m)T_{o,m-1})L_{m-1}g\,dy/s_o(y_m)$$

$$= -\int c_m DTL_{m-1}g + \frac{m-1}{2}\int \hat{s}_m TL_{m-1}g = \int(-D(c_mT) + \frac{m}{2}\hat{s}_mT)L_{m-1}g\,dy$$

where $T = T_{o,m}$. Integrating by parts gives (4.3). Using Lemma 3.1 in [3] (4.3) can also be shown to be valid when m of the $x_1,\dots,x_m,x$ are equal.

COROLLARY 4.1. *Suppose* $f \in \tilde{S}_m \cap U_m(g)$. *Fix* $x \in [a,b]$ *and let* $y_o,\dots,y_m$ *be a nondecreasing rearrangement of* $x_1,\dots,x_m,x$. *Then*

$$(4.4)\quad g(x) - f(x) = \prod_{k=1}^{m}(2s_k(x))\cdot\frac{m}{2}\int T_{o,m}(y)\tilde{L}_m g(y)\,dy/m!\ .$$

Moreover there is a ξ *in* $[a,b]$ *such that*

$$(4.5)\quad g(x) - f(x) = c(x)\prod_{k=1}^{m}(2s_k(s))\tilde{L}_m g(\xi)/m!$$

where $c(x) = \frac{m}{2}\int T_{o,m}$ *is bounded as in theorem* 3.1 .

Proof. We combine (4.1) and (4.3) and use the nonnegativity of $T_{o,m}$.

REFERENCES

1. Lanczos, C., *Discourse on Fourier series*, Oliver & Boyd, Edinburgh and London, 1966.
2. Lyche, T. and Winther, R., A stable recurrence relation for trigonometric B-splines, J. Approximation Theory, 25 (1979), 266-279.
3. Lyche, T., A Newton form for trigonometric Hermite interpolation, BIT 19 (1979), 229-235.
4. Schoenberg, I.J., On interpolation by spline functions and its minimal properties, in *On Approximation Theory*, Butzer, P.L. and J. Korevaar (Eds.),ISNM 5, Birkhäuser Verlag, Basel und Stuttgart, 1964, 109-129.

P.E. Koch
Institutt for informatikk
Universitetet i Oslo
Oslo 3, Norway

T. Lyche
Institutt for informatikk
Universitetet i Oslo
Oslo 3, Norway

BIRKHOFF INTERPOLATION: SOME APPLICATIONS OF COALESCENCE

G. G. Lorentz and S. D. Riemenschneider

The method of the coalescence of two rows of a matrix was first used implicitly by D. Ferguson [2, p. 14]. It was formally defined by Karlin and Karon [3], and extended and put on a broader basis by G. G. Lorentz [4,6]. Many applications of this method have already been given [3,4,6,7]. In this paper we give a few more applications; in particular, a simple proof of D. Ferguson's theorem describing the interpolation matrices which are complex regular.

1. THE METHOD OF COALESCENCE

Let E be a normal Pólya interpolation matrix $(e_{ik})_{i=0,k=0}^{m\ \ n}$. This means that e_{ik} are zeros or ones, with exactly $n+1$ ones and there are a total of at least $j+1$ ones in columns 0 thru j, for each $j = 0,1,\ldots,n$. The matrix is called a <u>Birkhoff matrix</u> if there are at least $j+2$ ones in columns 0 thru j, $j = 0,1,\ldots,n-1$. Corresponding to an interpolation matrix is a <u>Birkhoff interpolation problem</u>: Given numbers $\{C_{i,k}\}$ and a set of knots $X = \{x_1,\ldots,x_m\}$, does there exist a unique polynomial of degree n satisfying

$$(1.1)\quad P^{(k)}(x_i) = C_{i,k},\quad e_{ik} = 1.$$

If $g_k(x) = x^k/k!$, $k = 0,\ldots,n$, then the solution to this problem is equivalent to the question of whether the $(n+1)\times(n+1)$ determinant

$$(1.2)\quad D(E,X) = \det\left\{g_o^{(k)}(x_i),\ldots,g_n^{(k)}(x_i):e_{i,k}=1\right\}$$

is zero. (Here the rows are ordered according to the lexicographic ordering on (i,k)). If $D(E,X) \neq 0$, then such a polynomial exists and the pair (E,X) is called <u>regular</u>. If (E,X) is regular for each choice of X, then the matrix E is called <u>regular</u>; otherwise, E is called <u>singular</u>. The adjectives complex, real, or order are used to describe regularity when X is restricted to complex, real, or ordered $(x_1 < \ldots < x_m)$ numbers.

ISBN 0-12-213650-0

We shall need the notion of the (directed) coalescence [3] of row F_i to row F_j in $E(i \neq j)$ and also of the maximal coalescence [4] of row F_i in E. The coalescence of row F_i to row F_j in E can be described as follows: Let $\ell_1,\ldots,\ell_p$ denote the column positions of the ones in row F_i (always written in increasing order). The $(m-1) \times (n+1)$ matrix of coalescence, E', is obtained from E by omitting row F_i and placing the p ones of row F_i into positions $\ell_1',\ldots,\ell_p'$ of row F_j. (Shifting the ones from positions $\ell_1,\ldots,\ell_p$ to positions $\ell_1',\ldots,\ell_p'$ in row F_i is called the precoalescence of row F_i with respect to F_j). The positions ℓ_s' are determined according to the following rule:

(1.3) ℓ_1' is the first integer k such that $k \geq \ell_1$ and $e_{jk} = 0$; ℓ_s' is the first k such that $k \geq \ell_s$, $k > \ell_{s-1}'$ and $e_{j,k} = 0$.

Maximal coalescence of row F_i to F_j in E gives the matrix E^* obtained by omitting row F_i and placing ones in positions $\ell_1^*,\ldots,\ell_p^*$ of row F_j. To obtain ℓ_s^*, we first coalesce all the rows of E/F_i into a single row and then choose the positions according to the above rule. It turns out that

(1.4) $\ell_s' \leq \ell_s^* \qquad s = 1,\ldots,p.$

Associated with the coalescence of row F_i to row F_j in E is the coefficient of collision,

(1.5) $$\alpha_{ij} = \alpha(F_i \cup F_j) = \sum_{s=1}^{p} (\ell_s' - \ell_s)$$

and the coefficient of maximal collision

(1.6) $$\gamma_i = \sum_{s=1}^{p} (\ell_s^* - \ell_s).$$

The main importance of coalescence is expressed in the following theorem. Let X' be the knot set obtained from X by deleting x_i.

THEOREM 1.1. [3], [4], [8]. *The determinant* $D(E,X)$ *with* $X = (\ldots,x_i,\ldots, x_j,\ldots)$ *has, as a function of* x_i, *the expansion*

(1.7) $$D(E,X) = (-1)^{\sigma} \frac{C}{\alpha_{ij}!} (x_i - x_j)^{\alpha_{ij}} D(E',X') + \ldots + (-1)^{\sigma^*} \frac{C^*}{\gamma_i!} (x_i - x_j)^{\gamma_i} D(E^*,X').$$

The determinant $D(E',X')$ is obtained from $D(E,X)$ by replacing each row $g_0^{(\ell_s)}(x_i),\ldots,g_n^{(\ell_s)}(x_i)$ by the ℓ_s' derivatives at x_j, and then interchanging rows to order the rows of the matrix according to the lexicographic ordering induced by E'. Thus, $\sigma = \tau p + \sigma'$ where τ is the number of ones in the rows of E contained between rows F_i and F_j, and σ' is the number of interchanges which are required to bring the sequence of integers.

$$(1.8)\qquad \ell_1',\ldots,\ell_p',\bar{\ell}_1,\ldots,\bar{\ell}_q$$

into its natural order (here $\bar{\ell}_1 < \ldots < \bar{\ell}_q$ are the positions of ones in row F_j). If the rows are adjacent, then σ is just this number of permutations. Similar statements can be made for $D(E^*,X')$ and σ^*.

The constant $C = C(\ell_1,\ldots,\ell_p;\ell_1',\ldots,\ell_p') > 0$ (see [9]) is the number of different ways of shifting $(\ell_1,\ldots,\ell_p)$ to $(\ell_1',\ldots,\ell_p')$. A shift $\ell_s \to \ell_{s+1}$ is defined if $\ell_s + 1 \neq \ell_{s+1}$, and takes $(\ell_1,\ldots,\ell_p)$ to $(\ell_1,\ldots,\ell_{s-1},\ell_s+1,\ell_{s+1},\ldots,\ell_p)$. Exactly $\alpha_{i,j}$ shifts are required to transform $(\ell_1,\ldots,\ell_p)$ to $(\ell_1',\ldots,\ell_p')$. A representation for the important number $C(\ell_1,\ldots,\ell_p;\ell_1',\ldots,\ell_p')$ is one of our goals. The constant C^* reduces to this: $C^* = C(\ell_1,\ldots,\ell_p;\ell_1^*,\ldots,\ell_p^*)$.

For a detailed discussion of the method of coalescence consult [6], [7].

Theorem 1.1 has the following dual form:

THEOREM 1.1a. <u>As a function of</u> x_j, the determinant $D(E,X)$, $X = (\ldots, x_i,\ldots,x_j,\ldots)$ <u>has the expansion</u>

$$(1.9)\qquad D(E,X) = (-1)^{\bar{\sigma}} \frac{\bar{C}}{\alpha_{ij}!} (x_j-x_i)^{\alpha_{ij}} D(E',X') + \ldots + (-1)^{\bar{\sigma}} \frac{C^*}{\gamma_j!} (x_j-x_i)^{\gamma_j} D(\bar{E}^*,X^*)$$

Here $E',X',\alpha_{ij} = \alpha_{ji}$ are as before, but $\bar{C} = C(\bar{\ell}_1,\ldots,\bar{\ell}_q;\bar{\ell}_1',\ldots,\bar{\ell}_q')$ and $\bar{\sigma} = \tau p + \bar{\sigma}'$, where $\bar{\sigma}'$ is the number of interchanges which are required to bring the sequence

$$(1.10)\qquad \ell_1,\ldots,\ell_p;\bar{\ell}_1',\ldots,\bar{\ell}_q'$$

into its natural order. (Here $\bar{\ell}_1',\ldots,\bar{\ell}_q'$ are the positions of ones in the precoalesced row F_j).

With E also E' is a Pólya matrix. The determinant of a Pólya matrix is not identically zero. Comparing (1.7) and (1.9) we obtain the *duality formula*

$$(1.11)\quad C(\ell_1,\ldots,\ell_p;\ell_1',\ldots,\ell_p') = C(\bar{\ell}_1,\ldots,\bar{\ell}_q;\bar{\ell}_1',\ldots,\bar{\ell}_q').$$

2. THE NUMBER $C(\ell_1,\ldots,\ell_p;\ell_1',\ldots,\ell_p')$ AND SIGNS OF DETERMINANTS

Theorem 1.1 and the permutation numbers may be used to determine the signs of certain determinants arising in interpolation problems.

THEOREM 2.1. *Let E be a $2\times(n+1)$ Pólya matrix with $k_1,\ldots,k_p$ denoting the positions of ones in row 1 of E. If $X=\{0,x\}$, then*

$$(2.1)\quad D(E,X) = \frac{C}{\alpha!}(-1)^{p(p-1)/2+k_1+\ldots+k_p}x^{\alpha}$$

where $\alpha=\alpha(F_1\cup F_2)$ is the coefficient of collision for E and C is the constant of Theorem 1.1.

Proof. According to Theorem 1.1, $D(E,X)$ is a polynomial in x of total degree α and having a zero of order α at zero so that (2.1) holds for the positive constant C provided

$$\sigma \equiv p(p-1)/2 + k_1 + \ldots + k_p \pmod 2.$$

Let $\ell_1',\ldots,\ell_q'$ be positions of the zeros in row 1 of E. Then σ is the number of interchanges required to bring

$$k_1,\ldots,k_p,\ell_1',\ldots,\ell_q'$$

into the natural order. Since $p=n+1-q$, we have

$$\sigma = (p-\ell_1') + (p-(\ell_2'+1)) + \ldots + p - (\ell_q'+(q-1))$$
$$\equiv 1+2+\ldots+p-1+\sum_{j=1}^{n} j - \sum_{j=1}^{q}\ell_j' \pmod 2$$
$$\equiv p(p-1)/2 + \sum_{j=1}^{p} k_j \pmod 2. \quad \square$$

The first p rows of $D(E,X)$ contain only a single one. Indeed, the j-th row is the k_j-th derivative of $\{1,x,\ldots,x^n/n!\}$ at $x = 0$, and thus has a single non-zero entry 1 in the (k_j+1) column. This suggests the following

COROLLARY 2.2. *Let $M(E,X)$ be the minor of the determinant (2.1) obtained by omitting the first p rows and the columns $k_1 + 1,\ldots,k_p + 1$. Then*

$$(2.2) \quad M(E,X) = C\, x^{\alpha}/\alpha! > 0$$

(in particular, the sign of M does not depend on the number or distribution of ones in row 1, but only on the fact that E is a Pólya matrix).

Proof. Expanding $D(E,X)$ successively about its first p rows, we have $M(E,X) = (-1)^{\rho} D(E,X)$. The contribution of the row numbers to ρ is $1 + 2 + \ldots + p = p(p+1)/2$ while the contribution from the column numbers is $(k_1+1) + \ldots + (k_p+1) = \sum_{j=1}^{p} k_j + p$. Thus, $\rho = p(p+3)/2 + k_1 + \ldots + k_p$ and (2.2) follows from (2.1). □

Let $\ell_1',\ldots,\ell_q'$ be the positions of the zeros in row 1 of E, and $\ell_1,\ldots,\ell_q$ be the positions of ones in row 2 of E. Adopting the notation

$$(m)_j = m(m-1)\cdot \ldots \cdot(m-j+1); \quad (m)_o = 1; \quad \binom{m}{j} = \frac{(m)_j}{j!};$$

we have that

$$M(E,X) = \det\left\{(\ell_1')_{\ell_j}\frac{x^{\ell_1'-\ell_j}}{\ell_1'!},\ (\ell_2')_{\ell_j}\frac{x^{\ell_2'-\ell_j}}{\ell_2'!},\ \ldots,\ (\ell_q')_{\ell_j}\frac{x^{\ell_q'-\ell_j}}{\ell_q'!};\ j = 1,\ldots,q\right\}$$

$$= \frac{x^{\alpha}}{\ell_1'!\ldots\ell_q'!}\det\left\{(\ell_1')_{\ell_j},(\ell_2')_{\ell_j},\ldots,(\ell_q')_{\ell_j};\ j = 1,\ldots,q\right\}$$

where $\alpha = \sum_{j=1}^{q} (\ell_j'-\ell_j)$. Comparing (2.2) and (2.3) leads to two interesting consequences.

The first of them gives a formula for the "Lorentz-Zeller constant" C. The second has proved to be useful in certain interpolation problems ([1],[9]).

COROLLARY 2.3. *The number of different representations* $C = C(\ell_1,\ldots,\ell_p;\ell_1',\ldots,\ell_p')$ *for the multiple shift* Λ_α *of ones in positions* $(\ell_1,\ldots,\ell_p)$ *to positions* $(\ell_1',\ldots,\ell_p')$ *is*

$$C(\ell_1,\ldots,\ell_p;\ell_1',\ldots,\ell_p') = \frac{\alpha!}{\ell_1'!\ldots\ell_p'!}\ \det\left\{(\ell_1')_{\ell_j},\ldots,(\ell_p')_{\ell_j} : j=1,\ldots,q\right\}$$

where $\alpha = \sum_{j=1}^{p} (\ell_j'-\ell_j)$.

COROLLARY 2.4. [1], [10]. *For any positive integer* q *and integers* $\{\ell_j'\}_{j=1}^q$, $\{\ell_j\}_{j=1}^q$, *satisfying*

$$(2.5)\qquad \begin{cases} 0 \le \ell_1' < \ldots < \ell_q' \\ 0 \le \ell_1 < \ldots < \ell_q \\ \ell_j \le \ell_j' \quad j = 1,\ldots,q, \end{cases}$$

$$\det\left\{\binom{\ell_1'}{\ell_j},\ldots,\binom{\ell_q'}{\ell_j}\ ;\ j=1,\ldots,q\right\} = \frac{\ell_1!\ldots\ell_q!}{\ell_1'!\ldots\ell_q'!}\ \det\left\{(\ell_1')_{\ell_j},\ldots,(\ell_q')_{\ell_j}\ ; \ j=1,\ldots,q\right\}$$

is positive.

3. THE FUNCTION $\delta(E)$

In a Pólya matrix E, we single out a leading row, for example the first row, and denote by $\alpha_2,\ldots,\alpha_m$, γ the coefficients of collision of row 1 with rows $2,\ldots,m$, and the maximal coefficient of collision of row 1 in E, respectively. We put

$$(2.1)\qquad \delta = \delta(E) = \gamma - \sum_{i=2}^{m} \alpha_i .$$

Theorem 1.1 implies that for properly chosen knots $x_2,\ldots,x_m$, the determinant $D(E,X)$ is a polynomial of degree γ in x_1, which has roots of orders α_i at $x_i = 2,\ldots,m$. This shows that always $\delta \ge 0$.

If the Pólya matrix E decomposes vertically into Pólya matrices $E = E_1 \oplus \ldots \oplus E_\mu$, then

$$\alpha(E) = \sum_{\lambda=1}^{\mu} \alpha(E_\lambda), \quad \gamma(E) = \sum_{\lambda=1}^{\mu} \gamma(E_\lambda)$$

so that

$$(3.2) \quad \delta(E) = \sum_{\lambda=1}^{\mu} \delta(E_\lambda).$$

The value of $\delta(E)$ may depend upon the choice of the leading row in E. It is interesting to note that this is not the case when $m = 3$; since if δ_1 and δ_2 correspond to F_1 and F_2 as leading rows, then

$$(3.3) \quad \delta_1(E) = \delta_2(E).$$

This follows easily from equation (3.1.3) of Lorentz [6], for in his notation this relation is equivalent to

$$\alpha(F_1 \cup (F_2 \cup F_3)^\circ) + \alpha(F_2 \cup F_3) = \alpha(F_2 \cup (F_1 \cup F_3)^\circ) + \alpha(F_1 \cup F_3),$$

and both sides are equal to $\alpha(F_1 \cup F_2 \cup F_3)$. (Here $(F_i \cup F_j)^\circ$ is the row obtained from coalescing row F_i to F_j.)

EXAMPLE 3.1. For a two row matrix, $\gamma = \alpha_2$, hence $\delta = 0$.

EXAMPLE 3.2. Also for a Hermitian matrix E with rows F_i of lengths $p_i > 0$, $i = 1,\ldots,m$ one has $\delta = 0$. This follows by the easily established fact that for Hermite rows $\alpha_i = \alpha(F_1 \cup F_i) = p_1 p_i$, $i = 2,\ldots,m$, and $\gamma = p_1(p_2+\ldots+p_m)$.

EXAMPLE 3.3. Let E be a Birkhoff matrix of the following type. The (non-zero) rows F_i of E have Hermitian sequences of lengths $p_i \geq 0$, $i = 1,\ldots,m$, $m \geq 3$. In addition, F_1 has a sequence of ones for $k = n - r,\ldots,n - 1$, and F_2 has a single one in position $p_2 + s$; at least one of the numbers $r, s \geq 0$ is not zero. (If $r = 0$ or $s = 0$, then the corresponding group of ones is missing.) Then $s \geq 1$.

Since E is a Birkhoff matrix, we have

$$(3.4) \quad p_2 + s < n - r, \quad p_1 + \ldots + p_m + r = n.$$

Now

$$\alpha_i = p_1 p_i, \quad i = 3,\ldots,m$$

$$\alpha_2 = p_1p_2 + (p_1-s)_+ = p_1p_2 - s + \max(p_1,s).$$

The coalesced row $F' = (F_2 \cup \ldots \cup F_m)^0$ will have a Hermitian sequence of length $p_2 + \ldots + p_m$, and in addition a one at $k_0 = \max(p_2+\ldots+p_m, p_2+s)$. In coalescence of F' with F_1, the coefficient of collision of the first $p_2 + \ldots + p_m$ ones in F' with F_1 will be $p_1(p_2+\ldots+p_m)$, that of the one at k_0 with F_1 will be $n - k_0$, hence

$$\gamma = p_1(p_2+\ldots+p_m) + n - k_0,$$

$$\delta = \gamma - \alpha_2 - \ldots - \alpha_m = n + s - \max(p_1,s) - \max(p_2+\ldots+p_m, p_2+s).$$

Relations (3.4) yield $\delta \geq 1$ in all four possible cases.

Here are some applications of the function δ:

THEOREM 3.1. *If* $\delta(E)$ *is odd, then* E *is strongly real singular.*

Indeed, $(x_1-x_2)^{-\alpha_2} \ldots (x_1-x_m)^{-\alpha_m} D(E,X)$ is a polynomial in x_1 of odd degree, and not vanishing (for proper choice of knots) at $x_2,\ldots,x_m$. Hence it has a root x_1 different from $x_2,\ldots,x_m$. □

PROPOSITION 3.2. *A Pólya matrix* E *is complex regular if and only if* $\delta(E) = 0$.

Proof. For the polynomial $D(x_1) = D(E,X)$, with fixed complex knots $x_2,\ldots,x_m$ and variable x_1, we denote by g its degree, by $\beta_2,\ldots,\beta_m$ the multiplicities of its roots $x_2,\ldots,x_m$. By Theorem 1.1, $g \leq \gamma$, $\beta_i \geq \alpha_i$, $i = 2,\ldots,m$. Hence if $\delta = 0$, then $g \leq \beta_2 + \ldots + \beta_m$, and we must have $g = \beta_2 + \ldots + \beta_m$. This means that, for each choice of $x_2,\ldots,x_m$, $D(x_1)$ has no other zeros than $x_2,\ldots,x_m$, or that E is complex regular. Conversely, if $\delta > 0$, then for almost all choices of knots, $g = \gamma$, $\beta_i = \alpha_i$, $i = 2,\ldots,m$. For knots $x_2,\ldots,x_m$ of this type, $D(x_1)$ will have a zero different from them, and E will be complex singular. □

4. THE THEOREM OF FERGUSON

We shall give here a relatively simple proof of this theorem (actually, Ferguson's theorem is a special case of our Theorem 4.3). It is based upon a simplifying operation T for E; after finitely many applications, the problem is reduced to one of the simple cases: (a) the case of Example 3; (b) a two-row matrix. The necessity of using (at least) two cases of reduction explains perhaps the relative complexity of Ferguson's original proof [2].

For a Birkhoff matrix E with leading row F_1, we define an operation $T(E) = E_1$ as follows. In the rows $F_2,\ldots,F_m$ we select a one $e_{ik} = 1$ with largest possible k. If there are several k of this kind, we select a row with the maximal number of ones in it. Without loss of generality, let this be $e_{2k} = 1$. We replace this one by zero, and omit the last column of E (which consists of zeros), obtaining an $m \times n$ matrix $E' = T'(E)$. This E' is a Pólya -- (but not necessarily a Birkhoff) matrix. However, the Birkhoff condition is satisfied for all columns $\ell < k$ of E', and also for column k, if there is more than a single one in this column. It follows that all ones of rows $2,\ldots,m$ of E' are contained in the first matrix E_1 of the canonical decomposition $E' = E_1 \oplus \ldots E_\mu$ into Birkhoff matrices. Thus, all matrices E_λ, $\lambda \geq 2$ (if they exist) are one column matrices with a one in row 1, and with $\delta(E_\lambda) = 0$. From (3.2), $\delta(E_1) = \delta(E')$. We define $T(E) = E_1$.

For simplicity, we assume here and in Theorem 4.2 that the leading row F_1 has been so selected that $e_{10} = 1$.

LEMMA 4.1. *For a Birkhoff matrix* E, *and each selection of the leading row,*

(4.1) $\delta(TE) \leq \delta(E)$

Moreover, we have strict inequality in (4.1) *if after application of* T, *the matrix becomes a two row matrix.*

Proof. We show that

(4.2) $\delta(E') \leq \delta(E)$.

It is clear that the numbers $\alpha_i(E)$, $i = 3,\ldots,m$ are not affected by the operation T'.

Let $\ell_1 < \ldots < \ell_p$ be the positions of ones in row 1, $\ell_1' <\ldots <\ell_p'$ the positions of ones in precoalescence of row 1 with respect to row 2, let $\ell_1^* < \ldots < \ell_p^*$ be the positions of ones in the maximal precoalescence of row 1. None of the ℓ_j', ℓ_j^* is equal to k.

Since k is the highest non-zero column in $F_2 \cup \dots \cup F_m$, only those ℓ_j^* which satisfy $\ell_j^* > k$ will change if $e_{2k} = 1$ is removed. The row $(F_2 \cup \dots \cup F_m)^0$ has a sequence of ones, $[k,k_1]$, $k_1 \geq k$, followed by zeros, hence the $\ell_j^* > k$ are precisely the integers $k_1 + 1, \dots, n$. If $e_{2k} = 1$ is removed, these ℓ_j^* decrease by exactly one.

Similarly only $\ell_j' > k$ can change if $e_{2k} = 1$ is removed, and they can decrease by one or remain constant. Since $\ell_j^* \geq \ell_j'$, the sum $\gamma = \sum_{j=1}^{p} (\ell_j^* - \ell_j)$ will decrease by at least as much as $\alpha_2 = \sum_{j=1}^{p} (\ell_j' - \ell_j)$, and $\gamma - \alpha_2$ will decrease.

Suppose that TE is a two row matrix. Then E is a three row Birkhoff matrix where row F_2 contains only a single entry one, $e_{2k} = 1$, and in row F_3, $e_{30} = 1$ and $e_{3\ell} = 0$ for $\ell \geq k$. Let the ones in row F_1 of E be in positions $0 = \ell_1, \dots \ell_t, \dots, \ell_{t+r}, \dots, \ell_{t+r+s}$ where $\ell_{t+j} = k + j - 1$ $(j=1,\dots,r)$ if $r \neq 0$, and $\ell_{t+r+1} > k + r$. Then $\ell_{t+1}, \dots, \ell_{t+r}$ represents a maximal block of ones in row F_1 beginning in position k. Under the coalescence of row F_1 to row F_2 in E, this block moves to positions $[k+1,k+r]$ while all other ones remain in their original position. Thus,

$$\alpha_2 = r.$$

On the other hand, by the Birkhoff condition, the ones of E in columns $0,\dots,k-1$ are

$$n + 1 - (s+r+1) = n - (s+r) \geq k + 1$$

in number. Hence in the coalescence of row F_1 to F_3, these ones form the block $[0,\dots,n-(s+r)-1] (\supset [0,k])$ and $\ell_t, \dots, \ell_{t+r+s}$ go to positions $\ell_t'' \leq n - (s+r)$, $\ell_{t+1}'' = n - (s+r), \dots, \ell_{t+s+r}'' = n - 1$. However, in maximal coalescence $\ell_{t+s+r}^* = n$, $\ell_{t+s+r-1}^* = n - 1, \dots, \ell_t^* = n - (s+r)$. Therefore

$$\delta(E) = \gamma - \alpha_3 - r = \sum_{j=0}^{t+r+s} (\ell_j^* - \ell_j'') - r \geq 1 + s > 0,$$

and there is strict equality in (4.1). □

THEOREM 4.2. *For a Pólya matrix* E (i) $\delta(E) \geq 0$; (ii) *One has* $\delta(E) > 0$ *if and only if one of the matrices of the canonical decomposition of* E *is non-Hermitian, with at least three non-zero rows.*

Proof. Because of (3.2), we may assume that E is a Birkhoff matrix. We apply several times the operation T of Lemma 4.1. After finitely many steps, a Hermitian matrix results. Then $\delta = 0$. This gives (i).

To prove (ii), we assume that E is a Birkhoff matrix with $m \geq 3$ which has gaps. A gap is a maximal sequence of zeros followed by a one in a row. At each application of T, at most two gaps in E are destroyed: possibly in the row where $e_{ik} = 1$ has been replaced by zero, and in row 1, after canonical decomposition.

We can assume that at each step E has a non-zero first row and at least three rows. Then our process will stop at the moment when the gaps disappear. At the last application of T, the matrix E must be of the type of Example 3. For this matrix, $\delta \geq 1$, and we obtain (ii) again by Lemma 4.1. □

Combining Proposition 3.2 and Theorem 4.2, we have

THEOREM 4.3. *A Pólya matrix E is complex regular if and only if its canonical decomposition consists of matrices which are Hermitian, or have at most two non-zero rows.*

For Birkhoff matrices E, this has been given by D. Ferguson [2], the general case appears in [5].

REMARK 4.1. It is interesting to note that Lemma 4.1 and therefore Theorem 4.2 are valid also without the assumption that $e_{10} = 1$. To show this, one has to consider also the possibility that row F_1 disappears after an application of T. In this case, before the last application of T, the ones in row F_1 occupy positions $[n-r,\dots,n-1]$, $n - r \geq k + 1$. By the Birkhoff condition there cannot be any other ones in columns $n - r,\dots,n - 1$. Hence $\alpha_i = 0$, $i = 2,\dots,m$ while $\gamma = r > 0$. Therefore strict equality holds in (4.1) at this stage.

REMARK 4.2. One can use property (3.3) and remark 4.1 in order to avoid reduction to two rows in the proof of Theorem 4.2, thereby eliminating the need for the special argument in Lemma 4.1.

REFERENCES

1. Cavaretta Jr., A. S., A. Sharma, and R. S. Varga, Hermite-Birkhoff interpolation in the n-th roots of unity, Preprint.

2. Ferguson, D., The question of uniqueness for G. D. Birkhoff interpolation problems, J. Approximation Theory, 2 (1969), 1-28.

3. Karlin, S. and J. M. Karon, Poised and non-poised Hermite-Birkhoff interpolations, Indiana Univ. Math. J., 21 (1972), 1131-1170.

4. Lorentz, G. G., The Birkhoff interpolation problem: New methods and results, In: Linear Operators and Approximation II, Birkhauser Verlag, Basel, 1974 (ISNM25), 481-501.

5. Lorentz, G. G., Birkhoff Interpolation Problem, CNA Report - 103, The University of Texas at Austin, July 1975.

6. Lorentz, G. G., Coalescence of matrices, regularity and singularity of Birkhoff interpolation problems, J. Approximation Theory, 20 (1977), 178-190.

7. Lorentz, G. G. and S. D. Riemenschneider, Recent progress in Birkhoff interpolation, In: Approximation Theory and Functional Analysis, J. B. Prolla editor, North-Holland Publishing Co., (1979), 187-236.

8. Lorentz, G. G. and K. L. Zeller, Birkhoff interpolation problem: Coalescence of rows, Arch. Math. (Basel), 26 (1975), 189-192.

9. Riemenschneider, S. D., Birkhoff interpolation at the n-th roots of unity: convergence, Preprint.

10. Zia-Uddin, Note on an "alternant" with factorial elements, Proceedings Edin. Math. Soc., 3 (1933), 296-299.

G. G. Lorentz*
Department of Mathematics
The University of Texas
Austin, Texas
U. S. A.

S. D. Riemenschneider**
Department of Mathematics
University of Alberta
Edmonton, Alberta
Canada

* Supported in part by Grant MCS79-04689 of the National Science Foundation.

** Research supported by the National Sciences and Engineering Research Council of Canada grant A-7687.

n-WIDTHS OF OCTAHEDRA

Avraham A. Melkman

Exact values of the Kolmogorov n-widths of octahedra are obtained by drawing on a connection with two-graphs, Seidel [12]. Packing and covering provide a venue for estimates.

1. INTRODUCTION

This paper is concerned with the Kolmogorov n-widths of the octahedron $B_1^m = \{x \in R^m: \|x\|_1 \leq 1\}$ in ℓ_p^m

$$(1.1) \quad d_n(B_1^m, \ell_p^m) = \min_{X_n \subset R^m} \max_{\underline{x} \in B_1^m} \min_{y \in X_n} \|\underline{x} - \underline{y}\|_p$$

with $\|\underline{x}\|_p$ denoting the usual ℓ_p^m-norm. An equivalent problem is the best rank n-approximation of the identity

$$(1.2) \quad \delta_n(I; \ell_1^m, \ell_p^m) = \min_{\text{rank } P \leq n} \max_{\|x\|_1 \leq 1} \|\underline{x} - P\underline{x}\|_p$$

because e.g. [4], both expressions are equal to

$$(1.3) \quad \min_{X_n} \max_{k=1,\ldots,m} \min_{\underline{y} \in X_n} \|\underline{e}^k - \underline{y}\|_p$$

with $\underline{e}^k$ the k-th unit vector. Pinkus [11] contains much useful information about this and related problems, and in particular, the interesting observation that allowing $X_n \subset C^m$ may actually yield a smaller n-width. However, we will make this distinction only where strictly necessary because most results apply equally to both cases.

Section 2 establishes and exploits a connection with regular 2-graphs, Seidel [12], to derive many exact values. Section 3 surveys some estimates, in particular those of Kašin [5,6], Maiorov [9], and Höllig [3].

ISBN 0-12-213650-0

2. EXACT VALUES

This section is an extension of [10]; in particular, some proofs are omitted. Theorem 1 summarizes the general cases where exact values are known.

THEOREM 1. (i) *Stechkin* [15].

$$(2.1)\quad d_n(B_1^m, \ell_2^m) = \sqrt{1 - \frac{n}{m}}$$

and P *is an optimal rank n-approximation if and only if* P *is a projection (Hermitian, idempotent) with*

$$(2.2)\quad \|P\underline{e}^k\|^2 = P_{kk} = \frac{n}{m}\ , \ k = 1, \ldots, m.$$

(ii) *Pinkus* [11].

$$d_1(B_1^m, \ell_p^m) = [1 + (m-1)^{-q/p}]^{-1/q}\ ,\ \frac{1}{q} + \frac{1}{p} = 1\ ,\ p \leqq \infty\ .$$

(iii) *Pinkus* [11].

$$d_{m-1}(B_1^m, \ell_p^m) = m^{-1/q}\ ,\ p \leqq \infty\ .$$

REMARK. Condition (2.2) says that X_n is an optimal subspace if and only if the vectors $P\underline{e}^k$ form a spherical eutactic star, Seidel [13]. Thus when the latter is known, a corresponding optimal subspace can be identified. For an instance of this see section 3.

Further exact results will be based upon an examination of the following lower bound.

THEOREM 2. *For all* $2 \leqq p \leqq \infty$,

$$(2.3)\quad d_n(B_1^m, \ell_p^m) \geqq [1 + (m-1)^{1-q/2}(\tfrac{m}{n} - 1)^{-q/2}]^{-1/q}\ ,\ \frac{1}{p} + \frac{1}{q} = 1\ .$$

Proof. With arbitrary X_n, consider $E_p(e^k, X_n) = \min(\|\underline{e}^k - x\|_p : x \in X_n)$. Using duality and Cauchy-Schwartz on $\sum_{i \neq k} |y_i|^q$ yields

$$E_p(\underline{e}^k,X_n)^q = \max_{\underline{y}\perp\underline{X}_n} \frac{|y_k|^q}{|y_k|^q + \sum_{i\neq k} |y_i|^q}$$

$$\geq \max_{\underline{y}\perp\underline{X}_n} \left[1 + (m-1)^{1-q/2}\left(\sum_{i\neq k} \left|\frac{y_i}{y_k}\right|^2\right)^{q/2}\right]^{-1}$$

$$= \{1 + (m-1)^{1-q/2}[E_2(\underline{e}^k,X_n)^{-2} - 1]^{q/2}\}^{-1} .$$

The result follows since this expression is increasing in E_2 and, by Theorem 1, $\min_{X_n} \max_k E_2(\underline{e}^k,X_n) = (1-\frac{n}{m})^{1/2}$.

On examination of this proof, one sees that equality holds in (2.3) if and only if for the optimal X_n

(i) $\max_k E_2(\underline{e}^k,X_n) = (1-\frac{n}{m})^{1/2}$, i.e., X_n is optimal also for $d_n(B_1^m,\ell_2^m)$. Hence, by Theorem 1, $E_2(\underline{e}^k,X_n) = (1-\frac{n}{m})^{1/2}$ for all k.

(ii) Thus, equality in Cauchy-Schwartz (when $p>2$) yields for each k, a vector $y^k \perp X_n$ for which $|y_i^k| = 1$, $i \neq k$, $y_k^k = \gamma = \sqrt{(m-1)(\frac{m}{n}-1)}$. Now, $\|\underline{e}^k - [\gamma + (\frac{m-1}{\gamma})^{\frac{1}{p-1}}]\underline{y}^k\|_p =$ $[1 + (m-1)^{1-q/2}(\frac{m}{m-n}-1)^{-q/2}]^{-1/q}$.

Therefore, comparison with Theorem 2 shows that the vectors $\{\underline{y}^k\}_1^m$ span an (m-n)-dimensional optimal subspace for which equality is again attained in (2.3), for all p in fact. Hence by Theorem 1, the corresponding rank (m-n) matrix is a projection. Reversing this argument yields similar information about X_n, summarized in

THEOREM 3. *The lower bound* (2.3) *is attained for* $p>2$ *if and only if there exists a rank* n *projection* P *such that*

$$P_{kk} = \frac{n}{m},\ |P_{ik}|^2 = (1-\frac{n}{m})\ \frac{n}{m(m-1)},\ i\neq k,\ k=1,\ldots,m.$$

In that case, $X_n = PR^m$ *is an optimal subspace for* $d_n(B_1^m,\ell_p^m)$, $p\geq 2$, *while* $X_n^\perp$ *is optimal for* $d_{m-n}(B_1^m,\ell_p^m)$, $p\geq 2$, *again with equality in* (2.3).

In the above case, consider the lines spanned by the vectors $P\underline{e}^k$. Since $|(P\underline{e}^i, P\underline{e}^j)| = |P_{ij}|$ the angle between any two lines is the same. This set of equiangular lines in X_n is extremal in the sense that any set of n equiangular lines in R^n must have $|\text{inner product}| \geqq \sqrt{(1-\frac{n}{m})\frac{1}{m-1}}$, viz. Theorem 2; this is the content of Van Lint and Seidel [8, Lemma 6.1]. The existence of such extremal sets is equivalent to the existence of regular 2-graphs on n vertices, Seidel [12]. The above mentioned papers as well as Lemmens and Seidel [7], Delsarte, Goethals and Seidel [2], contain much relevant information which we now summarize.

I. <u>The real case</u>.

(a) Necessary for equality in (2.3) is $n = 1$ or $m \leqq \frac{1}{2}n(n+1)$ [7, Theorem 3.5], and since then there is equality for m-n as well, this can be sharpened to $n = 1$ or $n = m - 1$ or $\sqrt{2m + 1/4} - 1/2 \leqq n \leqq m - \sqrt{2m + 1/4} + 1/2$.

(b) Furthermore, from [12, Theorem 7.4] both $\sqrt{(m-1)\frac{n}{m-n}}$ and $\sqrt{(m-1)\frac{m-n}{n}}$ have to be odd integers unless $m = 2n$; in any case, it follows that m must be even.

(c) For $m = 2n$, $S = \sqrt{m-1}(2P - I)$ satisfies $S^2 = (m-1)I$ and hence S is a symmetric conference matrix, e.g., Seidel [12]. Necessary for its existence is $m \equiv 2 \pmod 4$, sufficient, e.g., $m = s^k + 1 \equiv 2 \pmod 4$, s prime. On the other hand, even if $m \equiv 0 \pmod 4$ we need not be far off. From [12, Theorem 12.3], if there exists a Hadamard matrix of order k, then there is equality with $m = k^2$ and $n = \frac{m}{2} - \sqrt{\frac{m}{2}}$ and so

$$(1 + \sqrt{m-1})^{-1} \leqq d_{m/2}(B_1^m, \ell_\infty^m) \leqq d_{m/2-\sqrt{m}/2}(B_1^m, \ell_\infty^m) = m^{-1/2} .$$

(d) In a similar vein, from [12, Theorem 9.7] which shows equality with $m = 2^{k-1}(2^k + 1)$, $n = \frac{2}{3}m - \frac{1}{6}(\sqrt{8m+1} + 1)$ and $m = 2^{k-1}(2^k - 1)$, $n = \frac{1}{3}m - \frac{1}{6}(\sqrt{8m+1} - 1)$ one gets $(1 + \sqrt{2m-2})^{-1} \leqq d_{(2/3)m}(B_1^m, \ell_\infty^m) \leqq (-1/2 + \sqrt{2m + 1/4})^{-1}$, $m = 2^{k-1}(2^k + 1)$;

$$(1 + \sqrt{\frac{m-1}{2}})^{-1} \leqq d_{(1/3)m}(B_1^m, \ell_\infty^m) \leqq (1/4 + \sqrt{\frac{m-1}{2} + \frac{9}{16}})^{-1},$$

$m = 2^{k-1}(2^k + 1)$.

These results point to the

CONJECTURE 1. For $n=\alpha m$, $\lim_{m\to\infty}\sqrt{\alpha m}\, d_{\alpha m}(B_1^m,\ell_\infty^m) = \sqrt{1-\alpha}$. For comparison, let us state another conjecture to be buttressed later on.

CONJECTURE 2. For $n=\sqrt{m}$, $\lim_{m\to\infty} m^{1/4}\, d^C_{\sqrt{m}}(B_1^m,\ell_\infty^m) = 1$. In these cases then, the lower bound (2.3) would be asymptotically sharp.

II. In the complex case, the necessary condition for equality is weakened to $\sqrt{m} \leqq n \leqq m-\sqrt{m}$, lending credence to Conjecture 2. Equality does prevail when $m = 2n$ and there exists a skew-Hadamard matrix of that order, [2], [10]. These are conjectured to exist for all $m \equiv 0 \pmod 4$ and known for infinite sequences of such m. Since by (c) above, equality cannot hold for such m in the real case, we get, e.g., $d_2^R(B_1^4,\ell_\infty^4) > d_2^C(B_1^4,\ell_\infty^4)$ as pointed out by Pinkus [11].

3. ESTIMATES

In the above, the rank n matrix P was constructed as the Grammian of m vectors $\underline{v}_i$ in R^n. In general, this technique can be, and has been, used to yield estimates. Namely, for any choice of $\{\underline{v}_i\}_1^m$ let

$$\gamma_p = \max_j \left[\sum_{i\neq j} \left| \frac{(\underline{v}_i,\underline{v}_j)}{(\underline{v}_j,\underline{v}_j)} \right|^p \right]^{1/p} .$$

Then $d_n(B_1^m,\ell_p^m) \leqq (1+\gamma_p^{-q})^{-1/q}$, $\frac{1}{p}+\frac{1}{q}=1$. Presumably, there is equality when $\{\underline{v}_i\}_1^m$ is chosen to minimize γ, but this has not been proven. Finding the minimum is itself not an easy task. For example, it is to be expected, but not proven, that the choice

$$\underline{v}_k = (\cos(k-1)\tfrac{\pi}{m}\, ,\ \sin(k-1)\tfrac{\pi}{m}),\ k=1,\dots,m$$

is optimal for $n=2$, yielding, e.g., $\gamma_2 = \sqrt{\frac{m}{2}-1}$, $\gamma_\infty = \cos\frac{\pi}{m}$ and $d_2(B_1^m,\ell_2^m) \leqq \sqrt{1-\frac{2}{m}}$ (compare 2.1), $d_2(B_1^m,\ell_\infty^m) \leqq (1+\sec\frac{\pi}{m})^{-1}$.

Lest one get the impression from this example and previous results that the same subspace is optimal for all p, whatever the value of n, let us mention two results from Seidel [13]: For $m = n(n-1)$ ($m = (1/2)n(n+1)$), the Grammian of the vectors $\underline{e}_i + \underline{e}_j$, $i < j$, $\underline{e}_i \in R^n$ ($\underline{e}_i - \underline{e}_j$, $i < j$, $\underline{e}_i \in R^{n+1}$), is optimal for $p = 2$ but for $p = \infty$ yields only the useless estimate $d_n \leqq 1/2$.

The case $p = \infty$ is particularly interesting; geometrically it corresponds to the problem of packing m equal spherical caps of largest possible radius on the unit hemisphere in R^n (with sewn equator). This point of view permits a simple estimate as follows.

Denote by $P(\theta)$ the maximum number of spherical caps of (spherical) radius θ which can be packed on the unit hemisphere in R^n. If θ is chosen such that $P(\theta) \geqq m$, then $d_n(B_1^m, \ell_\infty^m) \leqq (1 + \sec 2\theta)^{-1}$. For an estimate of $P(\theta)$ we follow Shannon [14] and Blachman [1]. Observe that, $P(\theta)$ being maximal, it is impossible to draw an additional circle of radius θ on the hemisphere and hence no point is at a distance $\geqq 2\theta$ from all centers of the caps. Thus, drawing concentric caps of radii 2θ yields a covering of the hemisphere; denoting the area of a cap of radius θ by $A(\theta)$,

$$P(\theta)A(2\theta) \geqq A(\tfrac{\pi}{2}) = 1/2 \, \frac{\pi^{n/2}}{\Gamma(n/2)}$$

or

$$P(\theta) \geqq \frac{\sqrt{\pi}}{2} \frac{\Gamma(\frac{n-1}{2})}{\Gamma(n/2)} \left[\int_0^{2\theta} \sin^{n-2}\psi \, d\psi \right]^{-1}$$

$$\geqq \frac{\sqrt{\pi}}{2} \frac{\Gamma(\frac{n-1}{2})}{\Gamma(n/2)} \frac{(n-1)\cos 2\theta}{(\sin 2\theta)^{n-1}} .$$

Taking for simplicity, $n = 2k + 1 \gg 1$ and setting $x = \cos 2\theta$

$$P(\theta) \geqq \sqrt{\pi k}\, x\,(1 - x^2)^{-k} = f(x) .$$

We wish to choose x such that $f(x) = m$. Since $f(x)$ is increasing and $f(\sqrt{\frac{\ell nm}{k}}) \geqq \sqrt{\pi}\, m \,\ell nm \geqq m$, it follows that the solution x^* to $f(x) = m$ satisfies

$$x^* = \cos 2\theta^* \leqq \left(\frac{\ell nm}{k}\right)^{1/2}$$

or

$$(3.1) \quad d_n(B_1^m, \ell_\infty^m) \leqq \left(\frac{\ell nm^2}{n}\right)^{1/2},$$

a result Kašin [5] obtained by different means (and with larger constant).

Better estimates for a more restricted range of values have relied on constructive methods, after the pattern set by Kašin [6],

$$d_n(B_1^m, \ell_\infty^m) \leqq \frac{C_\lambda}{\sqrt{n}} \text{ for } n \leqq m \leqq n^\lambda,$$

which, combined with (3.1), gives

$$(3.2) \quad d_n(B_1^m, \ell_\infty^m) \leqq C_\lambda \left(\frac{\ell n(m/n)}{n}\right)^{1/2}.$$

Thus, Höllig shows that for $n = p^2$, p prime, there exist at least $n^{\ell/2}$ vectors with $\cos 2\theta \leqq \frac{\ell-1}{\sqrt{n}}$, arbitrary, which entails

$$(3.3) \quad d_n(B_1^m, \ell_\infty^m) \leqq \frac{8}{\sqrt{n}} \frac{\ell nm}{\ell nn}.$$

Maiorov [9], on the other hand, considers complex vectors yielding the improvement that for prime n there exist n^ℓ vectors with $\cos 2\theta \leqq \frac{\ell-1}{\sqrt{n}}$. In particular, then

$$d_n(B_1^{n^2}, \ell_\infty^{n^2}) \leqq (1+\sqrt{n})^{-1}.$$

Comparison with the lower bound (2.3) justifies Conjecture 2. For general m,n one gets an estimate as in (3.3) with constant reduced from 8 to $2\sqrt{2}$.

REFERENCES

1. Blachman, N. M., On the capacity of a band-limited channel perturbed by statistically dependent interference, IRE Trans. IT, 8 (1962), 48-55.
2. Delsarte, P., J. M. Goethals and J. J. Seidel, Bounds for systems of lines and Jacobi polynomials, Phillips Res. Repts., 30 (1975), 91-105.
3. Höllig, K., Approximationszahlen von Sobolev-Einbettungen Math. Ann., 242 (1979), 273-281.
4. Hutton, C. V., J. S. Morrell and J. R. Retherford, Diagonal operators, approximation numbers and Kolmogorov diameters, J. Approx. Th., 16 (1976), 48-80.
5. Kašin, B. S., On Kolmogorov diameters of octahedra, Soviet Math. Dokl., 15 (1974), 304-307.
6. Kašin, B. S., On diameters of octahedra, Uspehi Mat. Nauk, 30 (1975), 251-252.
7. Lemmens, P. W. H. and J. J. Seidel, Equiangular lines, J. Alg., 24 (1973), 494-512.
8. von Lint, J. H. and J. J. Seidel, Equilateral point sets in elliptic geometry, Indag. Math., 28 (1966), 335-348.
9. Maiorov, V. E., Various widths of the class H_p^r in the space L_q, Math. USSR Izv., 13 (1979), 73-87.
10. Melkman, A. A., The distance of R^m from its axes and n-widths of octahedra, to appear.
11. Pinkus, A., Matrices and n-widths, Lin. Alg. Appl., 27 (1979), 245-278.
12. Seidel, J. J., A survey of 2-graphs, Proc. Intern. Coll. Teorie Combinatorie, Accad. Naz. Lincei., Roma 1976, Vol. I, 481-511.
13. Seidel, J. J., Eutactic stars, Coll. Math. Soc. J. Bolyai 18, Combinatorics, Keszthely 1976, 983-999.
14. Shannon, C. E., Probability of error for optimal codes in a Gaussian channel, Bell System Tech. J., 38 (1959), 611-656.
15. Stechkin, S. R., The best approximation of given classes of functions, Uspehi Mat. Nauk, 9 (1954), 133-134.

A. A. Melkman
Department of Mathematics
Ben-Gurion University
Beer-Sheva, Israel

It is a pleasure to acknowledge helpful conversations with Dr. Pinkus and Prof. Seidel.

THE APPROXIMATION OF MULTIPLE INTEGRALS BY USING INTERPOLATORY CUBATURE FORMULAE

I. P. Mysovskikh

This article is a survey of the theory of interpolatory cubature formulae which has been developed after 1970. The following are considered: lower bounds for the number of knots of cubature formulae which are exact for polynomials of some fixed degree, the connection between orthogonal polynomials and cubature formulae, the method of reproducing kernels, and invariant formulae.

1. INTRODUCTION

Our goal is to compute the integral

$$I(f) \overset{\text{def}}{=} \int_{\Omega} p(x)f(x)dx,$$

where $\Omega \subset \mathbb{R}^n$ is the domain of integration. Here, the symbol $\overset{\text{def}}{=}$ means "equal by definition". The integrand is given as a product of two functions. The first one, $p(x)$, is assumed to be fixed and is called the weight-function, the second one, $f(x)$, belongs to a sufficiently large class of functions, e.g., continuous and such that the integral $I(f)$ exists. It will be assumed that all moments of $p(x)$ exist, i.e., the integrals $I(x^\alpha)$, where $\alpha = (\alpha_1,\ldots,\alpha_n)$ is a vector with nonnegative integers as components and

$$x^\alpha \overset{\text{def}}{=} x_1^{\alpha_1}\ldots x_n^{\alpha_n}.$$

We shall also assume that the following conditions hold

(1) $\qquad p(x) \geq 0 \quad \text{for} \quad x \in \Omega, \text{ and } I(1) > 0.$

ISBN 0-12-213650-0

A cubature formulae for $I(f)$ will be the approximate equality

$$I(f) \cong \sum_{j=1}^{N} C_j f(x^{(j)}). \tag{2}$$

The sum on the right-hand side is called cubature sum, the points $x^{(j)}$ are called knots, and the numbers C_j are called coefficients. The interest in the approximation of multiple integrals by using cubature formulae has increased considerably in recent years. The books of S. L. Sobolev [51], A. H. Stroud [52] and some chapters of the books of V. I. Krylov [17], and P. J. Davis and P. Rabinowitz [3] were devoted to the construction of cubature formulae. A survey of results about finding approximate values of multiple integrals was given in the papers of S. Haber [10] and the author [32]. This article can be considered as a continuation of [32]. Questions which arose in [32] and which were answered in the seventies will be discussed here, as there are the specification of lower bounds for the number of knots in cubature formulae exact for polynomials of some fixed degree, new results for formulae with knots being roots of polynomials, in particular of orthogonal ones, an extension of the method of reproducing kernels and others. Finally the results on cubature formulae which are invariant with respect to some transformation group. Such formulae were introduced by S. L. Sobolev [50]. Today a number of papers are devoted to the construction and the analysis of this type of formulae. We note that in article [32] invariant formulae have not been considered.

Henceforth we will use the notation

$$M(r,s) \overset{\text{def}}{=} \frac{(r+s)!}{r!s!} .$$

By $\phi_j(x)$, $j = 1,2,3,\dots$, we denote the monomials x^α enumerated in such a way that the monomials of lower degree have smaller indices and monomials of the same degree are enumerated arbitrarily, for example, lexicographically. In particular, $\phi_1(x) = 1$. In such an enumeration, among $\phi_j(x)$, $j = 1,2,\dots,M(n,m)$, there are all exponents not higher than m.

2. LOWER BOUNDS FOR THE NUMBER OF KNOTS

If formula (2) is exact for polynomials of order not higher than m, then the number of its knots satisfies the inequality

$$N \geq \kappa \overset{\text{def}}{=} M(n,k), \tag{3}$$

where $k = [\frac{m}{2}]$ is the integer part of the number $\frac{m}{2}$. For $n = 2$ this inequality was used by J. Radon [41], for arbitrary $n \geq 2$ it was proved by A. H. Stroud [52]. The inequality holds also for the case, when the knots and coefficients of (2) are complex numbers. If the knots and the coefficients in (2) are real, then among the coefficients there are κ positive ones.

Equality in (3) is attained for $n = 1$, if (2) is a Gauss-quadrature formulae. For $n \geq 2$ and for common domains of integration equality in (3) will not be attained in general. In a number of papers certain domains of integration Ω are given for which cubature formulae (2) are constructed for prescribed m and n. For example, in [4] the case $m = 3$ and arbitrary $n \geq 2$ is studied.

THEOREM 1. *Let $\Omega \subset \mathbb{R}^n$ be given and numbers $C_j > 0$ and points $x^{(j)}$, $j = 1,2,\ldots,\kappa$, not lying on an algebraic hyperplane of order k, chosen in a special way. Then there exists a function p(x), continuous and non-negative in Ω, which can be taken as a weight-function, such that (2) is a formula exact for all polynomials of degree not higher than $2k+1$ and which defines the remaining parameters of the theorem.*

Theorem 1 was proved by V. A. Kuzmenkov [18]. The choice of the points $x^{(j)}$ mentioned in the theorem is described in [17, p.444]. We believe that the requirement of the special choice of knots is not necessary; it suffices to assume that they do not lie on an algebraic hyperplane of order k.

Let $\theta \overset{\text{def}}{=} (0,0,\ldots,0)$ denote the origin in $\mathbb{R}^n$. For notational simplicity we do not subscribe the dependence on n.

Henceforth we shall use such simplification and we shall not emphasize it any more. We assume that Ω and $p(x)$ satisfy central symmetry with respect to θ, i.e.,

$$(4) \qquad x \in \Omega \Rightarrow -x \in \Omega, \quad p(x) = p(-x) \quad (x \in \Omega).$$

THEOREM 2. *Let Ω and $p(x)$ satisfy conditions (1) and (4), and let formula (2) be exact for all polynomials of order not higher than $2k+1$. Then the number of knots satisfies the inequality*

$$(5) \qquad N \geq \begin{cases} 2(\kappa - \nu) - 1 & \text{for } k \text{ even,} \\ 2\nu & \text{for } k \text{ odd,} \end{cases}$$

where ν is the number of odd monomials of degree not higher than k in n variables. The equality in (5), for k even, holds if and only if θ is a knot of (2).

THEOREM 3. *Let Ω and $p(x)$ satisfy conditions (1) and (4), let formula (2) be exact for all polynomials of order not higher than $2k+1$, and let the number of its knots attain the lower bound given in (5). Then the cubature sum in (2) is centrally symmetric, i.e., the set of knots is centrally symmetric and the coefficients corresponding to symmetric knots are equal.*

Theorems 2 and 3 were obtained by H. M. Möller [27]. The case $k = 1$ has been derived in [32]. We remark that Theorem 2 for $k=3$, $n=2$ and for Ω and $p(x,y)$ of special kind was obtained by R. Franke [5].

THEOREM 4. *Let the conditions of Theorem 2 be satisfied. If θ is not a knot of formula (2), then*

$$(6) \qquad N \geq \begin{cases} 2(\kappa - \nu) & \text{for } k \text{ even,} \\ 2\nu & \text{for } k \text{ odd.} \end{cases}$$

If θ is one of the knots of the formula, then

$$(7) \qquad N \geq \begin{cases} 2(\kappa - \nu) - 1 & \text{for } k \text{ even,} \\ 2\nu + 1 & \text{for } k \text{ odd.} \end{cases}$$

Theorem 4 generalizes Theorem 2 and was obtained in [33].

To obtain inequalities (3),(5),(6), and (7) it is important that the domain of integration Ω has interior points. If this condition is not satisfied, other lower estimations for the number of knots can be obtained. Let, for example, the sphere

$$S_{n-1} \overset{\text{def}}{=} \{x \in \mathbb{R}^n \mid x_1^2 + x_2^2 + \ldots + x_n^2 = 1\}$$

be the domain of the integration such that the cubature formula has the form

$$(8) \qquad \int_{S_{n-1}} p(x)f(x)dS \cong \sum_{j=1}^{N} C_j f(x^{(j)}).$$

We assume that the integrals

$$\int_{S_{n-1}} p(x)x^\alpha dS$$

exist.

THEOREM 5. <u>If formula (8) is exact for all polynomials of degree not higher than m, and if the weight-function satisfies the condition</u>

$$(9) \qquad p(x) \geq 0 \quad \underline{\text{for}} \quad x \in S_{n-1}, \quad \underline{\text{and}} \quad \int_{S_{n-1}} p(x)dS > 0,$$

<u>then</u>

$$(10) \quad N \geq h(k,n-1) \overset{\text{def}}{=} (n+2k-1)\frac{(n+k-2)!}{k!(n-1)!}, \ \underline{\text{where}}\ k = [\tfrac{m}{2}].$$

<u>If the knots and the coefficients of the formulae are real, then at least $h(k,n-1)$ of its coefficients are positive.</u>

THEOREM 6. <u>If formula (8) is exact for polynomials of order not higher than $2k+1$, if $p(x)$ satisfies (9) and $p(x) = p(-x)$ for all $x \in S_{n-1}$ and if θ is not among the knots, then</u>

$$(11) \qquad N \geq 2M(n-1,k).$$

Theorems 5 and 6 are stated by the author in [36], the proofs are given in [37].

We denote the orthogonal polynomials of order $k+1$ for the domain $\Omega \subset \mathbb{R}^2$ and the weight-function $p(x,y)$ by

$$(12)\quad P_i^{k+1}(x,y) \stackrel{\text{def}}{=} x^{k+1-i}y^i + Q_i^{k+1}(x,y),\ Q_i^{k+1} \in \mathbb{P}_k,$$

$$i = 0,1,2,\ldots,k+1.$$

We set

$$(13)\quad \gamma_{ij}^{k+1} \stackrel{\text{def}}{=} I(P_i^{k+1}P_{j+1}^{k+1} - P_{i+1}^{k+1}P_j^{k+1}),\quad i,j = 0,1,2,\ldots,k.$$

J. Radon [41] has obtained a cubature formula for the double integral which is exact for polynomials of order not higher than 5 and which has 7 knots. The numbers γ_{01}^3, γ_{02}^3, γ_{12}^3 which are defined by the equations (13) for $k = 2$ play an important rôle in finding the knots for his formula. Radon assumes that not all of these numbers equal zero. The author proved in [17, p. 472] that under this assumption the number of 7 knots in Radon's formula is minimal. In [52, p. 120] it is incorrectly stated that the proof for the minimal number of 7 knots was given by J. Radon in [41].

We consider a matrix Γ of order $k+1$,

$$\Gamma = [\gamma_{ij}^{k+1}]_{i,j=0}^{k},$$

where the γ_{ij}^{k+1}s are defined by (13). As we can see from (13), the matrix Γ is skew-symmetric and, subsequently, the rank of it is even. The rank will be denoted by $2\alpha(k,I)$. H. M. Möller obtained in [29] the following specification of the estimation (3) for $n = 2$ and $m = 2k+1$,

$$N \geq M(2,k) + \alpha(k,I).$$

The specification of inequality (3) for $n > 2$ and $m = 2k+1$ is also given.

3. COMMON ROOTS OF POLYNOMIALS AND CUBATURE FORMULAE

The extension of cubature formulae of Gaussian type for the multivariate case is very important in the theory of interpolatory cubature formulae. The first paper on this subject was written by J. Radon [41]. Its details are given in [17]. In 1969 A. H. Stroud and the author independently obtained the following theorem for the case $n = 2$.

If two polynomials of degree k orthogonal with respect to Ω and weight-function $p(x,y)$ have exactly k^2 roots in common, finite and distinct, then these roots can be taken as knots of a cubature formula for an integral on Ω with the weight-function $p(x,y)$. This formula is exact for all polynomials of degree not higher than $2k-1$. This was generalized in papers of A. H. Stroud, R. Franke, and the author. For the corresponding results and references we refer to [32].

In 1973 C. Günther [9] stated a theorem about cubature formulae, for $n = 2$, where the knots are the common roots of orthogonal polynomials and the cubature sum includes values of the derivatives of the integrand at multiple common roots. This theorem differs from Theorem 2 in [32, p.86] by the fact that the terminology and some results from the theory of polynomial ideals are used.

In 1973 H. M. Möller's dissertation [27] was published, in which formulae were considered for arbitrary n, the theory of polynomial ideals was also used. This paper led to a general result which was achieved by using common roots of polynomials as knots of cubature formulae.

An ideal A in the ring of polynomials $\mathbb{P} \stackrel{\text{def}}{=} \mathbb{R}[x_1,x_2,\dots,x_n]$ can be defined as the collection of polynomials

$$f(x) = \sum_{j=1}^{s} a_j(x)\, f_j(x), \tag{14}$$

where $f_1,f_2,\dots,f_s$ are fixed polynomials and a_j are arbitrary polynomials from the ring $\mathbb{P}$. We say that the polynomials $f_1,f_2,\dots,f_s$ form a basis for the ideal A, and we write $A = (f_1,\dots,f_s)$.

A basis $f_1,\dots,f_s$ of the ideal A is called an H-basis, if for any $f \in A$ we can find a representation (14), where the polynomials $a_j(x)$ are such that the degree of a_jf_j is not greater than the degree of f, for all $j = 1,2,\dots,s$. The notation H-basis was introduced by F. S. Macaulay [25]. He also proved that any ideal has an H-basis. H. M. Möller introduced in [27] the notion of an H-basis under a different name (canonical basis) and proved its existence.

The set of the common roots of the polynomials of an ideal is called the zero-set of the ideal. The dimension of the zero-set (in the topological sense) will be called the dimension of the ideal. We shall deal with zero-dimensional ideals, i.e. with zero-sets which consist of a finite number of points.

Let V be a finite dimensional vector space, and let V' be the dual space of lineal functionals. If U is a subspace of V, then the collection of functionals

$$U_A \overset{\text{def}}{=} \{L \in V' \mid L(u) = 0, \text{ for all } u \in U\}$$

turns out to be a subspace of V'. It will be called the anihilator of the subspace U.

By $\mathbb{P}_d$ we denote the vector space of polynomials in n variables of degree not higher than d. Let the points

$$(15) \qquad x^{(1)}, x^{(2)}, \ldots, x^{(N)}$$

be fixed. We shall consider the collection

$$(16) \qquad U \overset{\text{def}}{=} \{f \in \mathbb{P}_d \mid f(x)^{(i)}) = 0 \ , \quad i = 1,2,\ldots,N\},$$

which turns out to be a subspace of $\mathbb{P}_d$.

THEOREM 7. _In order for the points_ (15) _to be the knots of a formula_ (2), _which is exact for polynomials of degree not higher than_ d, _it is necessary and sufficient that the following condition is satisfied_,

$$f \in U \Rightarrow I(f) = 0.$$

This theorem was proved by S. L. Sobolev [50]. If in the definition of the anihilator we take $V = \mathbb{P}_d$, and if U satisfies equality (16), then the condition of Theorem 7 is equivalent to $I \in U_A$. The connection between anihilator and cubature formulae shown in Theorem 7 turns out to be a very handy idea. It leads among others to the definition of generalized cubature formulae, see [27].

The vector space of functions defined on $\Omega \subset \mathbb{R}^n$ will be denoted by $\mathbb{F}$. We assume that the space of all polynomials with real coefficients is a subspace of $\mathbb{F}$. Let A be a zero-dimensional ideal in $\mathbb{P}$ and $L_1(f), L_2(f),\dots,L_N(f)$ be linear functionals defined on $\mathbb{F}$. The approximate equality

$$(17) \qquad I(f) \cong \sum_{j=1}^{N} C_j L_j(f)$$

is called generalized cubature formula of algebraic degree of precision d associated to the ideal A, if this equality is exact for all $f \in \mathbb{P}_d$ and not exact for all $f \in \mathbb{P}_{d+1}$ and if the linear span of functionals $L_1(f), L_2(f),\dots,L_N(f)$, considered as functionals on $\mathbb{P}_d$, coincides with the anihilator of $A \cap \mathbb{P}_d$, being a subspace of $\mathbb{P}_d$. A system of functionals $L_1(f), L_2(f),\dots,L_N(f)$ satisfying the properties above always exists.

It is clear that a generalized cubature formula (17) which is exact for all polynomials in $\mathbb{P}_d$ exists if and only if I belongs to the anihilator of the set $A \cap \mathbb{P}_d$ considered as a subspace of $\mathbb{P}_d$.

H. M. Möller introduced the notion of m-orthogonality. A polynomial f(x) is said to be m-orthogonal with respect to I, if $I(fg) = 0$ for an arbitrary g(x), such that $f \cdot g \in \mathbb{P}_m$.

THEOREM 8. Let $A = (f_1, f_2,\dots,f_s)$ be a zero-dimensional ideal and let $f_1, f_2,\dots,f_s$ be an H-basis of the ideal. Then the following are equivalent.

(a) The polynomials f_j are d-orthogonal with respect to I, for $j = 1,2,\dots,s$;

(b) There exists a generalized cubature formula (17) defined by the ideal A which is exact for polynomials $f \in \mathbb{P}_d$.

In the articles of H. M. Möller [28-30] the results from [27] are further developed.

The use of the theory of polynomial ideals in order to construct cubature formulae, was also considered in the papers of C. R. Morrow, T. N. L. Patterson [31], and H. J. Schmid [49]. We shall state one of the results from [49].

We set

$$H_i \stackrel{\text{def}}{=} P_i^k + \sum_{j=0}^{k-1} h_{ij} P_j^{k-1}, \qquad i = 0,1,2,\ldots,k,$$

where P_j^s is defined by (12) and the h_{ij}s are real constants.

THEOREM 9. *In order for a cubature formula to exist as an approximation of the integral on $\Omega \subset \mathbb{R}^2$ with weight* $p(x,y)$ *which is exact for all polynomials of degree not higher than* $2k-2$ *and for which the number of knots is equal to the lower bound* $\frac{1}{2}k(k+1)$, *it is necessary and sufficient that we can find real numbers* $\beta_{j\nu}$ *such that*

$$yH_{j-1} - xH_j = \sum_{\nu=0}^{k} \beta_{j\nu} H_\nu, \qquad j = 1,2,\ldots,k.$$

Theorem 9 leads us from the question of the existence of cubature formulae to the question of the existence of a solution of some nonlinear system in order to find the unknown parameters h_{ij}. It is remarkable that it follows from this theorem that a cubature formula for the square with weight 1 exists which is exact for all polynomials of degree not higher than 6 and which has 10 knots.

The construction of cubature formulae for concrete domains using orthogonal polynomials was discussed in articles of A. V. Gobran [7], W. Häcker [11], G. P. Ismatullaev [12, 13], A. K. Ponomarenko [40], G. G. Rasputin [43]. The point in the article of G. G. Rasputin [42] is to find a numerical solution for a system of nonlinear algebraic equations. It is necessary to solve such systems, if the knots of a cubature formula are chosen as common roots of polynomials.

4. THE METHOD OF REPRODUCING KERNEL

We apply the Schmidt orthonormalization to the system of monomials $\{\phi_i(x)\}_{i=1}^{\infty}$ with respect to the inner product

$$(\phi,\psi) \stackrel{\text{def}}{=} I(\phi\psi). \tag{18}$$

The system obtained will be denoted by $\{F_i(x)\}_{i=1}^{\infty}$, so that $(F_i,F_j) = \delta_{ij}$, $i = 1,2,\ldots$.

We set

$$K(x,y) \overset{\text{def}}{=} \sum_{i=1}^{\kappa} F_i(x)\, F_i(y),$$

where κ attains equality in (3). The polynomial $K(x,y)$ in $2n$ variables of degree $2k$ is a reproducing kernel in the space $\mathbb{P}_k$ with the inner product (18). We denote by V_k the set of common roots of all orthogonal polynomials of degree k with respect to the inner product (18).

Let $a^{(i)}$ be a point which does not belong to V_k. We denote by H_i the hypersurface defined by $K(a^{(i)},x) = 0$. Hence, if $a^{(i)} \notin V_k$, the order of H_i is equal to k. We choose n points

(19) $$a^{(1)}, a^{(2)}, \ldots, a^{(n)}$$

the following way. As $a^{(1)}$ we pick any point which does not belong to V_k. Suppose the points $a^{(1)}, a^{(2)}, \ldots, a^{(r-1)}$, $2 \le r \le n$, have been chosen. As $a^{(r)}$ we will pick any point from the intersection $\bigcap_{i=1}^{r-1} H_i$ which does not belong to V_k. We assume that infinity is not a common point of the hypersurfaces $H_1, H_2, \ldots, H_n$, defined by the points (19). Therefore the number of the points of the intersection is finite. Let us denote these points by

(20) $$x^{(1)}, x^{(2)}, \ldots, x^{(r)}.$$

The method of reproducing kernels works as follows. As knots for the cubature formula we choose the points in (19) and (20). The coefficients are chosen, such that the formula is exact for all polynomials in $\mathbb{P}_{2k}$ (if this is possible). The method was described by the author in 1968, see [32].

We state a theorem, proved by H. M. Möller [27].

THEOREM 10. *Let us assume that the points of intersection* (20) *are distinct and their number is equal to* k^n. *Then we can take the points* (19) *and* (20) *as knots of the formula*

$$I(f) \cong \sum_{i=1}^{n} \frac{1}{b_i} f(a^{(i)}) + \sum_{j=1}^{k^n} C_j f(x^{(j)}),$$

where $b_i = K(a^{(i)}, a^{(i)})$. *This formula is exact for all polynomials of degree not higher than* $2k$.

Theorem 10 was proved independently by G. N. Gegel [6].

H. M. Möller gave a modified method of reproducing kernels in the case of central symmetry, then Ω and $p(x)$ satisfy condition (4). We shall denote

$$\tilde{K}(x,y) \overset{\text{def}}{=} \sum_{j=1}^{\kappa}{}' F_j(x) F_j(y),$$

where $\sum'$ means that the summation is taken on j, such that the corresponding F_j have the same parity as k. $\tilde{K}(x,y)$ turns out to be a reproducing kernel in the linear space of polynomials having the same parity as k with the inner product (18). We construct the points in (19) the same way as before. However, we do assume that to the point $a^{(i)}$ there corresponds a hypersurface $\tilde{H}_i$ of order k which is determined by the equation $\tilde{K}(a^{(i)},x) = 0$. We shall further assume that infinity is not a common point of $\tilde{H}_1,\ldots,\tilde{H}_n$.

THEOREM 11. *Let* Ω *and* $p(x)$ *satisfy condition* (4). *We assume that the hypersurfaces* $\tilde{H}_1,\ldots,\tilde{H}_n$ *intersect in* k^n *distinct points* (20). *Then the points* (19) *and* (20) *can be taken as knots of the formula*

$$I(f) \cong \sum_{i=1}^{n} \frac{1}{2b_i} [f(a^{(i)}) + f(-a^{(i)})] + \sum_{j=1}^{k^n} C_j f(x^{(j)}),$$

where $b_i = \tilde{K}(a^{(i)}, a^{(i)})$. *This formula is exact for all polynomials of degree not higher than* $2k + 1$.

Theorems 10 and 11 turn out to be special cases of more general theorems given in [27], where generalized cubature formulae are considered.

The proof of Chakalov's theorem about the existence of interpolatory cubature formulae with positive coefficients and knots belonging to the domain of integration without the assumption of the boundedness of the domain of integration is shown in the paper [34] of the author. Here it is necessary to assume that the weight-function p(x) satisfies condition (1).

5. INVARIANT CUBATURE FORMULAE

Let O(n) be the group of all orthogonal transformations of $\mathbb{R}^n$ onto itself leaving the origin θ fixed. By G we denote a finite subgroup of O(n). More precisely, G is a transformation group of a regular polyhedron $\mathcal{U}$ with center θ.

A set $\Omega \subset \mathbb{R}^n$ is said to be invariant with respect to a group G, if $g(\Omega) = \Omega$ for any $g \in G$. The space $\mathbb{R}^n$, the sphere S_{n-1}, and the ball

$$B_n \overset{\text{def}}{=} \{x \in \mathbb{R}^n \mid x_1^2 + \ldots + x_n^2 \leq 1\}$$

are invariant with respect to G.

If G is a transformation group of a polyhedron $\mathcal{U}$ onto itself, then $\mathcal{U}$ is invariant.

The set of points ga, where $a \in \mathbb{R}^n$ is fixed, and g runs through all elements of a group G, is called an orbit or a G-orbit containing the point a; it is denoted by G(a). It is clear that a G-orbit is invariant with respect to G. The cardinality of an orbit depends on the point a.

A function $\phi(x)$, defined on $\mathbb{R}^n$, is said to be invariant with respect to the group G, if $\phi(g(x)) = \phi(x)$ for any $g \in G$. An example of an invariant function with respect to G is the function $\varphi(r)$ where φ is any real valued function defined on $[0,\infty)$, and $r = \sqrt{x_1^2 + \ldots + x_n^2}$.

A cubature formula (2) is said to be invariant with respect to G, if the domain of integration Ω and the weight-function p(x) are invariant with respect to G, and if the set of knots is a union of G-orbits, where the knots of one and the same orbit have the same coefficient.

Let us consider a sequence of functions $\{\psi_i(x)\}_{i=1}^{\infty}$ defined on $\Omega \subset \mathbb{R}^n$ and such that the integrals

$$\int_\Omega p(x)\ \psi_i(x)dx, \quad i = 1,2,\ldots$$

exist.

We denote by Ψ the real vector space spanned by the first μ functions of this sequence. We assume that the vector space Ψ is invariant with respect to G, i.e., $\psi(gx) \in \Psi$ for any $g \in G$, $\psi \in \Psi$.

THEOREM 12. _Let the formula (2) as well as the vector space Ψ be invariant with respect to G. In order for the formula (2) to be exact for all functions in Ψ it is necessary and sufficient that (2) is exact for all those functions from Ψ which are invariant with respect to G._

The terminology for invariant cubature formulae as well as Theorem 12 are due to S. L. Sobolev. Theorem 12 is presented in [50] where applications are also given together with a construction of invariant formulae for integrals on the sphere S_2 which are exact for polynomials of degree not higher than m.

As Ψ one can take the space $\mathbb{P}_m$ in order to use Theorem 12 for the construction of invariant formulae exact for all polynomials of degree not higher than m. The space $\mathbb{P}_m$ is invariant with respect to the group of all affine transformations of $\mathbb{R}^n$ onto itself and, subsequently, with respect to O(n) or any subgroup of it.

We state some of the results about polynomials, invariant with respect to G, which will be useful for the applications of Theorem 12. The finite group G considered so far has a set of n algebraically independent invariant forms such that every invariant is algebraically expressible in terms of them. If these forms have the smallest degree, we call them a set of basic invariant forms. For details see [2]. In particular, $x_1^2 + x_2^2 + \ldots + x_n^2$ will be one of the basic invariant forms.

THEOREM 13. *Let G be a finite subgroup of O(n), generated by reflections, and let $I_1(x),\dots,I_n(x)$ be a set of basic invariant forms of G with a degree $m_1,\dots,m_n$ respectively. Then any polynomial invariant with respect to G is a polynomial of the set of basic invariant forms, i.e., a linear combination with constant coefficients of the polynomials*

$$I_1^{\alpha_1}(x)\cdot I_2^{\alpha_2}(x)\cdot\ \dots\ \cdot I_n^{\alpha_n}(x) \tag{21}$$

where $\alpha_1,\alpha_2,\dots,\alpha_n$ are nonnegative integers. The order of G is equal to the product $m_1\cdot m_2\cdot\dots\cdot m_n$.

Theorem 13 was proved in a more general form in [1]. The polynomials (21) are linearly independent, since linear dependence among them would imply algebraic dependence in the set of basic invariant forms.

Let us consider invariant polynomials as functions on the sphere S_{n-1}. We assume that $I_1(x) = x_1^2 + \dots + x_n^2$. Then according to Theorem 13 any invariant polynomial considered as a function on S_{n-1} is a linear combination of the polynomials

$$I_2^{\alpha_2}(x)\cdot I_3^{\alpha_3}(x)\cdot\ \dots\ \cdot I_n^{\alpha_n}(x) \tag{22}$$

where $\alpha_2,\alpha_3,\dots,\alpha_n$ are nonnegative integers. One can prove that the polynomials (22) regarded as functions on S_{n-1} are linearly independent.

We denote by $\mathcal{O}_n$ the hyperoctahedron with vertices

$$(\pm 1,0,\dots,0),\ (0,\pm 1,\dots,0),\ \dots,\ (0,0,\dots,\pm 1).$$

Further we shall consider $k+1$ vertices of $\mathcal{O}_n (1 \le k+1 \le n)$

$$(0,0,\dots,0,\varepsilon_{r_i},0,\dots,0), \quad i = 1,2,\dots,k+1,$$

where $\varepsilon_{r_i}\in\{-1,1\}$ is a coordinate with number r_i, and where $r_i \neq r_j$ for $i \neq j$. The convex hull of these $k+1$ vertices is called a k-dimensional face of the hyperoctahedron. The number of k-dimensional faces is equal to $2^{k+1}\binom{n}{k+1}$. It is clear that a k-dimensional face is a regular k-dimensional simplex.

A regular simplex in $\mathbb{R}^n$ with vertices

$$a^{(r)} = (a_1^{(r)}, a_2^{(r)}, \ldots, a_n^{(r)}), \quad r = 1,2,\ldots,n+1, \tag{23}$$

where

$$a_i^{(r)} \overset{\text{def}}{=} \begin{cases} -\sqrt{\dfrac{n+1}{n(n-i+2)(n-i+1)}}, & i < r, \\ \sqrt{\dfrac{(n+1(n-r+1)}{n(n-r+2)}}, & i = r, \\ 0, & i > r, \end{cases} \tag{24}$$

will be denoted by C_n. The points (23) belong to the sphere S_{n-1}, and the center of C_n is θ. The convex hull of any $k+1$ vertices from (23), $1 \le k+1 \le n$, is called a k-dimensional face of C_n, and is a k-dimensional regular simplex. The number of k-dimensional faces of C_n is equal to $\binom{n}{k+1}$.

The cube

$$K_n \overset{\text{def}}{=} \{x \in \mathbb{R}^n \mid -1 \le x_i \le 1, \quad i = 1,2,\ldots,n\}$$

is a regular polyhedron in $\mathbb{R}^n$. It is known that for $n \ge 5$ in $\mathbb{R}^n$ there is no regular polyhedron but O_n, C_n, K_n.

The group of all orthogonal transformations of the hyperoctahedron O_n onto itself will be denoted by O_nG. This group is generated by reflections and its order is equal to $n!2^n$. The symmetric polynomials of order $2,4,\ldots,2n$ in the arguments $x_1^2, x_2^2, \ldots, x_n^2$, i.e.,

$$\sigma_2 \overset{\text{def}}{=} \sum_{i=1}^{n} x_i^2, \; \sigma_4 \overset{\text{def}}{=} \sum_{i<j} x_i^2 x_j^2, \ldots, \; \sigma_{2n} \overset{\text{def}}{=} x_1^2 \cdot x_2^2 \ldots x_n^2, \tag{25}$$

form a set of basic invariant forms of O_nG.

The group of all orthogonal transformations of the cube K_n onto itself, K_nG, is congruent with the group O_nG. Hence, we only consider the group O_nG.

We denote by C_nG the group of all orthogonal transformations of C_n onto itself. The group C_nG is generated by reflections and its order is $(n+1)!$. We set

$$L_r(x) \overset{\text{def}}{=} \sum_{i=1}^{n} a_i^{(r)} x_i, \quad r = 1,2,\ldots,n+1,$$

where $a_i^{(r)}$ are the coordinates of the vertex $a^{(r)}$ of C_n, defined by formula (24). A set of basic invariant forms of the group C_nG is given by

$$\pi_s(x) \overset{\text{def}}{=} \sum_{i=1}^{n+1} L_i^s(x), \quad s = 2,3,\ldots,n+1. \tag{26}$$

Let us consider an associated group C_nG^* which is obtained from C_nG by adding to its generators the transformations which are centrally symmetric with respect to θ. Since C_n is not centrally symmetric, C_nG^* is not a transformation group of C_n onto itself. The order C_nG^* equals $(n+1)!2$. The set of polynomials invariant with respect to C_nG^* is congruent to the set of even polynomials invariant with respect to the group C_nG.

By giving some examples we show how to construct cubature formulae for an integral on the sphere S_{n-1} which are invariant with respect to the group O_nG. From now on we assume the weight-function to be equal to 1.

We construct formulae which are exact for polynomials of degree not higher than m, where $m \leq 9$. According to Theorem 12 it suffices to require the formula to be exact for all invariant polynomials of degree not higher than m. Since O_nG is generated by reflections and since $\sigma_4, \sigma_6, \sigma_8$ (for $n \geq 4$) are in the set of basic invariant forms, we have 5 linear independent polynomials of degree not higher than 9 on S_{n-1} which are invariant with respect to O_nG, namely,

$$1, \sigma_4, \sigma_6, \sigma_8, \sigma_4^2. \tag{27}$$

The set of the centers of all k-dimensional faces of O_n (for a fixed k) turns out to be the most natural orbit of O_nG. Since $k \in \{0,1,\ldots,n-1\}$, we obtain n orbits. A k-dimensional face defined by the vertices

$$(\delta_{i1}, \delta_{i2}, \ldots, \delta_{i,k+1}, 0, 0, \ldots, 0), \quad i = 1,2,\ldots,k+1,$$

has the center

$$\frac{1}{k+1} (\overbrace{1,1,\ldots,1}^{k+1}, 0, \ldots, 0). \tag{28}$$

The centers of the remaining k-dimensional faces of $\mathcal{O}_n$ have coordinates which are obtained from the coordinates of the point (28) by all possible changes of sign and transposition.

Let us introduce a notation for the projection of the point (28) from θ onto the sphere S_{n-1} by setting

$$g^{(k)} \overset{\text{def}}{=} \frac{1}{\sqrt{k+1}} (\overbrace{1,1,\ldots,1}^{k+1},0,0,\ldots,0), \quad k = 0,1,2,\ldots,n-1.$$

As knots we shall take the orbits of $\mathcal{O}_n G(g^{(k)})$.

Let us construct a formula which is exact for all polynomials of degree not higher than 3. There is only one invariant polynomial (as a function on S_{n-1}) of degree not higher than 3, namely, the constant polynomial 1 - the first from the set given in (27). As knots we can only take the orbit $\mathcal{O}_n G(g^{(0)})$. Hence the coefficient of the formula is defined by the requirement that the formula is exact for constants. We obtain the cubature formula given in [52, p. 294]. The number of 2n knots attains the lower bound in Theorem 6. Choosing as knots orbits generated by $g^{(0)}$ and $g^{(1)}$, we obtain a formula exact for polynomials of degree not higher than 5. In this case the formula has to be exact for the first two polynomials from (27), i.e., 1 and σ_4. This defines the coefficients. The formula has $2n^2$ knots (for $n \neq 4$), it is known, see e.g. [52, p. 294]. In the same way we can construct a cubature formula exact for polynomials of degree not higher than 7. It suffices to require that the cubature formula is exact for the first three polynomials from (27). As knots we choose the orbits, generated by $g^{(0)}, g^{(1)}, g^{(2)}$. This will lead to a formula with $\frac{2}{3}n(2n^2 - 3n + 4)$ (for $n \neq 5$) knots. For n = 3 this formula is already known, see [52, p. 299].

It is obvious that each center of a k-dimensional face of $\mathcal{O}_n$ for $k = 0,1,2,\ldots,n-1$ belongs to at least one of the four hypersurfaces $x_1 = x_2$, $x_1 = -x_2$, $x_1 = 0$, $x_2 = 0$. Let ℓ be a straight line passing through the origin and through the center of M of a k-dimensional face of $\mathcal{O}_n$, and let L be the union of all the lines ℓ which we obtain when M runs through the centers of all k-dimensional faces, $k = 0,1,2,\ldots$ $\ldots,n-1$. An arbitrary line from L is contained in at least one of the mentioned hypersurfaces. From this follows that the polynomial $(x_1^2 - x_2^2)^2\, x_1^2\, x_2^2$ of degree 8 vanishes on any line from L. Since its integral on S_{n-1} is positive, no formula for the integral is exact for this polynomial, if as knots arbitrary points from the lines in L will be chosen, in particular projections of the centers of the k-dimensional faces from θ onto S_{n-1}.

Hence, in order to obtain a cubature formula whose algebraic degree of precision is greater than seven, we have to include points which do not belong to the lines from L. We can, for example, include orbits generated by the projection of an inner point of an edge of the hyperoctahedron distinct from its midpoint on S_{n-1} or by the projection of an inner point of a two-dimensional face different from its center on S_{n-1}, etc. In [37] a formula for S_{n-1} exact for all polynomials of degree not higher than 9 has been obtained.

Systematical use of the basic invariant forms in (25) for the construction of cubature formulae invariant with respect to O_nG is done in the papers of V. I. Lebedev [19-24] who obtained a great number of cubature formulae of high algebraic degree of precision for integrals on the sphere S_2.

Let us continue with the construction of formulae for an integral on the sphere S_{n-1} invariant with respect to the transformation group of the regular simplex C_n. From Theorem 13 it follows that any polynomial invariant with respect to C_nG, considered as a function on the sphere S_{n-1}, turns out to be one of the polynomials $\pi_3, \pi_4, \ldots, \pi_{n+1}$ which are defined by equation (26). Let us consider the following invariant polynomials of degree not higher than seven,

$$1,\ \pi_3,\ \pi_4,\ \pi_5,\ \pi_6,\ \pi_7,\ \pi_3^2,\ \pi_3\pi_4. \tag{29}$$

The even polynomials from (29)

(30) $$1,\ \pi_4,\ \pi_6,\ \pi_3^2$$

are invariant with respect to C_nG^* and have a degree not higher than 7.

The set of the centers of the k-dimensional faces of C_n (for fixed k) is a C_nG-orbit. The centers of the 0-dimensional faces are the vertices of C_n, denoted earlier by $a^{(j)}$. Let us denote by $b^{(j)}$ projections of the midpoint of an edge of C_n on S_{n-1}. If $b^{(1)}$ is a projection of the midpoint of an edge on S_{n-1}, connecting the vertices $a^{(1)}$ and $a^{(2)}$, then

(31) $$b^{(1)} = \left(\sqrt{\frac{n-1}{2n}},\ \sqrt{\frac{n+1}{2n}},0,0,\ldots,0\right).$$

The set of the centers of the k-dimensional faces of C_n (k fixed) is not an orbit of the group C_nG^*, if it is not centrally symmetric with respect to θ. But if this set is completed by its central symmetries, then we obtain an orbit of C_nG^*.

Let us construct a cubature formula for S_{n-1} invariant with respect to C_nG and exact for all polynomials of degree not higher than 2. From (29) it is seen that the constant function 1 is the only invariant polynomial of degree not higher than 2. Hence as knots we can choose the vertices of C_n. The coefficient of the formulae is $\mu(S_{n-1})/(n+1)$, where

$$\mu(S_{n-1}) \overset{\text{def}}{=} 2\pi^{n/2} / \Gamma\left(\frac{n}{2}\right).$$

The cubature formula has $n+1$ knots and is exact for polynomials of degree not higher than 2. The number of knots attains the lower bound given in Theorem 5. For $n = 3$ this formula was shown by S. L. Sobolev [50].

We derive a cubature formula invariant with respect to C_nG^* and exact for all polynomials of degree not higher than 5. It can be seen from (30) that there are two polynomials invariant with respect to C_nG^* of degree not higher than 5, namely, 1 and π_4. Hence as knots we can choose two orbits $C_nG^*(a^{(1)})$ and $C_nG^*(b^{(1)})$. Here the points $a^{(1)}$ and $b^{(1)}$ are defined by equations (23) and (31). The formula has the form

$$\int_{S_{n-1}} f(x)\,dS \cong A \sum_{j=1}^{n+1} [f(a^{(j)}) + f(-a^{(j)})] + B \sum_{j=1}^{n(n+1)/2} [f(b^{(j)}) + f(-b^{(j)})].$$

The requirement for the formula to be exact for 1 and π_4 leads us to a linear algebraic system in A and B. Solving this system we find

$$A = \frac{(7-n)n}{2(n+1)^2(n+2)}\,\mu(S_{n-1}), \quad B = \frac{2(n-1)^2}{n(n+1)^2(n+2)}\,\mu(S_{n-1}).$$

Hence we obtain a formula for $n \geq 4$ and $n \neq 7$ which has $(n+1)(n+2)$ knots and is exact for all polynomials of degree not higher than 5. The number of knots does not essentially exceed the lower bound $n(n+1)$ given in Theorem 6. For $n = 7$ the coefficient A vanishes such that the formula has 56 knots which agrees with the lower bound.

We note that the centers of all k-dimensional faces of C_n, $k = 0,1,2,\ldots,n-1$, lie in three hypersurfaces passing through the origin. From this it follows that in order to construct cubature formulae with an algebraic degree of precision greater than 5 it is necessary to use orbits different from the centers of the faces of C_n.

Other cubature formulae for S_{n-1} invariant with respect to the group C_nG^* can be found in paper [37] of the author.

In the papers [14, 16] of S. I. Konjaev cubature formulae for an integral on S_2 are constructed which are invariant with respect to the transformation group of the icosahedron onto itself. Here results about invariant polynomials are used.

The problem of constructing formulae for integrals on the sphere which are invariant with respect to a transformation group is discussed in the papers of A. D. McLaren [26], G. N. Salichov [44-47], G. N. Salichov and F. Sharipchodjaeva [48]. In these papers the same approach is used as in the paper of S. L. Sobolev [50]. The number of linear independent invariant polynomials on the sphere has to be determined. The explicit form of these polynomials is not used in the construction.

This approach has the advantage that it can be used even if it is difficult to obtain the invariant polynomials. Invariant cubature formulae cannot only be constructed for integrals on the sphere. For example, in [38] invariant formulae have been obtained for the ball, and in [15] for the space $\mathbb{R}_3$ with the weight-function $p(r)$, $r = \sqrt{x_1^2 + x_2^2 + x_3^2}$.

Let us conclude with the construction of cubature formulae for integrals on the simplex

$$T_n \overset{\text{def}}{=} \{x \in \mathbb{R}^n \mid x_i \geq 0, \quad i = 1,2,\ldots,n, x_1 + x_2 + \ldots + x_n \leq 1\}$$

with the weight-function $p(x) = 1$. Let us denote by T_nG the group of all affine transformations of T_n onto itself. The order of T_nG equals $(n+1)!$. Obviously it is possible to define the notion of a cubature formula for the integral on T_n which is invariant with respect to T_nG. For such a formula Theorem 12 holds.

The polynomials

$$(32) \quad \Pi_k(x) \overset{\text{def}}{=} \sum_{i=1}^{n} [(n+1)x_i - 1]^k + [(n+1)(1 - x_1 - \ldots - x_n) - 1]^k,$$
$$k = 2,3,\ldots,n+1,$$

are invariant with respect to the group T_nG. An arbitrary, invariant polynomial (with respect to T_nG) is a polynomial in the polynomials of (32).

By $T_{n,u}$ we shall denote the simplex in $\mathbb{R}^n$ with the vertices

$$(u,\ldots,u,\ldots,u)$$
$$(1 - nu,\ u,\ldots,u),$$
$$(u,\ 1 - nu,\ldots,u),$$
$$\ldots\ldots\ldots\ldots$$
$$(u,u,\ldots,1 - nu),$$

where u is a real number. The center of the simplex $T_{n,u}$ is $W \overset{\text{def}}{=} \left(\frac{1}{n+1}, \frac{1}{n+1}, \ldots, \frac{1}{n+1}\right)$, the center of T_n.

The set of the centers of the k-dimensional faces of the simplex $T_{n,u}$ (for fixed k) is a T_nG-orbit. It is not difficult to prove that the centers of all k-dimensional faces $T_{n,u}$ have exactly two distinct coordinates for all $k = 0,1,2,\ldots,n-1$ with the exception of the center of the zero-dimensional face $(u,u,\ldots,u)$. From this it follows that the centers of all k-dimensional faces of $T_{n,u}$, $k = 0,1,2,\ldots,n-1$, lie on the three hypersurfaces $x_1 = x_2$, $x_1 = x_3$, $x_2 = x_3$ (for $n \geq 3$) passing through the center of T_n.

Consequently, it follows for the construction of invariant formulae with an algebraic degree of precision higher than 5, that as knots we should take not only the centers of the faces, but also orbits which are distinct from them. The vertices and midpoints of the edges of the triangle $T_{2,u}$ lie on three straight lines $2x_1 + x_2 - 1 = 0$, $x_1 + 2x_2 - 1 = 0$, $x_1 - x_2 = 0$, passing through the center of the triangle, hence the stated assertions hold for $n = 2$.

The well-known cubature formula in [52, p. 308] for the simplex T_n is exact for all polynomials of a degree not higher than 3 and has $n+2$ knots. To construct it we take as knots the center W and the vertices of the simplex $T_{n,u}$, such that the formula depends on three parameters, two values for the coefficients and the parameter u. It is surprising that by a suitable choice of these three parameters we can construct formulae which are exact for the seven linear independent monomials $(n \geq 3)$

$$1,\ x_1,\ x_1^2,\ x_1x_2,\ x_1^3,\ x_1^2x_2,\ x_1x_2x_3.$$

Theorem 12 explains this fact. Since the formula is invariant with respect to T_nG, it will be exact for all polynomials of degree not higher than 3, if we require that it is exact for the three invariant polynomials 1, Π_2, and Π_3. In [39] a formula invariant with respect to $T_n G$ and exact for all polynomials of degree not higher than 7 was obtained. The number of knots is $M(n,3) + (n+2)(n+1) + 1$.

Also in [8] a cubature formula has been derived for the integral T_n invariant with respect to $T_n G$. This formula is exact for all polynomials of degree not higher than $2s+1$, and has $\binom{n+s+1}{s}$ knots, $s = 0,1,2,\ldots$. A conjecture is stated that for $n \geq 2s$ the number of knots is minimal. The explicit formula has been derived for the basis orthogonal polynomials for T_n and $p(x) = 1$.

REFERENCES

1. Chevalley, C., Invariants of finite groups generated by reflections, Amer. J. Math. 77 (1955), 778-782.
2. Coxeter, H. S. M., The product of the generators of a finite group by reflections, Duke Math. J. 18 (1951), 765-782.
3. Davis, P. J. and P. Rabinowitz, Methods of numerical integration, New York, Academic Press, 1975.
4. Flatto, L. and S. Haber, A quadrature formula of degree three, J. Approximation Theory 9 (1973), 44-52.
5. Franke, R., Minimal point cubatures of precision seven for symmetric planar regions, SIAM J. Numer. Anal. 10 (1973), 849-862.
6. Gegel, G. N., Construction of cubature formulae precise for polynomials of the 2nd degree (in Russ.), Vopros. Vycisl. i Prikl. Mat. 32 (1975), 5-9.
7. Gobran, A. V., Construction of cubature formulas for the approximation of double integrals (in Russ.), Metody Vycisl. Univ. Leningrad 10 (1976), 60-68.
8. Grundmann, A. and H. M. Möller, Invariant integration formulas for the n-simplex by combinatorial methods, SIAM J. Numer. Anal. 15 (1978), 282-290.
9. Günther, C., Zur Konstruktion mehrdimensionaler Integrationsformeln, Z. Angew. Math. Mech. 53 (1973), T194-T195.
10. Haber, S., Numerical evaluation of multiple integrals, SIAM Rev. 12 (1970), 481-526.
11. Häcker, W., On the construction of cubature formulae for a circle (in Russ.), Metody Vycisl. Univ. Leningrad 10 (1976), 73-80.

12. Ismatullaev, G. P., On some cubature formulae containing derived functions behind the integral sign (in Russ.), Vopros Vycisl. i Prikl. Mat., Tashkent 14 (1972), 117-129.

13. Ismatullaev, G. P., On the construction of cubature formulae for the hypersphere and surfaces of the sphere (in Russ.), Vopros. Vycisl. i Prikl. Mat., Tashkent 51 (1978), 191-203.

14. Konjaev, S. I., Quadrature formulae on the sphere invariant with regard to the group of the icosahedron (in (Russ.), Vopros. Vycisl. i Prikl. Mat., Tashkent 32 (1975), 69-76.

15. Konjaev, S. I., Quadrature formulae of the 9th degree invariant with regard to the group of the icosahedron (in Russ.), Dokl. Akad. Nauk, SSSR 233 No. 5 (1977), 784-787.

16. Konjaev, S. I., Quadratures of Gaussian type for the sphere invariant with regard to the group of the icosahedron with inversion (in Russ.), Mat. Zametki 25 No. 4 (1979), 629-634.

17. Krylov, V. I., Approximate calculation of multiple integrals (in Russ.), 2nd Edition, Nauka, Moskau 1967.

18. Kuzmenkov, V. A., Gaussian type cubature formulas and finite moment problem (in Russ.), Metody Vycisl. Univ. Leningrad 11 (1978), 21-42.

19. Lebedev, V. I., On quadratures on a sphere with highest possible degree of algebraic accuracy (in Russ.), Theory of cubature formulae and applications of functional analysis to problems of mathematical physics, ed. by S. L. Sobolev, Novosibirsk 1973, 31-35.

20. Lebedev, V. I., Values of the base-points and weights of Gauss-Markov quadrature formulae for a sphere from the 9th to the 17th order of accuracy invariant under the octahedron group with inversion, USSR Comp. Math. and Math. Phys. 15 (1975).

21. Lebedev, V. I., Evaluation of Gauss-Markov quadratures for domains and weights invariant under the octahedron group with inversion (in Russ.), Vopros. Vycisl. i Prikl. Mat., Tashkent 32 (1975), 77-84.

22. Lebedev, V. I., Quadratures on a sphere, USSR Comp. Math. and Math. Phys. 16 (1976), 10-24.

23. Lebedev, V. I., A type of quadrature formulas of increased algebraic accuracy for the sphere (in Russ.), Dokl. Akad. Nauk, SSSR 231 No. 1 (1976), 32-34.

24. Lebedev, V. I., Quadrature formulae for the sphere of degree 25 to 29 (in Russ.), Sibirsk Mat. Z. 18 No. 1 (1977), 132-142.

25. Macauley, F. S., Algebraic theory of modular systems, Cambridge tracts in mathematics and mathematical physics No. 19, Cambridge University Press 1916.

26. McLaren, A. D., Optimal numerical integration on a sphere, Math. Comp. 17 (1963), 361-383.

27. Möller, H. M., Polynomideale und Kubaturformeln, Dissertation in der Abteilung Mathematik der Universität Dortmund, Dortmund 1973.

28. Möller, H. M., Mehrdimensionale Hermite-Interpolation und numerische Integration, Math. Z. 148 (1976), 107-118.

29. Möller, H. M., Kubaturformeln mit minimaler Knotenzahl, Numer. Math. 25 (1976), 185-200.

30. Möller, H. M., Hermite interpolation in several variables using ideal-theoretic methods, Lecture Notes in Math. 571 (1977), 155-163.

31. Morrow, C. R. and T. N. L. Patterson, Construction of algebraic cubature rules using polynomial ideal theory, SIAM J. Numer. Anal. 15 (1978), 953-976.

32. Mysovskikh, I. P., Interpolatory cubature formulae (in Russ.), Theory of cubature formulae and applications of functional analysis to problems of mathematical physics, ed. by S. L. Sobolev, Novosibirsk 1973, 73-90.

33. Mysovskikh, U. P., On the construction of cubature formulae (in Russ.), Vopros. Vycisl. i Prikl. Mat., Tashkent 32 (1975), 85-98.

34. Mysovskikh, I. P., On Chaklov's theorem, USSR Comp. Math. and Math. Phys. 15 (1975), 221-227.

35. Mysovskikh, I. P., Orthogonal polynomials of several variables, Metody Vycisl. Univ. Leningrad 10 (1976), 26-35.

36. Mysovskikh, I. P., On the evaluation of integrals over the surface of a sphere, Soviet Math. Dokl. 18 No. 4 (1977), 925-929.

37. Mysovskikh, I. P., On cubature formulas invariant with respect to groups of transformations (in Russ.), Metody Vycisl. Univ. Leningrad 11 (1978), 3-21.

38. Mysovskikh, I. P., Invariant cubature formulae (in Russ.), Theory of cubature formulae and applications of functional analysis to problems of mathematical physics, Papers of a seminar of S. L. Sobolev, Novosibirsk 1978, No. 1, 69-78.

39. Mysovskikh, I. P., On a cubature formula for the simplex (in Russ.), Vopros. Vycisl. i Prikl. Mat., Tashkent 51 (1978), 74-90.

40. Ponomarenko, A. K., On the construction of cubature formulae for the square (in Russ.), Vopros. Vycisl. i Prikl. Mat., Tashkent 38 (1970), 93-99.

41. Radon, J., Zur mechanischen Kubatur, Monatsh. Math. 52 (1948), 286-300.

42. Rasputin, G. G., On the numerical solution of systems of nonlinear algebraic equations (in Russ.), Metody Vycisl. Univ. Leningrad 11 (1978), 57-81.

43. Rasputin, G. G., Some cubature formulae for the triangle and the square (in Russ.), Vopros. Vycisl. i Prikl. Mat., Tashkent 51 (1978), 119-128.

44. Salichov, G. N., Groups of regular polyhedrons and cubature formulae on hyperspheres (in Russ.), Vopros. Vycisl. i Prikl. Mat., Tashkent 14 (1972), 139-147.

45. Salichov, G. N., On the theory of cubature formulae on spheres (in Russ.), Theory of cubature formulae and applications of functional analysis to problems of mathematical physics, ed. by S. L. Sobolev, Novosibirsk 1973, 22-27.

46. Salichov, G. N., The group of a regular 600-hedron and cubature formulae of highest algebraic accuracy on the hypersphere (in Russ.), Vopros. Vycisl. i Prikl. Mat., Tashkent 32 (1975), 139-154.

47. Salichov, G. N., Cubature formulae for the hypersphere, invariant with regard to the group of regular 600-hedron (in Russ.), Dokl. Akad. Nauk, SSSR 223 No. 5 (1975), 1075-1078.

48. Salichov, G. N., and F. Sharipchodjaeva, On cubature formulae on the sphere of the five-dimensional space invariant with regard to the group 32-hedron (in Russ.), Vopros. Vycisl. i Prikl. Mat., Tashkent 51 (1978), 204-222.

49. Schmid, H. J., On cubature formulae with a minimal number of knots, Numer. Math. 31 (1978), 281-297.

50. Sobolev, S. L., Formulas of mechanical cubature on the surface of a sphere (in Russ.), Sibirsk Mat. Ž. 3 No. 5 (1962), 769-796.

51. Sobolev, S. L., Introduction to the theory of cubature formulae (in Russ.), Nauka, Moscow 1974.

52. Stroud, A. H., Approximate calculation of multiple integrals. Prentice-Hall, Englewood Cliffs, N. J, 1971.

Prof. Dr. I. P. Mysovskikh
Mat.-Mech. Fak.
Leningr. Gosudarstvenn. Univ.
Bibliotečnaja Pl., 2
Staryi Petergof
Leningrad, 198904
USSR

И. П. Мысовских
Математико-Механический Фак.
Ленингр. Госуд. Университета
Библиотечная Площадь, 2
Старый Петергоф
Ленинград, 198904
СССР

INTERPOLATORY CUBATURE FORMULAE AND REAL IDEALS

H. J. Schmid

This is a follow-up to I. P. Mysovskikh's survey on interpolatory cubature formulae.

1. INTRODUCTION

It seems appropriate to continue the survey by giving a general characterization of this important type of cubature formulae. The paper is based on the preceeding article and uses I. P. Mysovskikh's notation. We refer to equations and theorems of his article by using a preposed [M].

The ideal-theoretical approach by H. M. Möller, as pointed out in [M, sec. 3], will be refined by introducing real ideals. Real ideals satisfying some additional properties are necessary and sufficient for the existence of interpolatory cubature formulae. As a consequence, properties known from the one-dimensional case can be regained in several dimensions. For instance, the knots of an interpolatory cubature formula are the pairwise distinct common zeros of certain orthogonal polynomials. These polynomials have only common simple real zeros. Finally, attention will be drawn to several applications in the two-dimensional case, when the number of knots attains the lower bounds given in [M, (3) and (5)]. In particular, the existence of such formulae will be discussed for special two-dimensional product integrals.

Let σ be a positive Radon measure on R^n and let Ω be a measurable subset of R^n. Let us denote by $P(\Omega)$ the restriction of the real ring $P = R[x_1, x_2, \ldots, x_n]$ to Ω. We shall consider a positive linear functional

$$(1) \quad I: \ P(\Omega) \to R: \ P \to I(P) = \int_\Omega P(x)\,d\sigma(x).$$

ISBN 0-12-213650-0

We want to approximate (1) by means of point evaluations in the form

$$(2) \quad K(P) = \sum_{j=1}^{N} C_j(x^{(j)}) , C_j > 0 , x^{(j)} \in R^n .$$

The cubature sum (2) is called a cubature formula of degree m, if $I(P) = K(P)$ for all $P \in P_m$ and if there is at least one Q in P_{m+1} such that $I(Q) \neq K(Q)$. We shall always assume that

$$N \leqq \dim P_m .$$

This seems to be reasonable due to Chakalov's theorem, see reference [34] in [M].

A cubature formula of degree m is called interpolatory, if there are N linear independent polynomials $U_1, U_2, \ldots, U_N$ in P_m such that

$$(3) \quad \det [U_i(x^{(j)})_{i,j=1,2,\ldots,N}] \neq 0$$

holds. If (2) is a formula of degree m which is not interpolatory, it is well known that there are $C_j^* \geqq 0$ - some coefficients will vanish - such that

$$K^*(P) = \sum_{j=1}^{N} C_j^* P(x^{(j)})$$

is an interpolatory cubature formula of degree m, see [1] and [11]. Hence the essential and most interesting part of all cubature formulae are the interpolatory ones.

If N is minimal for a fixed degree of exactness and all knots $x^{(j)}$ are in the closed convex hull of Ω, then (2) will be called a Gaussian formula. In particular, minimal formulae are interpolatory.

2. REAL IDEALS

In this section, we want to motivate the use of real ideals in order to characterize interpolatory cubature rules. It will be seen that the polynomials vanishing at the knots of a formula generate an ideal with a special property. Such ideals are called real. They have been characterized in a very useful way which allows us to obtain a general characterization theorem.

Let us enlarge the definition of m-orthogonality with respect to I, as it has been defined in [M, sec. 3]. A set $\{R_1, R_2, \ldots, R_s\}$ of polynomials in P is called _strictly m-orthogonal_ (with respect to I), if all R_i are m-orthogonal and there is at least one R_i which is not (m+1)-orthogonal.

Furthermore, we need the notion of fundamentality. A subset $\{R_1, R_2, \ldots, R_s\}$ of P_m is called _fundamental of degree m_, whenever the monic terms R_i^* of degree m of R_i span all monomials of degree m.

Let A be an ideal in P generated by the polynomials $R_1, R_2, \ldots, R_s$, i.e., $A = (R_1, R_2, \ldots, R_s)$. (Recall that (0) is the ideal generated by the zero function.) The set

$$NG(A) = \{y \in R^n : \; Q(y) = 0 \text{ for all } Q \in A\}$$

is called the _real zero set of_ A. If y is in NG(A), it is called a _simple root_, if there are n linear independent polynomials $Q_1, Q_2, \ldots, Q_n$ in A such that

$$\det\left[\left(\frac{\partial}{\partial x_i} Q_j(y)\right)_{i,j=1,2,\ldots,n}\right] \neq 0$$

holds.

The following necessary conditions for the existence of interpolatory cubature formulae are easy to derive.

LEMMA 1. _Let (2) be an interpolatory cubature formula of degree_ m _with_ $N \leq \dim P_m$ _knots and positive coefficients. Then the following assertions hold._

(i) _There are exactly_ $t = \dim P_{m+1} - N$ _linear independent polynomials_ $R_1, R_2, \ldots, R_t$ _in_ P_{m+1} _vanishing at the knots_ $x^{(j)}$, i.e., _if_ $A = (R_1, R_2, \ldots, R_t)$, _then_ $x^{(j)} \in NG(A)$.

(ii) _The space_ P_{m+1} _can be represented as_ $P_{m+1} = R \oplus U$, _where_ $R = \text{span}\,\{R_1, R_2, \ldots, R_t\}$ _and_ $U = \text{span}\,\{U_1, U_2, \ldots, U_N\}$. _Here,_ U _is chosen such that_ (3) _holds._

(iii) _The_ R_is _are fundamental of degree_ m+1 _and strictly m-orthogonal._

(iv) _The following conditions hold,_ $A \cap U = (0)$, _and for all_

$U \in \mathcal{U}$ we have $I(U^2 - R^*) > 0$, where $R^* \in A$ is chosen such that $U^2 - R^* \in P_m$.

It is rather surprising that these conditions are sufficient, too. This can be proved by the theory of ideals. The next lemma shows that the notion of a real ideal is necessarily connected to cubature formulae.

LEMMA 2. *Let $A = (R_1, R_2, \ldots, R_t)$ be the uniquely determined ideal which belongs to an interpolatory formula (2) with $N \leqq \dim P_m$ and positive coefficients. Then the following condition holds,*

(5) $Q \in P$ *vanishes at* NG(A), *if and only if* $Q \in A$.

A proof of the two lemmas is given in [10].

Hence the knots of an interpolatory formula belong to the zero set of a very special ideal. Note that for an ideal A in P one direction of (5) is trivial, while the other is not satisfied in general. Ideals satisfying property (5) are well-known in the theory of ideals, they are called *real ideals*. The following theorem, credited to D. W. Dubois and G. Efroymson [2] and J. J. Risler [7], characterizes real ideals and is essential for the proof of the main theorem.

THEOREM 1. *Let A be an ideal in the real ring P. Then A is real, if and only if the following is true. If*

$$\sum_{i=1}^{\mu} Q_i^2 \in A \, , \; Q_i \in P \, , \; \mu \in N \, ,$$

then $Q_i \in A$, $i = 1, 2, \ldots, \mu$.

We need one further theorem for the proof of the main theorem, its proof goes back to I. P. Mysovskikh [6]. In this theorem an upper bound for the pairwise distinct common zeros of an ideal is given which is generated by a fundamental set of polynomials.

THEOREM 2. Let $R_1, R_2, \ldots, R_t$ be linear independent polynomials in $\mathcal{P}_{m+1}$ which are fundamental of degree m+1. Then the following holds,

(i) $|NG(A)| \leq \dim \mathcal{P}_{m+1} - t$,

(ii) $|NG(A)| = \dim \mathcal{P}_{m+1} - t$, if and only if no $Q \in \mathcal{P}_{m+1} \setminus \mathcal{R}$ vanishes at $NG(A)$. Here $\mathcal{R}$ is the linear space spanned by $R_1, R_2, \ldots, R_t$.

For a proof, we refer to [6] or [10].

3. CHARACTERIZATION OF INTERPOLATORY CUBATURE FORMULAE

Since the conditions (i) to (iii) in Lemma 1 are necessary and easy to achieve, if (iv) holds, we state our main theorem as follows.

THEOREM 3. Let $R_1, R_2, \ldots, R_t$ be linear independent polynomials in $\mathcal{P}_{m+1}$ which are fundamental of degree m+1 and strictly m-orthogonal. Let $A = (R_1, R_2, \ldots, R_t)$ and $\mathcal{R} = \operatorname{span}\{R_1, R_2, \ldots, R_t\}$. Then the following are equivalent.

(6) An interpolatory cubature formula (2) of degree m exists with knots $x^{(j)} \in NG(A)$, $j = 1, 2, \ldots, N$, $N = \dim \mathcal{P}_{m+1} - t \leq \dim \mathcal{P}_m$.

(7) Let $\mathcal{U}$ be a complement of $\mathcal{R}$ in $\mathcal{P}_{m+1}$. Then the following conditions hold,

(i) $A \cap \mathcal{U} = (0)$,

(ii) for all $U \in \mathcal{U}$ we have $I(U^2 - R^*) > 0$, where $R^* \in A$ is chosen such that $U^2 - R^* \in \mathcal{P}_m$.

If such a formula (6) exists, we obtain in addition; A is real, $NG(A) = \{x^{(1)}, x^{(2)}, \ldots, x^{(N)}\}$, all zeros in $NG(A)$ are simple, $R_1, R_2, \ldots, R_t$ do not have common complex zeros. The coefficients can be computed via the formula

$$(8) \quad (C_j)^{-1} = \sum_{i=1}^{N} \frac{U_i^2(x^{(j)})}{I(U_i^2 - R_i^*)}, \quad j = 1, 2, \ldots, N,$$

<u>where</u> $U_1, U_2, \ldots, U_N$ <u>is an orthogonalized basis of</u> $\mathcal{U}$ <u>and where the</u> $R_i^* \in \mathcal{A}$ <u>are chosen such that</u> $U_i^2 - R_i^* \in \mathcal{P}_m$.

Since the terms $I(U_i^2 - R_i^*)$ have to be computed in order to check condition (ii) of the theorem, they are available. Formula (8) is a generalization of the formula by I. P. Mysovskikh developed for minimal formulae of even degree attaining the bound given in [M, (3)], see reference [32] in [M].

Theorem 3, which has been derived in [10] by using Theorems 1 and 2, gives new information about the structure of interpolatory formulae with positive coefficients. Many properties known for the one-dimensional case are generalized successfully. Hence one is tempted to try the classical approach to Gaussian quadrature formulae in higher dimensions for strictly positive linear functionals as well. Lower bounds are known, it remains to show that they will be attained. This, however, fails badly. In the next section we shall illustrate this for two-dimensional product integrals with an even weight-function.

One important property cannot be settled by the theorem. In the one-dimensional case for the most important quadrature formulae it is possible to decide whether the knots are in the domain of integration. It seems to be a very hard task to prove corresponding results in higher dimensions.

4. APPLICATIONS

The main aim in this final section is to show how Theorem 3 can be used in order to obtain interpolatory cubature formulae, in particular, minimal or Gaussian ones. In [M, sec. 2], examples are given for minimal cubature formulae which have a prescribed number of knots. We shall give further examples.

We restrict ourselves to simple two-dimensional integrals, though the results hold for a more general class. Let $n = 2$ be fixed throughout this section and let

$$(9) \quad I: \ \mathcal{P}([-1,1]^2) \to \mathbb{R}: \quad P \to I(P) = \int_{-1}^{+1}\int_{-1}^{+1} P(x,y)p(x)p(y)\,dxdy$$

be a strictly positive linear functional with an even weight-function p(u) on [-1,1]. The lower bounds [M, (3) and (5)] have the form

$$(10)\quad N \geqq \begin{cases} \frac{1}{2}k(k+1), & \text{if } m = 2k-2, \\ \frac{1}{2}k(k+1) + [\frac{k}{2}], & \text{if } m = 2k-1. \end{cases}$$

Trying the one-dimensional approach, one is tempted to use Theorem 3 in order to characterize those (minimal) formulae attaining the bound in (10). For m, even this leads to [M, Theorem 9]. This theorem follows directly from Theorem 3. The only condition which has to be checked is (ii). It can be shown that (ii) is equivalent to the condition stated in [M, Theorem 9]. The special form of (9) allows us to state this theorem in a simple way. Here we denote, as in [M, sec. 2, (12)], by P_i^k, $i = 0,1,\dots,k$, the monic polynomials of degree k orthogonal to $\mathcal{P}_{k-1}$ with respect to I. The moments will be denoted by

$$M_i^k = I(P_i^k P_i^k), \quad i = 0,1,\dots,k, \quad k = 0,1,\dots .$$

THEOREM 4. *A minimal formula of degree* $m = 2k-2$ *for the integral* (9) *exists if and only if the following non-linear system can be solved.*

$$(11)\quad \begin{aligned} M_i^k &= \sum_{\nu=0}^{k-1} \frac{1}{M^{k-1}} (\gamma_{i+\nu}^2 - \gamma_{i+\nu-1}\gamma_{i+\nu+1}), \quad i = 1,2,\dots,k-1, \\ 0 &= \sum_{\nu=0}^{k-1} \frac{1}{M_\nu^{k-1}} (\gamma_{i+\nu}\gamma_{\nu+j-1} - \gamma_{\nu+j}\gamma_{i+\nu+1}), \\ &\qquad i = 1,2,\dots,k-1, \\ &\qquad j = 0,1,\dots,i-2. \end{aligned}$$

If $\gamma_0,\gamma_1,\dots,\gamma_{2k-1}$ *is a solution of* (11), *then the real ideal characterizing the formula is spanned by*

$$(12)\quad R_i = P_i^k + \sum_{\nu=0}^{k-1} \frac{1}{M_\nu^{k-1}} \gamma_{i+\nu} P_\nu^{k-1}, \quad i = 0,1,\dots,k.$$

The knots of the formula are the $\frac{1}{2}k(k+1)$ *pairwise distinct common zeros of* $R_0,R_1,\dots,R_k$.

For a proof of Theorem 4, see [10]. A weaker form of Theorem 4 has been proved by C. R. Morrow and T. N. L. Patterson [5]. Their arguments do not involve the theory of real ideals as, e.g., in Theorem 1, stated above, and it seems not at all clear that the polynomials R_i in (12) have the prescribed number of pairwise distinct zeros.

In [5], an example is given showing that (9) with the weight-function $p(u) = \frac{2}{\pi}\sqrt{1 - u^2}$ possesses Gaussian formulae for each even degree of exactness.

If m is odd, H. M. Möller [4] has given a theorem characterizing those formulae attaining the bound in (10). Here the same difficulty occurs. The number of pairwise distinct common zeros of the orthogonal polynomials under consideration must attain the bound given in (10). This is guaranteed by property (ii) of Theorem 3, which has to be added to give a complete description.

In [8] we derived the following theorem. To formulate it, we denote by $\dot{\mathcal{P}}_k$ the linear space spanned by the monic polynomials of degree k orthogonal to $\mathcal{P}_{k-1}$ in the variables x and y.

THEOREM 5. *A minimal cubature formula of degree* $m = 2k - 1$ *exists for the integral* (9), *if and only if the following assertions hold.*

There are $s = k + 1 - [\frac{k}{2}]$ *linear independent polynomials* $R_1, R_2, \ldots, R_s$ *in* $\dot{\mathcal{P}}_k$ *spanning a linear space* $\mathcal{R}$, *and an ideal* $\mathcal{A}$ *such that the following conditions are satisfied.*

(i) *The set* $\{xR_i, yR_i,\ i = 1,2,\ldots,s\}$ *is fundamental of degree* $k + 1$,

(ii) *Whenever* $Q_0 = yQ_1 - xQ_2 \in \mathcal{P}_k$, $Q_1, Q_2 \in \dot{\mathcal{P}}_k$, *then* Q_0, Q_1, *and* $Q_2 \in \mathcal{R}$ (*there are* $k - [\frac{k}{2}]$ *linear independent equations of this type*),

(iii) *Whenever* $Q = x^2Q_1 - xyQ^2 + y^2Q_3 \in \mathcal{P}_k$, $Q_1, Q_2, Q_3 \in \mathcal{R}$, *then* $Q \in \mathcal{R}$ (*there are* $2k - 3[\frac{k}{2}]$ *linear independent equations of this type*),

(iv) *If* $\mathcal{U}$ *is a complement to* $\mathcal{R}$ *in* $\dot{\mathcal{P}}_k$, *spanned by* $U_1, U_2, \ldots, U_{[\frac{k}{2}]}$, *then*

$$I(U_i^2 - H_i^*) > 0 \ , \quad i = 1,2,\ldots,[\tfrac{k}{2}] \ ,$$

__where__ $H_i^* \in A$ __is__ __chosen__ __such__ __that__

$$U_i^2 - H_i^* \in P_{2k-1} \ ,$$

__and__

$$I(U_iU_j - H_{ij}^*) < I(U_i^2 - H_i^*)I(U_j^2 - H_j^*) \ , \quad i \neq j, \ i = 1,2,\ldots,[\tfrac{k}{2}] \ ,$$

__where__ $H_{ij}^* \in A$ __is__ __chosen__ __such__ __that__ $U_iU_j - H_{ij}^* \in P_{2k-1}$.

The conditions (i)-(iii) are equivalent to the solvability of a quadratic system of equations. This approach has been derived in [10], and can be extended in order to obtain minimal and non-minimal formulae. For details, see [9].

Let us conclude with some results obtained in cooperation with N. Kroll, J. Linden. In [3], all formulae for integrals of the form (9) were characterized for $m \leqq 9$ by using Theorem 4 and Theorem 5. No difficulty occurred for $m \leqq 7$. In the case $m = 8$, the solution of the non-linear system could be achieved only by extensive numerical work. No formula was obtained having all its knots inside the domain of integration. The case $m = 9$, however, turned out to be very informative. Necessary and sufficient conditions for the moments were obtained to guarantee the existence of cubature formulae of degree 9 attaining the lower bound in (10) for even weight-functions $p(u)$. For the weight-function $p(u) = \frac{1}{2}$, these conditions are satisfied; the corresponding formula is due to H. M. Möller [24]. The weight-function $p(u) = \frac{\sqrt{3}}{2}(1 - u^2)$, on the other hand, does not satisfy these conditions, showing that the lower bound in (10) is not attained for all even weight-functions. This is quite disappointing, the more as the domain of integration and the weight-function seem to be quite regular.

However, for the weight-function $p(u) = \frac{2}{\pi}\sqrt{1 - u^2}$, it was proven in [3] that for all $m \in N$ a Gaussian formula exists attaining the lower bound in (10). Thus, the example of C. R. Morrow and T. N. L. Patterson - given for m even and

$m = 2k - 1$, k even - could be completed. This is the first example of a sequence in the degree of exactness of Gaussian formulae for classical integrals.

REFERENCES

1. Davis, P. J., A construction of nonnegative approximate quadratures, Math. Comp., 21 (1967), 578-587.
2. Dubois, D. W. and G. Efroymson, Algebraic theory of real varieties I, Studies and essays presented to Yu-Why Chen on his sixtieth birthday, Academica Sirica, Taipei 1970, 107-135.
3. Kroll, N., J. Linden and H. J. Schmid, Minimale Kubaturformeln für Integrale über dem Einheitsquadrat, Preprint of the SFB 72, Bonn 1980.
4. Möller, H. M., Kubaturformeln mit minimaler Knotenzahl, Numer. Math., 25 (1976), 242-245.
5. Morrow, C. R. and T. N. L. Patterson, Construction of algebraic cubature rules using polynomial ideal theory, SIAM J. Numer. Anal., 15 (1978), 953-976.
6. Mysovskikh, I. P., Numerical characteristics of orthogonal polynomials, Vestnik Leningrad Univ. Math., 3 (1976), 323-332.
7. Risler, J. J., Une charactérization des idéaux des variétés algébraic réelles, Note aux CRAS Paris 271 (1970), 1171-1173.
8. Schmid, H. J., On Gaussian cubature formulae of degree 2k - 1, Numerische Integration, ed. by G. Hämmerlin, ISNM 45, Birkhäuser Verlag, Basel, Boston, Stuttgart 1979, 252-263.
9. Schmid, H. J., Construction of cubature formulae using real ideals, Multivariate Approximation Theory, ed. by W. Schempp and K. Zeller, ISNM 51, Birkhäuser Verlag, Basel, Boston, Stuttgart 1979, 359-377.
10. Schmid, H. J., Interpolatorische Kubaturformeln und reelle Ideale, to appear in Math. Z., 1980.
11. Wilson, M. W., A general algorithm for nonnegative quadrature formulas, Math. Comp., 23 (1969), 253-258.

H. J. Schmid
Mathematisches Institut
Universität Erlangen-Nürnberg
Bismarckstr. 1 1/2
D-8520 Erlangen

STRONG UNIQUENESS OF BEST APPROXIMATIONS AND WEAK CHEBYSHEV SYSTEMS

G. Nürnberger

This paper contains a local version of Haar's Theorem, i.e. a complete characterization of those functions in $C_o(T)$ which have a strongly unique best approximation from an n-dimensional weak Chebyshev subspace in $C_o(T)$, where T is a locally compact subset of the real line.

1. INTRODUCTION

An extended version of the classical Haar's Theorem (Haar [3]), due to Newman, Shapiro [6] and Ault, Deutsch, Morris, Olson [1], says that if $G=\text{span}\{g_1,\ldots,g_n\}$ is an n-dimensional subspace of $C_o(T)$, then <u>each</u> $f\in C_o(T)$ has a strongly unique best approximation from G if and only if for each distinct points $t_1,\ldots,t_n$ in T $\det(g_i(t_j))\neq 0$. (A function $g_o\in G$ is called <u>strongly unique best approximation</u> of $f\in C_o(T)$ from G, if there exists a constant $K_f>0$ such that for each $g\in G$ $\|f-g\| \geq \|f-g_o\| + K_f\|g-g_o\|$.)

Considering this result it is natural to ask, which functions $f\in C_o(T)$ have a strongly unique best approximation, if strong uniqueness of best approximations may not be given for all $f\in C_o(T)$. In this context we prove the following characterization theorem:

A function $f\in C_o(T)\setminus G$, where $T\subset R$, has a strongly unique best approximation g_o from a weak Chebyshev subspace $G=\text{span}\{g_1,\ldots,g_n\}$ if and only if $f-g_o$ has at least n+1 alternating extreme points and $\det(g_i(t_j))\neq 0$ for certain sets $\{t_1,\ldots,t_n\}$ of extreme points of $f-g_o$.

As a consequence we obtain a characterization for an important class of weak Chebyshev subspaces in C[a,b], namely the spline functions of degree N with k fixed knots, from which partial results of Schaback [8], Schumaker [9]

ISBN 0-12-213650-0

and Nürnberger [7] follow.

2. THE CHARACTERIZATION THEOREM

For a locally compact Hausdorffspace T let $C_o(T)$ be the space of all real-valued continuous functions f on T vanishing at infinity, i.e. for each $\varepsilon>0$ the set $\{t\in T:\ |f(t)|\geq\varepsilon\}$ is compact, endowed with the norm $\|f\| = \sup\{|f(t)|:\ t\in T\}$ for each $f\in C_o(T)$. If T is compact then we denote $C_o(T)$ by $C(T)$.

In the following we consider n-dimensional subspaces $G=\text{span}\{g_1,\ldots,g_n\}$ of $C_o(T)$, where T is a locally compact subset of the real line containing at least n+1 points. Such a space G is called <u>weak Chebyshev</u>, if for each basis $\{g_1,\ldots,g_n\}$ of G there exists an $\varepsilon=\pm1$ such that for each points $t_1<\ldots<t_n$ in T $\varepsilon\det(g_i(t_j))\geq 0$.

For a function $f\in C_o(T)$ we use the following notation: We call points $t_1<\ldots<t_r$ in T <u>alternating extreme points</u> of f, if $\varepsilon(-1)^i f(t_i)=\|f\|$, $i=1,\ldots,r$, $\varepsilon=\pm1$. If A and B are subsets of T then we say $A<B$, if for each $a\in A$ and each $b\in B$ $a<b$. We call subsets $T_1<\ldots<T_r$ of T <u>alternation sets</u> of f, if $\varepsilon(-1)^i f(t_i)=\|f\|$, $t_i\in T_i$, $i=1,\ldots,r$, $\varepsilon=\pm1$, and say T_i, $1\leq i\leq r$, is a <u>positive</u> (respectively <u>negative</u>) <u>alternation</u> set of f, if for each $t_i\in T_i$ $f(t_i)=\|f\|$ (respectively $f(t_i)=-\|f\|$).

Furthermore for points $s_1,\ldots,s_n$ in T we set

$$D(s_1,\ldots,s_n) = \begin{vmatrix} g_1(s_1) & \ldots & g_1(s_n) \\ \vdots & & \\ g_n(s_1) & \ldots & g_n(s_n) \end{vmatrix} .$$

By $D(s_1,\ldots,s_{n-1})$ we denote the function in G, defined by $D(s_1,\ldots,s_{n-1})(t) = D(s_1,\ldots,s_{n-1},t)$ for each $t\in T$.

For proving the characterization theorem we need the following result, which is a consequence of Corollary 3.7 in Nürnberger [7] for normed linear spaces E, specialized to $E=C_o(T)$.

THEOREM 1. Let G be a linear subspace of $C_o(T)$, $f \in C_o(T) \setminus \bar{G}$ and $g_o \in G$. Then the following statements are equivalent:

(1) g_o is a strongly unique best approximation of f.

(2) For each $g \in G$, $g \neq 0$, $\min\{(f-g_o)g(t) : t \in M(f-g_o)\} < 0$, where $M(f-g_o) = \{t \in T : |(f-g_o)(t)| = \|f-g_o\|\}$.

Furthermore we need the following extension of the classical Alternation Theorem of Tschebyscheff [11], which has been proved by Jones, Karlovitz [4] for $T=[a,b]$ and in the final form by Deutsch, Nürnberger, Singer [2].

THEOREM 2 (Deutsch, Nürnberger, Singer [2]). Let G be an n-dimensional subspace of $C_o(T)$, where T is a locally compact subset of the real line. Then the following statements are equivalent:

(1) G is weak Chebyshev.

(2) For each $f \in C_o(T)$ there exists a best approximation $g_o \in G$ such that $f-g_o$ has at least $n+1$ alternating extreme points.

The main result is the following characterization concerning strongly unique best approximations from weak Chebyshev subspaces in $C_o(T)$.

THEOREM 3 (Local version of Haar's Theorem). Let $G=\mathrm{span}\{g_1,\ldots,g_n\}$ be an n-dimensional weak Chebyshev subspace of $C_o(T)$, where T is a locally compact subset of the real line, $f \in C_o(T) \setminus G$ and $g_o \in G$. Then the following statements (1) and (2) are equivalent:

(1) g_o is a strongly unique best approximation of f.

(2) There exist $n+m$ alternation sets $T_1,\ldots,T_{n+m}$ of $f-g_o$, where $m \geq 1$, with the following properties: (a) There exist n alternating extreme points $t_1,\ldots,t_n$ of $f-g_o$ in $\cup\{T_i : 1 \leq i \leq n+m\}$ with $D(t_1,\ldots,t_n) \neq 0$. (b) For each $n-1$ distinct integers $k_1,\ldots,k_{n-1} \in \{1,\ldots,n+m\}$, for which there exists an integer k_p from the linearly ordered complement $\{k_n,\ldots,k_{n+m}\}$ and points $t_{k_i} \in T_{k_i}$, $i=1,\ldots,n-1,p$, with $D(t_{k_1},\ldots,t_{k_{n-1}},t_{k_p}) \neq 0$, there exists an integer $k_q \in \{k_n,\ldots,k_{n+m}\}$ and points $s_{k_i} \in T_{k_i}$, $i=1,\ldots,n-1,q$, with $D(s_{k_1},\ldots,s_{k_{n-1}},s_{k_q}) \neq 0$, where q is odd, if p is even, or conversely.

Before proving Theorem 3 at the end of this section we state some corollaries which are easy consequences of this theorem. For the statement of these results we always assume that the assumption of Theorem 3 is given.

COROLLARY 4. *Let* $n=1$, *i.e.* $G=\text{span}\{g_1\}$, *then the following statements are equivalent*:

(1) g_o *is a strongly unique best approximation of* f.

(2) *There exist two alternating extreme points* t_1, t_2 *of* $f-g_o$ *such that* $g_1(t_1)\neq 0$ *and* $g_1(t_2)\neq 0$.

COROLLARY 5. *We consider the following statements*:

(1) g_o *is a strongly unique best approximation of* f.

(2) *There exist* $n+1$ *alternation sets* $T_1,\dots,T_{n+1}$ *of* $f-g_o$ *such that for each* n *distinct integers* $k_1,\dots,k_n\in\{1,\dots,n+1\}$ *there exist points* $t_{k_i}\in T_{k_i}$, $i=1,\dots,n$, *with* $D(t_{k_1},\dots,t_{k_n})\neq 0$.

Then (2) *implies* (1).

REMARK 6. Condition (1) in Corollary 5 does not imply condition (2), as can be seen by the following example: Let $G=\text{span}\{g_1,g_2\}$ be the two-dimensional weak Chebyshev subspace of $C(T)$, where $T=\{1,2,3,4\}$, defined by $g_1=(1,1,0,0)$ and $g_2=(0,0,1,1)$, $f=(1,-1,1,-1)$ and $g_o=0$, then by Theorem 3 g_o is a strongly unique best approximation of f, but condition (2) in Corollary 5 is not satisfied.

CORALLARY 7. *If* $f-g_o$ *has exactly* $n+1$ *maximal alternation sets* $T_1,\dots,T_{n+1}$ *then the following statements are equivalent*:

(1) g_o *is a strongly unique best approximation* of f.

(2) *For each* n *distinct integers* $k_1,\dots,k_n\in\{1,\dots,n+1\}$ *there exist points* $t_{k_i}\in T_{k_i}$, $i=1,\dots,n$, *with* $D(t_{k_1},\dots,t_{k_n})\neq 0$.

Now we give some examples concerning the preceeding results. By c_o we denote the space of all real sequences $f=(\beta_N)$ converging to zero, endowed with the norm $\|f\| = \sup\{|\beta_N|: N=1,2,\dots\}$. The Banachspace c_o is a space of type $C_o(T)$, where T are the natural numbers.

EXAMPLES 8. (i) Let $0=t_o<t_1<\dots<t_{n+1}\leq 1$ be $n+2$ points in $[0,1]$, $G=\text{span}\{g_1,\dots,g_n\}$ be the n-dimensional weak Chebyshev subspace of $C[0,1]$, defined by $g_i(t)=t^i$, $i=1,\dots,n$. Furthermore let f_1 be defined by $(-1)^i f_1(t_i)=1$, $i=1,\dots,n+1$,

$f_1(0)=0$ and linear elsewhere. Then by Corollary 5 or Corollary 7 $g_o=0$ is a strongly unique best approximation of f_1. Moreover, let f_2 be defined by $(-1)^i f_2(t_i)=1$, $i=0,1,\dots,n$, $f_2(1)=0$ and linear elsewhere. Then g_o is a best approximation of f, but by Corollary 7 not a strongly unique best approximation of f.

(ii) Let $G=\text{span}\{g_1,\dots,g_n\}$ be the n-dimensional weak Chebyshev subspace of c_o, defined by $g_i = (\beta_N^{(i)})$, where $\beta_N^{(i)}=1$, if $N=i$; and 0, if $N\neq i$. Then no f in $c_o \setminus G$ has a strongly unique best approximation, since condition (2b) in Theorem 3 is violated, although the function $f = (\beta_N)$, where $\beta_N=(-1)^{N+1}$, if $1\leq N\leq n+1$; and $\beta_N=0$ otherwise; together with $g_o=0$ fulfills condition (2a) in Theorem 3.

(iii) Let $G=\text{span}\{g_1,\dots,g_n\}$be the n-dimensional weak Chebyshev subspace of c_o, defined by $g_i = (\beta_N^{(i)})$, where $\beta_N^{(i)}=1$, if $N=2i-1$ or $N=2i$; and $\beta_N^{(i)}=0$ otherwise; $f = (\beta_N)$, where $\beta_N=(-1)^{N+1}$, if $1\leq N\leq 2n$; if $1\leq N\leq 2n$; and $\beta_N=0$ otherwise; and $g_o=0$. Then by Theorem 3 g_o is a strongly unique best approximation of f.

(iv) Now we give a similar example as in (iii) for C[0,1]. Let $0=t_1<t_2<\dots<t_{2n-1}<t_{2n}=1$ be 2n points in [0,1], $G=\text{span}\{g_1,\dots,g_n\}$ be the n-dimensional weak Chebyshev subspace of C[0,1], defined by $g_i(t)=1$, if $t\in\{t_{2i-1},t_{2i}\}$; $g_i(t)=0$, if $t\in\{t_1,\dots,t_{2n}\}\setminus\{t_{2i-1},t_{2i}\}$; and linear elsewhere; $f(t)=(-1)^{i+1}$, if $t=t_i$, $1\leq i\leq 2n$; and linear elsewhere; and $g_o=0$. Then by Theorem 3 g_o is a strongly unique best approximation of f.

(v) Concerning condition (2b) in Theorem 3 we remark that in general the points $s_{k_1},\dots,s_{k_{n-1}}$ cannot be chosen as $t_{k_1},\dots,t_{k_{n-1}}$, as the following example shows:

Let $G=\text{span}\{g_1,g_2\}$ be the two-dimensional weak Chebyshev subspace of C(T), where $T=\{1,2,3,4\}$, defined by $g_1=(1,0,0,-1)$ and $g_2=(0,1,1,0)$; $f=(1,-1,1,1)$ and $g_o=0$. Then by Theorem 3 g_o is a strongly unique best approximation of f, but for $t_{k_1}=3$ and $t_{k_p}=1$ from condition (2b) in Theorem 3 the only possible choice of s_{k_1} and s_{k_q} is $s_{k_1}=4$ and $s_{k_q}=2$.

Now we consider an important class of weak Chebyshev subspaces, namely spaces of spline functions, and apply the preceeding results:

Let $a=x_o<x_1<\ldots<x_K<x_{K+1}=b$ be K fixed knots in the compact interval [a,b]. The space $S_{N,K}$ of the usual <u>polynomial splines</u> of degree N with K fixed knots is defined by $S_{N,K} = S_{N,K}(x_1,\ldots,x_K) =$ span $\{1,t,\ldots,t^N,(t-x_1)_+^N,\ldots,(t-x_K)_+^N\}$, where the function $(t-x_i)^N$ is defined to be O, if $t\leq x_i$, and to be $(t-x_i)^N$, if $t>x_i$.

It is well-known that the space $S_{N,K}$ is an (N+K+1)-dimensional weak Chebyshev subspace of C[a,b] and that each function g in $S_{N,K}$ lies in $C^{N-1}[a,b]$ and that the restriction of g to any interval $[x_i,x_{i+1}]$, i=O,1,...,K, represents a polynomial of degree N. Furthermore it is easy to verify that in general a function in C[a,b] has more than one best approximation from $S_{N,K}$.

Karlin [5] has shown that for
$S_{N,K}$ = span $\{1,t,\ldots,t^N,(t-x_1)_+^N,\ldots,(t-x_K)_+^N\}$
$D(t_1,\ldots,t_{N+K+1})\neq 0$ if and only if
$t_i<x_i<t_{i+N+1}$, i=1,...,K, $(a<t_1)$.

Using this result it is not too difficult to verify that from Theorem 3 we obtain the following corollary. (The details of the verification will appear elsewhere.)

COROLLARY 9. <u>For</u> $f\in C[a,b]\setminus S_{N,K}$ <u>and</u> $g_o\in S_{N,K}$ <u>the following statements are equivalent</u>:

(1) g_o <u>is a strongly unique best approximation of</u> f.

(2) $f-g_o$ <u>has at least</u> N+K+2 <u>alternating extreme points in</u> [a,b] <u>and</u> $f-g_o$ <u>has at least</u> j+1 <u>alternating extreme points in each interval of the form</u>

$$[a,x_j)\ ,\ (x_{K-j+1},b]\ ,\ j=1,\ldots,K,$$
$$(x_i,x_{i+j+N})\ ,\ j\geq 1\quad (\text{if } K>N+1)\ .$$

We remark, that Corollary 9 is also true for the more general class of Chebyshevian spline functions, since the above mentioned result of Karlin [5] holds for this class. Strauß [10] has shown that condition (2) in Corollary 9 implies that g_o is a unique best approximation of f.

Finally at the end of this section we will prove Theorem 3. Since the complete proof is very long, here we only give a sketch of the arguments. The complete proof will appear elsewhere.

<u>Sketch of the proof of Theorem 3</u>. We give an outline of the arguments for the case that the appearing alternation sets consist of a single point:

(1) $\Rightarrow$ (2) : We assume that (1) holds. Since g_o is a unique best approximation of f, by Theorem 2 $f-g_o$ has at least n+1 alternation sets. Let $T_1,\ldots,T_{n+m}$, where $m\geq 1$, be all alternation sets of $f-g_o$.

(1) $\Rightarrow$ (2a) : Assume that (2a) fails. First of all we assume that

(3) there exist n-1 alternating extreme points $t_{\bar{k}_1},\ldots,t_{\bar{k}_{n-1}}$ such that the function $D(t_{\bar{k}_1},\ldots,t_{\bar{k}_{n-1}})\neq 0$.

Now we choose points $t_{k_1},\ldots,t_{k_{n-1}}$ by induction as follows: Let t_{k_1} be the minimum of points $t_{\bar{k}_1}$ as above. If $t_{k_1},\ldots,t_{k_i}$ are already chosen, then let $t_{k_{i+1}}$ be the minimum of points $t_{\bar{k}_{i+1}}$ with the property that there exist points $t_{\bar{k}_{i+2}},\ldots,t_{\bar{k}_{n-1}}$ such that $t_{k_1},\ldots,t_{k_i},t_{\bar{k}_{i+1}},\ldots,t_{\bar{k}_{n-1}}$ are alternating extreme points of $f-g_o$ and the function $D(t_{k_1},\ldots,t_{k_i},t_{\bar{k}_{i+1}},\ldots,t_{\bar{k}_{n-1}})\neq 0$.

We set $g = D(t_{k_1},\ldots,t_{k_{n-1}})$, $g\neq 0$. Since G is weak Chebyshev there exists an integer $\bar{\varepsilon}=\pm 1$ such that

$\bar{\varepsilon}(-1)^i g(t)\geq 0$, $t\in[t_{k_i},t_{k_{i+1}}]\cap T$, $i=0,1,\ldots,n-1$, where $t_{k_o}=-\infty$ and $t_{k_n}=\infty$.

It is easy to verify, using this property and the fact that (2a) fails, that there exists an integer $\varepsilon=\pm 1$ such that $\min\{\varepsilon(f-g_o)g(t): t\in M(f-g_o)\}\geq 0$, which by Theorem 1 is a contradiction to (1).

If (3) fails then again by similar arguments as above we construct a function $g\in G$, $g\neq 0$, such that $\min\{\varepsilon(f-g_o)g(t): t\in M(f-g_o)\}\geq 0$, which is a contradiction to (1). Roughly speaking the function g is constructed as follows:

First of all we may assume that

(4) there exists an integer i, $1\leq i\leq n+m$, and points $s_2,\ldots,s_{n-1}\in T$ such that the function $D(t_i,s_2,\ldots,s_{n-1})\neq 0$, where $T_i=\{t_i\}$.

Because, since $g_1,\ldots,g_n$ are linearly independent, there exist points $u_1,\ldots,u_{n-1}\in T$ such that the function $D(u_1,\ldots,u_{n-1})\neq 0$. Now, if (4) fails then for each i, $1\leq i\leq n+m$, we have $D(u_1,\ldots,u_{n-1})(t_i)=0$, where $T_i=\{t_i\}$. Then $g = D(u_1,\ldots,u_{n-1})$ has the desired property.

Now by (4) there exist j ($1\leq j\leq n-2$) alternating extreme points $t_{k_1},\ldots,t_{k_j}$ of $f-g_o$ and points $s_{j+1},\ldots,s_{n-1}\in T$ such that the function $D(t_{k_1},\ldots,t_{k_j},s_{j+1},\ldots,s_{n-1})\neq 0$. Then it can be shown that, if we choose j to be maximal and the points $s_{j+1},\ldots,s_{n-1}$ properly, the function $g = D(t_{k_1},\ldots,t_{k_j},s_{j+1},\ldots,s_{n-1})$ has the desired property. (We omit the details here.)

(1) ⇒ (2b) : Assume that (2b) fails. Then there exist n-1 points $t_{k_1}<\ldots<t_{k_{n-1}}$, where $T_{k_i}=\{t_{k_i}\}$, such that $D(t_{k_1},\ldots,t_{k_{n-1}})\neq 0$ and for each odd (respectively even) index q and each integer k_q from the linearly ordered complement $\{k_n,\ldots,k_{n+m}\}$ $D(t_{k_1},\ldots,t_{k_{n-1}},t_{k_q})=0$, where $T_{k_q}=\{t_{k_q}\}$. We set $g = D(t_{k_1},\ldots,t_{k_{n-1}})$, $g\neq 0$, and again as above it is easy to verify that there exists an integer $\varepsilon=\pm 1$ such that $\min\{\varepsilon(f-g_o)g(t): t\in M(f-g_o)\} \geq 0$, which is a contradiction to (1).

(2) ⇒ (1) : Assume that (2) holds, but not (1). Since (1) fails, it follows from Theorem 1 that there exists a function $g\in G$, $g\neq 0$, such that for each $t\in M(f-g_o) \supset \cup\{T_i: 1\leq i\leq n+m\}$ we have $(f-g_o)g(t) \geq 0$. This implies that

(5) for each i, $1\leq i\leq n+m$, and each $t_i\in T_i$ $g(t_i)\geq 0$ (respectively $g(t_i)\leq 0$), if T_i is a positive (respectively negative) alternation set of $f-g_o$.

In the following we show that (5) leads to a contradiction. According to (2a) there exist n alternating extreme points $u_1,\ldots,u_n$ such that $D(u_1,\ldots,u_n)\neq 0$. Then by (5) $\varepsilon(-1)^{i+1}g(u_i)\geq 0$, $i=1,\ldots,n$, $\varepsilon=\pm 1$. We assume that $\varepsilon=1$, since the other case follows analogously.

Since $D(u_1,\ldots,u_n)\neq 0$, there exists a basis $\{g_1,\ldots,g_n\}$ of G such that $(-1)^{i+1}g_i(u_j)=\delta_{ij}$, $i=1,\ldots,n$, $j=1,\ldots,n$.
Then $g=a_1g_1+\ldots+a_ng_n$, where $a_1\geq 0,\ldots,a_n\geq 0$.
Now we show that

(6) for each $s\in(-\infty,u_1)\cap T$ and each m, $1\leq m\leq n$, $g_m(s)\geq 0$; for each $s\in(u_n,\infty)\cap T$ and each m, $1\leq m\leq n$, $g_m(s)\geq 0$ (respectively $g_m(s)\leq 0$), if n is odd (respectively even); and for each i, $1\leq i\leq n-1$, and each $s\in(u_i,u_{i+1})\cap T$ $g_1(s)\geq 0,\ldots,g_i(s)\geq 0$, $g_{i+1}(s)\leq 0,\ldots,g_n(s)\leq 0$ (respectively $g_1(s)\leq 0,\ldots,g_i(s)\leq 0$, $g_{i+1}(s)\geq 0,\ldots,g_n(s)\geq 0$), if i is odd (respectively even).

To do this, we set $D=D(u_1,\ldots,u_n)$ and for each $s\in(-\infty,u_1)\cap T$ and each m, $1\leq m\leq n$, $D_m=D(s,u_1,\ldots,u_{m-1},u_{m+1},\ldots,u_n)$. Since G is weak Chebyshev $DD_m\geq 0$, which implies that $g_m(s)\geq 0$.
Analogously we can show the second statement in (6).
Furthermore for each i, $1\leq i\leq n-1$, each $s\in(u_i,u_{i+1})\cap T$ and each m, $1\leq m\leq n$, we set
$D_{i,m}=D(u_1,\ldots,u_i,s,u_{i+1},\ldots,u_{m-1},u_{m+1},\ldots,u_n)$, where the considered points shall be linearly ordered. Since G is weak Chebyshev $DD_{i,m}\geq 0$, which implies that
$g_1(s)\geq 0,\ldots,g_i(s)\geq 0$, $g_{i+1}(s)\leq 0,\ldots,g_n(s)\leq 0$
(respectively $g_1(s)\leq 0,\ldots,g_i(s)\leq 0$, $g_{i+1}(s)\geq 0,\ldots,g_n(s)\geq 0$), if i is odd (respectively even).
Now we define the following sets:
We denote by $S_{i,1}<\ldots<S_{i,m(i)}$ the negative (respectively positive) alternation sets of $f-g_o$ which are contained in (u_{i-1},u_i), if i is odd (respectively even) and by
$\bar{S}_{i,1}<\ldots<\bar{S}_{i,\bar{m}(i)}$ the negative (respectively positive) alternation sets of $f-g_o$ which are contained in (u_i,u_{i+1}), if i is odd (respectively even), where $u_o=-\infty$ and $u_{n+1}=\infty$ (provided such sets exist).
Furthermore for each i, $1\leq i\leq n$, we set

$S_i = \cup\{S_{i,k(i)}: 1\leq k(i)\leq m(i)\}$ and

$\bar{S}_i = \cup\{\bar{S}_{i,\bar{k}(i)}: 1\leq\bar{k}(i)\leq\bar{m}(i)\}$.

We may assume that

(7) for each set $\{v_1,\ldots,v_n\}$ of n alternating extreme points of $f-g_o$, where $v_1<\ldots<v_n$, $\{v_1,\ldots,v_n\}\subset[u_1,u_n]$ and $\{v_1,\ldots,v_n\}\neq\{u_1,\ldots,u_n\}$, we have $D(v_1,\ldots,v_n)=0$.

Otherwise we can replace the set $\{u_1,\dots,u_n\}$ by a set of n alternating extreme points, which we again denote by $\{u_1,\dots,u_n\}$ and which has property (7).

Now by using (7) we show that

(8) for each i, $1 \leq i \leq n-1$, and each $\bar{s}_i \in \bar{S}_i$ (respectively $s_{i+1} \in S_{i+1}$) $g_{i+1}(\bar{s}_i)=\dots=g_n(\bar{s}_i)=0$ (respectively $g_1(s_{i+1})=\dots=g_i(s_{i+1})=0$).

<u>Sketch of the proof of (8).</u> To show (8) we prove by induction on j that

(9) for each $j \geq 0$, each i, $1 \leq i \leq n-1$, and each $\bar{s}_i \in \bar{S}_i$ (respectively $s_{i+1} \in S_{i+1}$) $g_{i+1}(\bar{s}_i)=\dots=g_{i+j+1}(\bar{s}_i)=0$ (respectively $g_{i-j}(s_{i+1})=\dots=g_i(s_{i+1})=0$).

First let j=0. We have $g_{i+1}(\bar{s}_i)=0$ (respectively $g_i(s_{i+1})=0$), otherwise $D(u_1,\dots,u_i,\bar{s}_i,u_{i+2},\dots,u_n) \neq 0$ (respectively $D(u_1,\dots,u_{i-1},s_{i+1},u_{i+1},\dots,u_n) \neq 0$), contradicting (7). This shows that (9) holds for j=0.

Now we assume that (9) holds for j and show that then (9) holds for j+1. To do this let i=1 and assume that $g_{j+3}(\bar{s}_1) \neq 0$. (The other cases follow analogously.) We show that this assumption leads to a contradiction.

We show that

(10) for each p, $2 \leq p \leq j+2$, and each $s \in [(-\infty,\bar{s}_1) \cup (u_{j+3},\infty)] \cap T$ $g_p(s)=0$.

If $g_p(s) \neq 0$ then, since $g_{j+3}(\bar{s}_1) \neq 0$, the product of $D(u_1,\bar{s}_1,u_2,\dots,u_{j+2},u_{j+4},\dots,u_n)$ with $D(s,u_1,\bar{s}_1,u_2,\dots,u_{p-1},u_{p+1},\dots,u_{j+2},u_{j+4},\dots,u_n)$, where the considered points shall be linearly ordered, is less than zero, contradicting weak Chebyshev.

Now first of all we assume that

(11) there exists an integer p, $2 \leq p \leq j+2$, such that for each k, $2 \leq k \leq j+2$, and each $s_k \in S_k$ (respectively $\bar{s}_k \in \bar{S}_k$) $g_p(s_k)=0$ (respectively $g_p(\bar{s}_k)=0$).

In this case we denote by $T_{k_1},\dots,T_{k_{n-1}}$ the alternation sets of $f-g_o$ to which the points $u_1,\dots,u_{p-1},u_{p+1},\dots,u_n$ belong such that $T_{k_1} < \dots < T_{k_{n-1}}$ and by T_{k_p} the alternation set of $f-g_o$ to which the point u_p belongs. Then by using (10) and (11) it is easy to verify that condition (2b) is not satisfied, which is a contradiction.

If (11) fails then by using the fact that G is weak Chebyshev it can be shown that we get a contradiction to (7). (We omit the details here.) This shows (8).

Now by using (5), (6) and (8) we finally show that we get a contradiction to (2b).

First recall that the function g from above has the representation $g=a_1g_1+\dots+a_ng_n$, where $a_1\geq 0,\dots,a_n\geq 0$. Let j be an integer with $a_j\neq 0$. Such an integer exists, since $g\neq 0$. We denote by $T_{k_1},\dots,T_{k_{n-1}}$ the alternation sets of $f-g_o$ to which the points $u_1,\dots,u_{j-1},u_{j+1},\dots,u_n$ belong such that $T_{k_1}<\dots<T_{k_{n-1}}$. Let $\{k_n,\dots,k_{n+m}\}$ be the linearly ordered complement of $\{k_1,\dots,k_{n-1}\}$ with respect to $\{1,\dots,n+m\}$ and denote the alternation set of $f-g_o$ to which the point u_j belongs by T_{k_p}, where $k_p\in\{k_n,\dots,k_{n+m}\}$.

Then there exist points $t_{k_i}\in T_{k_i}$, $i=1,\dots,n-1,p$, such that $D(t_{k_1},\dots,t_{k_{n-1}},t_{k_p})\neq 0$, namely $u_1,\dots,u_n$.

Now we show that

(12) for each i, $1\leq i\leq n$, and each $\bar{s}_i\in\bar{S}_i$ (respectively $s_i\in S_i$) $g_j(\bar{s}_i)=0$ (respectively $g_j(s_i)=0$).

Let i=1. (The other cases follow analogously.) By (5) $g(\bar{s}_1) = a_1(\bar{s}_1)+\dots+a_ng_n(\bar{s}_1)\leq 0$, by (6) $g_1(\bar{s}_1)\geq 0$ and by (8) $g_2(\bar{s}_1)=\dots=g_n(\bar{s}_1)=0$. Since $a_1\geq 0,\dots,a_n\geq 0$ and $a_j\neq 0$, this implies that $g_j(\bar{s}_1)=0$.

Furthermore by (5) $g(s_1) = a_1g_1(s_1)+\dots+a_ng_n(s_1)\leq 0$ and by (6) $g_1(s_1)\geq 0,\dots,g_n(s_1)\geq 0$. Again, since $a_1\geq 0,\dots,a_n\geq 0$ and $a_j\neq 0$, this implies that $g_j(s_1)=0$.

Now by using (12) it immediately follows that there does <u>not</u> exist an integer $k_q\in\{k_n,\dots,k_{n+m}\}$ and points $s_{k_i}\in T_{k_i}$, $i=1,\dots,n-1,q$, such that $D(s_{k_q},s_{k_1},\dots,s_{k_{n-1}})\neq 0$, where q is odd, if p is even or conversely.

This is a contradiction to (2b) and completes the proof of Theorem 3.

REFERENCES

1 Ault D.A., Deutsch F.R., Morris P.D., Olson J.E., Interpolating subspaces in approximation theory, J. Approximation Theory 3 (1970), 164-182

2 Deutsch F., Nürnberger G., Singer I., Weak Chebyshev subspaces and alternation, to appear in Pacific J.Math.

3 Haar A., Die Minkowskische Geometrie und die Annäherung an stetige Funktionen, Math. Annalen 78 (1918), 294-311

4 Jones R.C., Karlovitz L.A., Equioscillation under non-uniqueness in the approximation of continuous functions, J. Approximation Theory 3 (1970), 138-145

5 Karlin S., Total positivity, Standford University Press, Stanford, California (1968)

6 Newman D.J., Shapiro H.S., Some theorems on Chebyshev approximation, Duke Math. J. 30 (1963), 673-684

7 Nürnberger G., Unicity and strong unicity in approximation theory, J. Approximation Theory 26 (1979), 54-70

8 Schaback R., On alternation numbers in nonlinear Chebyshev approximation, J Approximation Theory 23 (1978), 379-391

9 Schumaker L.L., Uniform approximation by Tschebyscheffian spline functions, J. Math. a. Mech. 18 (1968), 369-378

10 Strauß H., Eindeutigkeit bei der gleichmäßigen Approximation mit Tschebyscheffschen Splinefunktionen, J. Approximation Theory 15 (1975), 78-82

11 Tschebyscheff P.L., Sur les questions de minima qui se rattachant à la réprésentation approximative des fonctions, Oeuvres, Bd. I, St. Petersburg (1891), 273-378

G. Nürnberger
Institut für Angewandte Mathematik
Universität Erlangen-Nürnberg
Martensstraße 3
8520 Erlangen
West Germany

ON THE CONNECTION BETWEEN RATIONAL UNIFORM APPROXIMATION AND POLYNOMIAL L_p APPROXIMATION OF FUNCTIONS

V. A. Popov

In this note we shall consider the connection between rational uniform approximation of real valued functions on the finite interval and L_p-polynomial approximation of it and its derivatives.

1. INTRODUCTION

P. Turan was the first to note that there should exist a connection between the rational uniform approximation and the polynomial L_p-approximation of functions. This question was examined in [1], where some estimations are given of the best rational uniform approximation by means of the integral modulus of continuity of the derivatives of the function. Here we shall give a direct connection between rational uniform approximation and polynomial L_p-approximation. As a corollary we shall improve the results of [1] and we shall give some new results of this type.

By H_n we shall denote the set of all algebraical polynomials of n-th degree; by R_n, the set of all rational functions of n-th degree, i.e., $r \in R_n$ if $r(x) = (a_k x^k + \dots + a_0)/(b_m x^m + \dots + b_0)$, $k,m \leqq n$.

The best L_p-approximation, $1 \leqq p \leqq \infty$, of functions $f \in L_p(a,b)$ by means of algebraical polynomials of n-th degree, respectively by means of rational functions of n-th degree, is given by

$$E_n(f)_{L_p} = \inf\{\|f - p\|_{L_p} : p \in H_n\},$$

$$R_n(f)_{L_p} = \inf\{\|f - p\|_{L_p} : p \in R_n\},$$

$$\|f\|_{L_p} = \{\int_a^b |f(x)|^p dx\}^{1/p} .$$

ISBN 0-12-213650-0

In the case $p = \infty$ we shall write briefly $R_n(f)$. We shall use the k-th moduli of continuity of f in L_p:

$$\omega_k(f;\delta)_{L_p} = \inf_{0<h<\delta} \left\{ \int_a^{b-kh} |\Delta_h^k f(x)|^p dx \right\}^{1/p}$$

where

$$\Delta_h^k f(x) = \sum_{m=0}^{k} (-1)^{k+m} \binom{k}{m} f(x+mh) .$$

If f is a function with bounded variation in [a,b], we shall denote by $V_a^b f$ the variation of the function f in the interval [a,b].

In [2], we prove the following:

THEOREM A. *We have* (*for* $n > r \geq 1$):

$$\sup_{V_a^b f^{(r)} \leq M} R_n(f) \leq c(r) \frac{M(b-a)^r}{n^{r+1}}$$

where c(r) *is a constant, depending only on* r.

From Theorem A we obtain immediately:

THEOREM B. *If the function* f *has* r+1-*th derivative* $f^{(r+1)} \in L_p(a,b)$, $1 \leq p \leq \infty$, *then for* $1 \leq r < n$

$$R_n(f)_{L_p} \leq (b-a)^{1/p} R_n(f) \leq c(r)(b-a)^{r+1} \frac{\|f^{(r+1)}\|_{L_p}}{n^{r+1}} .$$

2. MAIN RESULTS

THEOREM 1. *If* f *has a derivative* $f' \in L_p$, $p > 1$, *then*

$$R_n(f) \leq c(p) n^{-1} E_n(f')_{L_p}$$

where c(p) *is a constant, depending only on* p.

In the case $p = 1$, we have only

THEOREM 2. *If* f *has a derivative* $f'' \in L_1$, *then*

$$R_n(f) \leqq c(b-a)n^{-2}E_n(f'')_{L_1}$$

where c *is an absolute constant.*

The proof of Theorem 2 is very simple if we use Theorem B for $r = 1$ and $p = 1$.

Let $p \in H_n$ be the algebraical polynomial of the best L_1 approximation of f'', i.e.,

$$\|f'' - p\|_{L_1} = E_n(f'')_{L_1}$$

and let $Q \in H_{n+2}$ be such that $Q'' = p$. Then for $n \geqq 2$, using Theorem B for $r = 1$, $p = 1$, we have:

$$R_{3n}(f) \leqq R_n(f-Q) + R_{2n}(Q) = R_n(f-Q)$$

$$\leqq c(1)(b-a)\frac{\|(f-Q)''\|_{L_1}}{n^2} = c(1)(b-a)n^{-2}E_n(f'')_{L_1}$$

and it proves Theorem 2.

Obviously the proof of Theorem 1 is the same if we use the following theorem:

THEOREM 3. *If* f *has a derivative* $f' \in L_p$, $p > 1$, *then*

$$R_n(f) \leqq c'(p)n^{-1}\|f'\|_{L_p}$$

where the constant $c'(p)$ *depends only on* p.

The proof of Theorem 3 can be done by the method of [2]. In the next section we give the full proof of Theorem 3. Some preliminary lemmas are needed.

From Theorems 1 and 2, using Jackson's theorem for algebraical polynomials (see, for example, [3]):

$$E_n(f)_{L_p} = 0(\omega_k(f;n^{-1})_{L_p})$$

we obtain:

COROLLARY 1. *If* $f' \in L_p[a,b]$, $p > 1$ *or if* $f'' \in L_1$, *we have*

$$R_n(f) = 0\left(\frac{\omega_k(f';n^{-1})_{L_p}}{n}\right), \quad p > 1,$$

or, respectively,

$$R_n(f) = 0\left(\frac{\omega_k(f'';n^{-1})_{L_1}}{n^2}\right).$$

Remark 1. Let us mention, that at the conference in Bonn, Ju. A. Brudnyi said to the author that he had obtained Corollary 1 (and as a consequence, Theorem 3) in another way (see [4], where Corollary 1 is given without proof).

Remark 2. Corollary 1 improves the results from [1].

Remark 3. In [5], P. Petrushev improves Theorem A as follows:

If $V_a^b f^{(r)} < \infty$, then $R_n(f) = o(n^{-r-1})$, $r \geq 1$.

Corollary 1 shows what this "little o" is in the case when $f^{(r)}$ is absolutely continuous, $r \geq 1$.

3. PROOF OF THEOREM 3

We shall prove here Theorem 3. We shall follow the scheme of the proof of Theorem A given in [2].

We set:

$$\phi_{n,A} = \sup_{\substack{\|f'\|_{L_p[0,1]} \leq 1 \\ f(0)=0}} \; \inf_{\substack{g \in R_n \\ \|g\|_{C(-\infty,\infty)} \leq A}} \|f - g\|_{C[0,1]}.$$

LEMMA 1. We have ($1/p + 1/q = 1$):

$$\sup_{\substack{\|f'\|_{L_p[a,b]} \leq M \\ f(a)=\alpha}} \inf_{r \in \tilde{R}_n} \|f - r\|_{C[a,b]} \leq M(b-a)^{1/q}\phi_{n,A}$$

where $\tilde{R}_n = \{r \in R_n: \|r\|_{C(-\infty,\infty)} \leq |\alpha| + M(b-a)^{1/q}A\}$.

Proof. Let $[a,b] = [0,b]$ and let us set $g(x) = M^{-1}b^{-1/q} \cdot (f(bx) - \alpha)$. Then $g(0) = 0$ and

$$\|g'\|_{L_p[0,1]} = M^{-1}b^{1/p}\left\{\int_0^1 |f'(bx)|^p dx\right\}^{1/p}$$

$$= M^{-1}\left\{\int_0^b |f'(x)|^p dx\right\}^{1/p} \leq 1 .$$

Consequently, there exists $\tilde{r} \in R_n$ such that

$$\|g - \tilde{r}\|_{C[0,1]} \leq \phi_{n,A}, \quad \|\tilde{r}\|_{C(-\infty,\infty)} \leq A .$$

The second condition gives us:

$$\phi_{n,A} \geq \max_{x \in [0,1]} \left| \frac{f(bx)}{Mb^{1/q}} - \frac{\alpha}{Mb^{1/q}} - \tilde{r}(x) \right|$$

$$= M^{-1}b^{-1/q} \max_{x \in [0,b]} \left| f(x) - \alpha - \frac{\tilde{r}(\frac{x}{b})}{M^{-1}b^{-1/q}} \right| ,$$

i.e., for $r(x) = \alpha + Mb^{1/q}\, r(\frac{x}{b}) \in R_n$, we have:

$$\|f - r\|_{C[0,b]} \leq Mb^{1/q}\phi_{n,A}$$

$$\|r\|_{C(-\infty,\infty)} \leq |\alpha| + Mb^{1/q}\|r\|_{C(-\infty,\infty)} \leq |\alpha| + Mb^{1/q}A .$$

LEMMA 2. Let $\|f'\|_{L_p[0,1]} \leq 1$. Then there exists $x_0 \in [0,1]$ such that ($1/p + 1/q = 1$):

$$\|f'\|_{L_p[0,x_0]}\, x_0^{1/q} \leq 1/2; \quad \|f'\|_{L_p[x_0,1]}\, (1 - x_0)^{1/q} \leq 1 .$$

The proof of Lemma 2 is based on

LEMMA 3. *Let* $a_i^{1/p} b_i^{1/p} = 1/2$, $i = 1,2, a_i, b_i \geqq 0$, $p \geqq 1$, $a_1 + a_2 = A \leqq 1$. *Then* $b_1 + b_2 \geqq 1$.

Proof. In fact, let us find $\min(b_1 + b_2) = \min 2^{-q}(a_1^{-q/p} + a_2^{-q/p})$ at the condition $a_1 + a_2 = A \leq 1$. Obviously, we must have $a_1 = a_2 = A/2$ and $\min(b_1 + b_2) = 2^{-q} 2^{q/p} A^{-q/p} = A^{-q/p} \geqq 1$.

Proof of Lemma 2. Let $x_0 \in [0,1]$ be such that $\|f'\|_{L_p[0,x_0]} x_0^{1/2} = 1/2$ (if such x_0 do not exist, the lemma follows for this f). Let us assume that $\|f'\|_{L_p[x_0,1]} (1 - x_0)^{1/q} > 1/2$. In such case, there exists $x' \in [0,1]$ so that $\|f'\|_{L_p[x_0,x']} (x' - x_0)^{1/q} > 1/2$. Setting $a_1 = \|f'\|^p_{L_p[0,x_0]}$, $a_2 = \|f'\|^p_{L_p[x_0,x_1]}$, $b_1 = x_0$, $b_2 = x_1 - x_0$, we have $a_1 + a_2 = \int_0^{x_1} |f'(x)|^p dx \leqq 1$ and by Lemma 2, $x_1 \geqq 1$, which contradicts $x_1 < 1$. This contradiction proves the lemma.

LEMMA 4. *If* $\|f'\|_{L_p[0,1]} \leqq 1$, $p > 1$, *then*

$$|f(x) - f(y)| \leqq |x - y|^{1/q}, \quad 1/p + 1/q = 1 .$$

Proof.

$$|f(x) - f(y)| \leqq \int_y^x |f'(t)| dt \leqq \left\{\int_y^x |f'(x)|^p dx\right\}^{1/p} |x - y|^{1/q}$$

$$\leqq |x - y|^{1/q} .$$

LEMMA 5. *For every* $r > 0$ *there exists a constant* $d(r)$ *depending only on* r *such that there exists* $\sigma_n \in R_m$, $m \leqq d(r) \log^2 n$, *for which*

$$|\sigma_n(x)| \leqq n^{-2} \text{ for } -1 \leqq x \leqq -n^{-r},$$

$$|1 - \sigma_n(x)| \leqq n^{-2} \text{ for } n^{-r} \leqq x \leqq 1,$$

$$0 \leqq \sigma_n(x) \leqq 1 \text{ for every } x.$$

Proof (see [6], Lemma 4). From the results of A. A. Gončar ([7], Lemmas 1-3), it follows that there exists a constant $d(r)$, depending only on r, such that for every natural $n > 1$, there exists an even number N, $N \leq d(r)\log^2 n$, such that for the rational function $\tilde{\sigma}_n(x) = (1 + s_n(x))^{-1}$, $s_n(x) = p_n(x)/p_n(-x)$, $p_n(x) = \prod_{i=1}^{N} (x - \delta^{i/N})$, $\delta = n^{-r}$, the following inequalities hold:

$$|\tilde{\sigma}_n(x)| \leq n^{-2} \text{ for } -1 \leq x \leq -n^{-r},$$

$$|1 - \tilde{\sigma}_n(x)| \leq n^{-2} \text{ for } n^{-r} \leq x \leq 1,$$

$$0 \leq \tilde{\delta}_n(x) \leq 1 \text{ for } |x| \leq n^{-r}.$$

Since N is an even number it is easy to see that $s_n(x) \geq 0$ for $|x| \geq 1$ and therefore $0 \leq \tilde{\sigma}_n(x) \leq 1$ for $|x| \leq 1$. Then

$$\sigma_n(x) = \frac{1 - \frac{1}{2}n^{-2}}{1 + \frac{1}{2}n^{-2}} \tilde{\sigma}_n(x) + \frac{1}{2} n^{-2}$$

satisfies the conditions of the lemma.

Let us set $\phi_n = \phi_{n,1}$.

LEMMA 6 (the main lemma). *There exists a natural number $n_o(p)$, depending only on p such that for $n > n_0(p)$ we have*

$$\phi_n \leq \frac{\psi(k)}{n}\left(1 + c(p)\frac{\log^2 n}{n^2}\right)^2$$

if $\phi_k \leq \psi(k)/k$, $\psi(k) \geq 1$, for $k = [\frac{n}{2}(1 - d(q)\frac{\log^2 n}{n})]$, where $c(p)$ is a constant, depending only on p and $d(q)$ is a constant from Lemma 5. $[x]$ as usual, denotes the integer part of x, $1/p + 1/q = 1$.

Proof. Let $f(0) = 1$. From Lemma 2, it follows that there exists $x_0 \in [0,1]$ such that ($\|f'\|_{L_p[0,1]} \leq 1$):

$$\|f'\|_{L_p[0,1]} x_0^{1/q} \leq 1/2; \quad \|f'\|_{L_p[x_0,1]}(1 - x_0)^{1/q} \leq 1/2 .$$

Let us set $\Delta_1 = [0,x_0]$, $\Delta_2 = [x_0,1]$. From Lemma 1, it follows that there exists two rational functions $r_i \in R_k$, $i = 1,2$, such that

$$(1)\quad \begin{aligned} &\| f - r_i\|_{C(\Delta_i)} \leq \phi_k/2, \quad i = 1,2; \\ &\| r_1\|_{C(-\infty,\infty)} \leq 1, \quad \| r_2\|_{C(-\infty,\infty)} \leq |f(x_0)| + 1/2 \leq 1 \end{aligned}$$

(since $|f(x_0)| = |\int_0^{x_0} f'(t)dt| \leq \{\int_0^{x_0} |f'(t)^p dt\}^{1/p} x_0^{1/q} \leq 1/2)$.

Let us denote $\Delta_1' = [0,x_0+\delta_k]$, $\Delta_2' = [x_0-\delta_k,1]$, where $\delta_k = k^{-2q}$, and let T_i, $i = 1,2$, be the linear transformations of the interval Δ_i' into the interval Δ_i, $i = 1,2$. Then

$$(2)\quad |T_i x - x| \leq \delta_k \text{ for } x \in \Delta_i', \ i = 1,2 .$$

We have

$$(3)\quad \begin{aligned} \| f(x) - r_i(T_i x)\|_{C[\Delta_i']} &\leq \max_{x\in\Delta_i'} |f(x) - f(T_i x)| + \\ &\max_{x\in\Delta_i'} |f(T_i x) - r_i(T_i x)| \\ &\leq \omega(f;\delta_k) + \max_{x\in\Delta_i} |f(x) - r_i(x)| \\ &\leq \omega(f;\delta_k) + \phi_k/2; \quad i = 1,2 . \end{aligned}$$

Let us consider the rational function $g \in R_N$, where $N \leq 2k + d(q)\log^2 k$:

$$g(x) = r_1(T_1 x) + \sigma_k(x - x_0)\{r_2(T_2 x) - r_1(T_1 x)\}$$

where σ_k and $d(q)$ are from Lemma 5.

Let $x \in [0,1]$. If $x < x_0-\delta_k$, then using (3) and Lemma 5, we obtain

$$\begin{aligned} |f(x)-g(x)| &\leq |f(x)-r_1(T_1 x)| + \\ &\quad + k^{-2}(\|r_1\|_{C(-\infty,\infty)} + \|r_2\|_{C(-\infty,\infty)}) \\ &\leq \omega(f;\delta_k) + \phi_k/2 + 2k^{-2} \end{aligned}$$

or using Lemma 4:

(4) $\quad |f(x)-g(x)| \leqq \phi_k/2 + 3k^{-2}.$

Similarly if $x \in [x_0-\delta_k,1]$, we again obtain (4).

If $x \in [x_0-{}_k, x_0+{}_k]$, using that $0 \leqq \alpha = \sigma_k(x-x_0) \leqq 1$ and (3), we get:

$$\begin{aligned} |f(x)-g(x)| &= |(1-\alpha)(f(x)-r_1(T_1x)) + \alpha(f(x)-r_2(T_2x))| \\ &\leqq (1-\alpha)\|f(x)-r_1(T_1x)\|_{C[\Delta_1']} + \alpha\|f(x)-r_2(T_2x)\|_{C[\Delta_2']} \\ &\leqq \phi_k/2 + k^{-2}. \end{aligned}$$

Consequently, (4) holds for every $x \in [0,1]$. On the other hand, Lemma 5 and (1) give us:

$$\|g\|_{C(-\infty,\infty)} = \sup_x \{(1-\sigma_k(x-x_0))|r_1(T_1x)| + \sigma_k(x-x_0)|r_2(T_2x)|\} \leqq 1 . \tag{5}$$

The inequalities (4) and (5) give us:

$$\begin{aligned} &\|f-g\|_{C[0,1]} \leqq \phi_k/2 + 3k^{-2}; \\ &\|g\|_{C(-\infty,\infty)} \leqq 1,\ g \in R_N,\ N \leqq 2k + d(g)\log^2 k . \end{aligned} \tag{6}$$

There exists a constant $c(p)$ and a natural number $n_0(p) > 2$ depending only on p, $1/p + 1/q = 1$, such that for $n > n_0(p)$ and $k = \left[\frac{n}{2}\left(1 - d(q)\frac{\log^2 n}{n}\right)\right]$ we have:

$$1 \leqq k \leqq n/2;\ k^{-1} \leqq \frac{2}{n}\left(1 + c(p)\frac{\log^2 n}{n}\right) ;$$

$$1 + 6/k \leqq (1 + c(p)/n) .$$

Then for $n > n_0(p)$, we have, if $\phi_k \leqq \frac{\psi(k)}{k}$, $\psi(k) \geqq 1$, $k = \left[\frac{n}{2}\left(1 - d(q)\frac{\log^2 n}{n}\right)\right]$:

$$\phi_k/2 + 3k^{-2} \leqq \frac{\psi(k)}{2k}(1 + \tfrac{6}{k}) \leqq \frac{\psi(k)}{n}\left(1 + c(p)\frac{\log^2 n}{n}\right)^2 . \tag{7}$$

Since under the same conditions, $2k + d(q)\log^2 k \leqq n - d(q)\log^2 n + d(q)\log^2 k \leqq n$, from (6) and (7) follows the lemma.

<u>Proof of Theorem 3</u>. We shall prove a stronger result of $\phi_n = \phi_{n,1} \leqq \frac{c_1(p)}{n}$, where $c_1(p)$ is a constant, depending only on p. In view of Lemma 1, this is sufficient. We define the function $y(x)$ by $y(x) = \left[\frac{x}{2}\left(1 - d(q)\frac{\log^2 x}{x}\right)\right]$ and $y^s(x) = y(y^{s-1}(x))$, $y^0(x) = x$.

Let $N_0(p)$ be so large that $N_0(p) \geqq n_0(p)$, where $n_0(p)$ is from Lemma 6 and that for $n > N_0(p)$ we have
$\frac{\log^2 n}{n} \leqq \frac{2}{3}\frac{\log^2 y(n)}{y(n)}$, i.e.,

$$(8) \quad \frac{\log^2 n}{n} \leqq \frac{2}{3}\frac{\left[\log^2 \frac{n}{2}(1 - d(q)\frac{\log^2 n}{n})\right]}{\left[\frac{n}{2}(1 - d(q)\frac{\log^2 n}{n})\right]}.$$

Obviously, there exists a constant $c'(p)$ depending only on p such that for $n \leqq N_0(p)$ we have;

$$(9) \quad \phi_{n,1} \leqq \frac{c'(p)}{n}, \quad c'(p) \geqq 1.$$

Let $n > N_0(p)$. Then there exists s_0 such that $y^{s_0}(n) \leqq N_0$, $y^{s_0-1}(n) > N_0$. Using Lemma 6 and (9), we obtain:

$$\phi_{y^{s_0}(n)} \leqq \frac{c'(p)}{y^{s_0}(n)}$$

$$\phi_{y^{s_0-1}(n)} \leqq \frac{c'(p)}{y^{s_0-1}(n)}\left(1 + c(p)\frac{\log^2 y^{s_0-1}(n)}{y^{s_0-1}(n)}\right)^2$$

$$\cdots$$

$$\phi_n \leqq \frac{c'(p)}{n}\prod_{s=0}^{s_0-1}\left(1 + c(p)\frac{\log^2 y^s(n)}{y^s(n)}\right)^2.$$

Since $y^{s_0-1}(n) > N_0(p)$, we have from (8):

$$\frac{\log^2 y^{i-1}(n)}{y^{i-1}(n)} \leqq \frac{2}{3}\frac{\log^2 y^i(n)}{y^i(n)}.$$

Therefore

$$\prod_{s=0}^{s_0-1} \left(1 + c(p)\frac{\log^2 y^s(n)}{y^s(n)}\right) \leqq c''(p)$$

which proves Theorem 3.

REFERENCES

1. Popov, V. A. and J. Szabados, A remark on the rational approximation of functions, C. R. Acad. Bulgare Sci., 28 (1975), 1303-1306.
2. Popov, V. A., Uniform rational approximation of the class V_r and its applications, Acta Math. Acad. Sci. Hung., 29(1-2) (1977), 119-129.
3. Brudnii, J. A., Approximation of functions of n-th variables by quasi-polynomials, Isv. Aca. Nauk SSSR, seria math., 34 (1970), 564-583.
4. Brudnii, J. A., Rational approximation and imbedding theorems, Dokl. Acad.. Nauk SSSR, 247 (1979), 269-272.
5. Petrushev, P. P., Uniform rational approximation of the functions of the class V_r, Mat. Sbornik, 108(150) (1979), 418-432.
6. Popov, V. A. and P. P. Petrushev, The exact order of the best uniform approximation of convex functions by rational functions, Mat. Sbornik, 103(145) (1977), 285-292.
7. Gončar, A. A., Estimates of the growth of rational functions and some applications, Mat. Sbornik, 72(114) (1967), 489-503.

V. A. Popov
Mathematical Institut of
Bulgarian Acad. of Sci., Sofia
BULGARIA

BEST CHEBYSHEV APPROXIMATION BY SMOOTH FUNCTIONS

U. Sattes

A characterization is given for the best approximation of a continuous function by smooth functions having a uniformly bounded r-th derivative.

Let C(I) be the space of all real continuous functions on the interval $I = [-1,+1]$ under the Chebyshev norm $\|\cdot\|_\infty$, and let $L_\infty(I)$ be the space of all (equivalence classes of) real, Lebesgue-measurable, essentially bounded functions on I. Here $\|\cdot\|_\infty$ denotes the essential supremum.

Furthermore, for $r \in \mathbb{N}$, let

$$L_\infty^r(I) = \{f: f, f', \ldots, f^{(r-1)} \text{ absolutely continuous, } f^{(r)} \in L_\infty(I)\}$$

and

$$W_r = \{ g \in L_\infty^r(I) : \| g^{(r)} \|_\infty \leq 1 \}.$$

For $r = 1$, N. P. Korneichuk stated in 1961 (see [1], p. 225) a characterization for the best approximation of a continuous function f by W_1, namely :

$g^* \in W_1$ is an element of best approximation of f in W_1,

i. e. $\| f - g^* \|_\infty = \inf (\| f - g \|_\infty : g \in W_1)$,

iff

there exists an interval $(\alpha,\beta) \subset I$, such that g^* is linear on (α,β) with angular coefficient $\gamma = \pm 1$ and

$$(g^* - f)(\alpha) = (f - g^*)(\beta) = \gamma \cdot \| f - g^* \|_\infty .$$

It is the aim of the paper to prove a corresponding result for $r \geq 2$.

Let us start with the remark, that the existence of elements of best approximation of $f \in C(I)$ in W_r is given simply by the boundedly compactness of that class.

ISBN 0-12-213650-0

The characterization and uniqueness theorem reads as follows :

THEOREM . *Let* $f \in C(I)$, $g^* \in W_r$, $r \geq 2$. *Then* g^* *is an element of best approximation of* f *in* W_r ,

iff

there exists a subinterval $(\alpha,\beta) \subset I$ *and an* $N \in \mathbb{N}$, $N \geq r+1$, *such that*

(i) $g^*|_{(\alpha,\beta)}$ *is a perfect spline with exactly* $N-r-1$ *knots*

$$y_o := \alpha < y_1 < y_2 < \ldots < y_{N-r-1} < \beta =: y_{N-r} .$$

(ii) *There exist* $\alpha = \xi_1 < \xi_2 < \ldots < \xi_N = \beta$ *such that*

$$(f - g^*)(\xi_i) = \varepsilon(-1)^i \| f - g^* \|_\infty, \ \varepsilon = \pm 1, \ 1 \leq i \leq N.$$

(iii) $\xi_{i+1} < y_i < \xi_{i+r}$, $1 \leq i \leq N-r-1$.

(iv)
$$g^{*(r)}(x) \equiv \begin{cases} -\varepsilon \ , \text{ a.e. on } (y_o,y_1), \text{ if } r \text{ is even} \\ +\varepsilon \ , \text{ a.e. on } (y_o,y_1), \text{ if } r \text{ is odd} . \end{cases}$$

Furthermore, g^* *is uniquely determined on* $[\alpha,\beta]$.

For the proof, we need the following lemma, which has some interest in its own:

LEMMA . *Let* $r, k \in \mathbb{N}$, $k \geq r \geq 2$ *and* $-1 < x_1 < \ldots < x_k < 1$. *Furthermore, let* $B_1, \ldots , B_{k+1-r}$ *be sets of non-zero measure in* I *with*

$$B_i < B_{i+1} \qquad , 1 \leq i \leq k-r$$

$$(\text{ i.e. } \sup B_i \leq \inf B_{i+1} \qquad , 1 \leq i \leq k-r)$$

and
$$x_i < B_i < x_{i+r-1} \qquad , 1 \leq i \leq k+1-r.$$

Finally, let $j \in \mathbb{N}$, $1 \leq j \leq k+1-r$ *and* $p_j \in L_\infty(I)$ *with*

$$p_j(x) > 0 \text{ a.e. on } B_j \text{ and } p_j(x) \equiv 0 \text{ elsewhere;}$$

define

$$u_j(x) := \int_{-1}^{+1} p_j(t)\, g_r(x;t)\, dt \qquad , x \in I ,$$

where
$$g_r(x;t) = (x - t)_+^{r-1}/(r-1)! \quad , \quad x,t \in I.$$

Then $G := \text{span}\{1, x, \ldots, x^{r-1}, u_1, \ldots, u_{k+1-r}\}$ *is a weak Chebyshev space of dimension* $k+1$ *, and there is a function* p^* *in* G, *which has simple zeros exactly in* $x_1,\ldots,x_k$.

Recall, that an n-dimensional subspace H of C(I) is a weak Chebyshev space, if each $h \in H$ has at most n-1 sign changes, i.e., there do not exist points $-1 \le x_o < \ldots < x_n \le 1$, such that $h(x_i)\cdot h(x_{i+1}) < 0$ for $0 \le i \le n-1$.

Recall further the following version of Rolle's theorem, see [2] :

Let $[a,b]$ be a non-degenerate compact interval, let $f \in C([a,b])\setminus\{0\}$ be absolutely continuous , such that $f(a) = f(b) = 0$. Then f' has at least one sign change in (a,b), i.e., there exist sets A, B of positive measure in (a,b) with $A < B$ and $\varepsilon = \pm 1$, such that

$\varepsilon\cdot f'(x) > 0$ a.e. on A and $\varepsilon\cdot f'(x) < 0$ a.e. on B.

If f' is continuous on (a,b) , then it has at least one zero in this interval.

Proof of the lemma.

If G is not a weak Chebyshev space, there is a $g \in G$ having at least $k+1$ sign changes. Applying Rolle's theorem r times gives that $g^{(r)}$ has at least $k+1-r$ sign changes in $(-1,+1)$, which is impossible.

To find a non-trivial function $g \in G$, which vanishes in $x_1, x_2,\ldots, x_k$, consider the linear system of equations

$$\sum_{i=o}^{k} a_i u_i(x_j) = 0 \quad , 1 \le j \le k$$

$$\sum_{i=o}^{k} a_i u_i(-1) = 1 .$$

The system has either a unique solution or the corresponding homogeneous one has a non-trivial solution, furnishing in both cases a non-trivial function g in G vanishing at the prescribed points.

Fix such a g.

We know there is an interval $[x_i, x_{i+1}]$, $1 \le i \le k-1$, such that $f \not\equiv 0$ on $[x_i, x_{i+1}]$; for else g would be a polynomial of degree $r-1$, having at most $r-1 < k$ zeros.

It is no restriction to assume, that in any interval $[x_i, x_{i+1}]$: $\varepsilon_i g(x) \geq 0$, $\varepsilon_i = \pm 1$, $1 \leq i \leq k-1$.

Contrary to the assertion let us assume that g has a non-simple zero in at least one of the roots $x_1, \ldots, x_k$. We distinguish two cases:

I. g has no zero-intervals $[x_i, x_{i+1}]$, $1 \leq i \leq k-1$ (but possibly zero-intervals $[-1, x_1]$, $[x_k, +1]$ or $[a,b] \subsetneqq [x_i, x_{i+1}]$).

II. g has zero-intervals $[x_i, x_{i+1}]$ for some i, $1 \leq i \leq k-1$.

ad I. By assumption and Rolle's theorem, g' has at least one sign change in each (x_i, x_{i+1}), $1 \leq i \leq k-1$, with corresponding s.c. point $z_i \in (x_i, x_{i+1})$, i.e.

$$f'(z_i) = 0 \text{ and } f'(x) \not\equiv 0 \text{ on } (x_i, z_i) \text{ and } (z_i, x_{i+1})$$

Furthermore there is a smallest index i_o, $1 \leq i_o \leq k$, such that $f'(x_{i_o}) = 0$.

If $i_o = 1$, then g'' has at least $k-1$ sign changes in (x_1, z_1), (z_1, z_2), $\ldots$, (z_{k-2}, z_{k-1}), again by Rolle's theorem; similarly for $i_o = k$.

If $1 < i_o < k$, then g'' has at least $k-1$ sign changes in $(z_1, z_2), \ldots, (z_{i_o -1}, x_{i_o})$, $(x_{i_o}, z_{i_o}), \ldots, (z_{k-2}, z_{k-1})$.

Repeated application of Rolle's theorem yields at least $k+1-r$ sign changes of $g^{(r)}$, a contradiction.

ad II.

case 1: There exist indices i_o, j_o, $1 \leq i_o < j_o \leq k$, such that: x_{i_o}, x_{j_o} are zeros of multiplicity r of g and $g(x) \not\equiv 0$ on (x_i, x_{i+1}), $i_o \leq i \leq j_o - 1$.

By Rolle's theorem, g' has at least $j_o - i_o$ sign

changes in (x_{i_o},x_{j_o}) with corresponding s.c. points $z_i^1 \in (x_i,x_{i+1})$, $i_o \leq i \leq j_o-1$.

Having seen that $g^{(j)}$, $1 \leq j \leq r-1$, has at least j_o-i_o+j-1 sign changes in (x_{i_o},x_{j_o}) with corresponding s.c. points z_i^j, we find that $g^{(j+1)}$ has at least j_o-i_o+j-2 sign changes in $(z_{i_o}^j, z_{j_o-1}^j)$ and at least one in each $(x_{i_o}, z_{i_o}^j)$ and $(z_{j_o-1}^j, x_{j_o})$, such that finally holds:

$g^{(r)}$ has at least j_o-i_o+r-1 sign changes in (x_{i_o},x_{j_o}), which is impossible, for (x_{i_o},x_{j_o}) carries at most the j_o-i_o+r-2 (or less) sets $B_{max(1;i_o+2-r)},\ldots, B_{j_o-1}$.

<u>case 2</u>: There is an index i_o, $2 \leq i_o \leq k-1$, such that x_{i_o} is a zero of multiplicity r of g and $g(x) \not\equiv 0$ on (x_i,x_{i+1}) for $i_o \leq i \leq k-1$ or $1 \leq i \leq i_o-1$.

Without restriction, let $g(x) \equiv 0$ on $[0,x_{i_o}]$. Then g' has at least $k-i_o$ sign changes in (x_{i_o},x_k) with s.c. point $z_{i_o}^1 \in (x_{i_o},x_{i_o+1})$.

Consequently, g'' has at least $k-i_o-1$ sign changes in $(z_{i_o}^1,x_k)$, as well as one in $(x_{i_o},z_{i_o}^1)$, which leads to $k-i_o$ sign changes of $g^{(r)}$ in (x_{i_o},x_k). But (x_{i_o},x_k) carries at most the $k-i_o$ (or less) sets $B_{max(1;i_o+2-r)},\ldots B_{k+1-r}$.

Thus the lemma is proved.

<u>Proof</u> of the theorem.

" $\Rightarrow$ " Let $g^* \in W_r$ with $\|f - g^*\|_\infty = \inf(\|f-g\|_\infty : g \in W_r)$.

As in the proof of Chebyshev's alternation theorem, the

existence of at least $r+1$ alternation points of $f - g^*$ is shown.

We then divide I in closed blocks $d_1 < d_2 < \ldots < d_{k+1}$ ($k \geq r$) of extremal points of $f - g^*$, without restriction as follows :

d_1 contains just E^+ - points (of $f - g^*$), i.e.,

if $x \in d_1$, then $(f - g^*)(x) = + \| f - g^* \|_\infty$;

$\xi_1^2 := \max d_1$ is E^+-point, $\overline{d}_1 := [-1, \xi_1^2]$ contains no E^--point.

d_j contains just $\begin{Bmatrix} E^- \\ E^+ \end{Bmatrix}$ - points, if j is $\begin{Bmatrix} \text{even} \\ \text{odd} \end{Bmatrix}$, $1 \leq j \leq k$;

$\xi_j^1 := \min d_j$ and $\xi_j^2 := \max d_j$ are $\begin{Bmatrix} E^- \\ E^+ \end{Bmatrix}$ - points and

$\overline{d}_j := [\xi_j^1, \xi_j^2]$ contains no $\begin{Bmatrix} E^+ \\ E^- \end{Bmatrix}$ - point.

d_{k+1} contains just $\begin{Bmatrix} E^+ \\ E^- \end{Bmatrix}$ - points, if k is $\begin{Bmatrix} \text{odd} \\ \text{even} \end{Bmatrix}$;

$\xi_{k+1}^1 := \min d_{k+1}$ is $\begin{Bmatrix} E^+ \\ E^- \end{Bmatrix}$ - point and $\overline{d}_{k+1} := [\xi_{k+1}^1, 1]$

contains no $\begin{Bmatrix} E^- \\ E^+ \end{Bmatrix}$ - point

and $I \setminus \bigcup_{j=1}^{k+1} \overline{d}_j$ contains no extremal points of $f - g^*$.

Consider the following assumption:

(A) There are sets $B_i \subset (\xi_i^2, \xi_{i+r}^1)$, $1 \leq i \leq k+1-r$ of positive measure, such that

$(-1)^{i+r-1} \cdot g^{*(r)}(x) < 1$ a.e. on B_i, $1 \leq i \leq k+1-r$

and $B_i < B_{i+1}$, $1 \leq i \leq k-r$.

If (A) holds, we may choose $x_i \in (\xi_i^2, \xi_{i+1}^1), 1 \leq i \leq k$, with the property, that $x_i < B_i < x_{i+r-1}$, $1 \leq i \leq k+1-r$.

Let

$$p_i(x) := \begin{cases} (-1)^{r+i-1} - g^{*(r)}(x) & , \text{ on } B_i, \\ 0 & , \text{ else }, \end{cases} \quad 1 \leq i \leq k+1-r ;$$

then by the lemma, we find a function $p^* \in L_\infty^r(I)$, having exactly k simple zeros in $x_1, \ldots, x_k$.

We may assume: $p^*(x) > 0$ on $\overline{d}_1$. (Otherwise we take $p^{**} = -p^*$.)

Then we obtain by construction for $\delta > 0$ sufficiently small:

$$g^* + \delta p^* \in W_r \quad \text{and} \quad \| f - (g^* + \delta p^*) \|_\infty < \| f - g^* \|_\infty ,$$

a contradiction.

Thus, the assumption (A) does not hold, and we get the conditions of the theorem by induction.

Let $k = r$; then $(-1)^r g^{*(r)}(x) \equiv 1$ a.e. on (ξ_1^2, ξ_{r+1}^1), such that conditions (i) - (iv) hold with $N = r+1 = k+1$.

Let $k = r+1$; then

$$(-1)^r g^{*(r)}(x) \equiv 1 \qquad \text{a.e. on } (\xi_1^2, \xi_{r+1}^1) ,$$

giving conditions (i) - (iv) with $N = r+1 = k$; or

$$(-1)^{r+1} g^{*(r)}(x) \equiv 1 \qquad \text{a.e. on } (\xi_2^2, \xi_{r+2}^1) ,$$

giving conditions (i) - (iv) with $N = r+1 = k$; or

we find $y_1 \in (\xi_2^2, \xi_{r+1}^1)$ such that

$$(-1)^r g^{*(r)}(x) \equiv 1 \qquad \text{a.e. on } (\xi_1^2, y_1) \quad \text{and}$$

$$(-1)^{r+1} g^{*(r)}(x) \equiv 1 \qquad \text{a.e. on } (y_1, \xi_{r+2}^1) ,$$

giving conditions (i) - (iv) with $N = r+2 = k+1$.

Now, if $k > r$ whatever, then there either exist

$$\xi_1^2 =: y_o < y_1 < \ldots < y_{k-r} < y_{k+1-r} := \xi_{k+1}^1$$

with

$$y_j \in (\xi_{j+1}^2, \xi_{j+r}^1) \qquad , \quad 1 \leq j \leq k-r ,$$

such that

$$(-1)^{i+r} g^{*(r)}(x) \equiv 1 \quad \text{a.e. on } (y_i, y_{i+1}),\ 0 \le i \le k-r,$$

giving conditions (i) - (iv) with $N = k+1 > r+1$; or else, we can restrict to $[-1, \xi_k^1]$ or $[\xi_2^2, +1]$, and find (i) - (iv) fulfilled for some $N \le k$.

„ ⇐ " Conversely, let (i) - (iv) be fulfilled for some $g^* \in W_r$, which is not an element of best approximation.

Then by the Kolmogorov criterion for convex sets , we find some function $g^{**} \in W_r$ with the property, that

$$(\ f - g^*\ )(\xi_i) \cdot (\ g^{**} - g^*\ )(\xi_i) \quad 0,\ 1 \le i \le N.$$

We may assume ξ_1 to be an E^+-point of $f - g^*$, giving

$$(-1)^{i+1}\ (\ g^{**} - g^*\ )(\xi_i) > 0 \qquad ,1 \le i \le N.$$

Repeated application of the mean-value theorem now gives, setting $h := g^{**} - g^*$:
There are

$$\eta_i^{(1)} \in (\xi_i, \xi_{i+1})\ ,\ \ni\ (-1)^i \cdot h'(\eta_i^{(1)}) > 0 \qquad ,1 \le i \le N-1\ ,$$

and for $1 \le j \le r-2$, there are

$$\eta_i^{(j+1)} \in (\eta_i^{(j)}, \eta_{i+1}^{(j)}) \subset (\xi_i, \xi_{i+j+1})\ ,\ \ni\ (-1)^{i+j} h^{(j+1)}(\eta_i^{(j+1)}) > 0$$

$$,1 \le i \le N-j-1.$$

By the absolute continuity of $h^{(r-1)}$, we find sets of positive measure

$$B_i < B_{i+1} \qquad ,1 \le i \le N-r-1,$$

such that

$$B_i \subset (\xi_i, \xi_{i+1}) \qquad \text{and}$$

$$(-1)^{i+r+1} h^{(r)}(x) > 0 \quad \text{a.e. on } B_i \qquad ,1 \le i \le N-r.$$

An inductive argument will now lead to a contradiction.

If $B_1 < y_1$, then we have

(+) $$1 \geq g^{**(r)}(x) = g^{*(r)}(x) + h^{(r)}(x) > 1 \quad \text{a.e. on } B_1, \text{ if } r \text{ is even,}$$

or

(++) $$-1 \leq g^{**(r)}(x) = g^{*(r)}(x) + h^{(r)}(x) < -1 \quad \text{a.e. on } B_1, \text{ if } r \text{ is odd ;}$$

a contradiction, or else $B_2 > y_1$.

Let $1 < i_o \leq N-r-1$ and no contradiction have been obtained on $B_1, \ldots, B_{i_o-1}$; then $B_{i_o} > y_{i_o-1}$.

If $B_{i_o} < y_{i_o}$, then we get a contradiction on B_{i_o} via (+) or (++); else we have $B_{i_o+1} > y_{i_o}$.

Having reached the index $N-r-1$ without obtaining a contradiction, we finally get

$$y_{N-r-1} < B_{N-r} < y_{N-r} = \beta.$$

But

$$(-1)^{N+1} \cdot h^{(r)}(x) > 0 \qquad \text{on } B_{N-r},$$

yielding (+) on B_{N-r}, if N is odd, or (++) on B_{N-r}, if N is even, showing altogether the sufficiency of the conditions (i) - (iv).

The uniqueness of g^* on $[\alpha,\beta]$ is proved as follows:

Let $g^*, g^{**} \in W_r$ be two elements of best approximation of f in W_r, and let g^* satisfy (i) - (iv) on $[\alpha,\beta]$. Without restriction, let ξ_1 be an E^+-point of $f - g^*$.

Then we have :

$$(-1)^{i+1}(f - g^{**})(\xi_i) \leq (-1)^{i+1}(f - g^*)(\xi_i), \quad ,1 \leq i \leq N,$$

or, setting $h := g^{**} - g^*$:

$$(-1)^{i+1}\, h(\xi_i) \geq 0 \qquad ,1 \leq i \leq N.$$

We claim

$$h(x) \equiv 0 \quad \text{on} \quad [\xi_1, \xi_N].$$

For $1 \leq i \leq N-1$, we have either

$$h(x) \equiv 0 \quad \text{on} \quad [\xi_i, \xi_{i+1}]$$

or

$$h' \text{ has a sign change on } (\xi_i, \xi_{i+1}).$$

If h has no zero-interval, then we get a contradiction as in part „ ⇐ " of the proof of the theorem. Let therefore h have a zero-interval.

<u>case 1</u> : There are indices $i_o, j_o \in \{1, \ldots, N\}$, $i_o < j_o$, such that ξ_{i_o} and ξ_{j_o} are zeros of multiplicity r of h, and $h(x) \not\equiv 0$ on (ξ_i, ξ_{i+1}), $1 \leq i \leq j_o-1$.

Then, as in the proof of the lemma, $h^{(r)}$ has at least j_o-i_o+r-1 sign changes in (ξ_{i_o}, ξ_{j_o}). But (ξ_{i_o}, ξ_{j_o}) contains at most j_o-i_o+r-2 knots y_i of g such that we can find a knot-interval Y with the property:

$$h^{(r)}(x) > 0 \quad \text{on a set} \quad C_1 \subset Y \quad \text{of positive measure}$$

and

$$h^{(r)}(x) < 0 \quad \text{on a set} \quad C_2 \subset Y \quad \text{of positive measure,}$$

yielding contradiction (+) or (++) on C_1 or C_2, respectively.

<u>case 2</u> : h has just one zero-interval $[\xi_{i_o}, \xi_{j_o}]$, and for $i \leq i_o - 1$, $i \geq j_o$ there holds $h(x) \not\equiv 0$ on $[\xi_i, \xi_{i+1}]$.

Then we again get a contradiction as in the lemma, corresponding to the first case above.

This completes the proof of the theorem.

REFERENCES

1 Korneichuk, N. P., Extremal Problems in Approximation Theory, Nauka, Moscow, 1976 (in Russian).

2 Schumaker, L. L., Toward a constructive theory of generalized spline functions; in Böhmer, Meinardus and Schempp, Spline Functions, Springer, Berlin-Heidelberg-New York, p. 265-331, 1975.

U. Sattes
Institut für Angewandte Mathematik
Universität Erlangen-Nürnberg
D - 852o Erlangen
Martensstr. 3

APPROXIMATIONS IN THE HARDY SPACE $\mathcal{H}^1(D)$ WITH RESPECT TO THE NORM TOPOLOGY

Walter Schempp

The purpose of this note is to investigate Patil type approximations of functions $f \in \mathcal{H}^1(D)$ with respect to the Hardy norm $\|?\|_{\mathcal{H}^1(D)}$. Since there does not exist any continuous projector of $L^1(\mathbf{T})$ onto $\mathcal{H}^1(\mathbf{T})$ by Newman's theorem, norm continuous Toeplitz operators on $\mathcal{H}^1(\mathbf{T})$ with symbols merely belonging to $L^\infty(\mathbf{T})$ are not available. Our main tool are Toeplitz operators $\mathcal{H}^1(\mathbf{T}) \to \mathcal{H}^1(\mathbf{T})$ with symbols that are smooth enough to multiply the (strong) dual BMO($\mathbf{T}$) (= vector space of all complex-valued functions on $\mathbf{T}$ with bounded mean oscillation) of the Hardy subspace $\mathcal{H}^1(\mathbf{T})$ of $L^1(\mathbf{T})$.

1. INTRODUCTION

Let $D = \{z \in \mathbb{C} \mid |z|<1\}$ denote the open unit disc in the complex plane $\mathbb{C}$ and ∂D the unit circle bounding D. In the sequel we shall identify ∂D with the one-dimensional compact torus group $\mathbf{T}$. Let λ denote the Haar measure of $\mathbf{T}$ with total mass 1. The corresponding complex Lebesgue spaces will be denoted by $L^p(\mathbf{T})$ and their norms by $\|?\|_p$ $(p \in [1,\infty])$.

Let $\mathcal{H}(D)$ be the complex vector space of all functions that are holomorphic in D. For $f \in \mathcal{H}(D)$ and $r \in [0,1[$ define the dilation $f_r : \mathbf{T} \ni w \rightsquigarrow f(rw) \in \mathbb{C}$ of f. Fix an exponent $p \in [1,\infty[$. Then $\mathcal{H}^p(D)$ is the vector subspace of $\mathcal{H}(D)$ that consists of all functions $f \in \mathcal{H}(D)$ such that

$$\|f\|_{\mathcal{H}^p(D)} = \sup_{r\in[0,1[} \|f_r\|_p = \sup_{r\in[0,1[} \left(\int_{\mathbf{T}} |f(rw)|^p d\lambda(w)\right)^{1/p} < +\infty$$

ISBN 0-12-213650-0

holds. Furthermore, $\mathcal{H}^{\infty}(D)$ is the vector subspace of $\mathcal{H}(D)$ of all bounded functions $f \in \mathcal{H}(D)$ with norm $f \rightsquigarrow \|f\|_{\infty} = \sup_{z \in D} |f(z)|$. The family $(\mathcal{H}^p(D))_{p \in [1,\infty]}$ of complex Banach spaces (with norms $\|?\|_{\mathcal{H}^p(D)}$) is referred to as the scale of Hardy spaces modelled on the disc D. It is well-known that for each function $f \in \mathcal{H}^p(D)$ the radial limit $\lim_{r \to 1-} f_r(w) = \tilde{f}(w)$ exists for λ-almost all points $w \in \mathbf{T}$, that $\|f\|_{\mathcal{H}^p(D)} = \|\tilde{f}\|_p$ and that Egoroff's theorem implies $\lim_{r \to 1-} \|f_r - \tilde{f}\|_p = 0$ when $p \in [1,\infty[$. Thus, the $\mathbb{C}$-linear mapping $f \rightsquigarrow \tilde{f}$ which assigns to each $f \in \mathcal{H}^p(D)$ its boundary-value function class $\tilde{f} \in L^p(\mathbf{T})$ is an isometry between $\mathcal{H}^p(D)$ and a closed vector subspace $\mathcal{H}^p(\mathbf{T})$ of the complex Lebesgue space $L^p(\mathbf{T})$. We may identify the complex Banach spaces $\mathcal{H}^p(D)$ and $\mathcal{H}^p(\mathbf{T})$, $p \in [1,\infty]$, by way of this linear isometry. Therefore $\mathcal{H}^p(\mathbf{T})$ is called the Hardy subspace of $L^p(\mathbf{T})$. For details, the reader is referred to Zygmund [10].

Let $\mathfrak{D}'(\mathbf{T})$ be the vector space of complex distributions on $\mathbf{T}$. It will be convenient to consider $\mathfrak{D}'(\mathbf{T})$ as the topological antidual of the vector space $\mathfrak{D}(\mathbf{T})$ over $\mathbb{C}$ of all infinitely differentiable complex-valued functions on $\mathbf{T}$ (cf. Schwartz [7]) when $\mathfrak{D}(\mathbf{T})$ is equipped with its canonical $\mathcal{C}^{\infty}$-topology, i.e., with the topology of uniform convergence of the functions and of all their derivatives. If $\mathfrak{D}'_0(\mathbf{T})$ denotes the vector subspace of $\mathfrak{D}'(\mathbf{T})$ formed by the distributions $T \in \mathfrak{D}'(\mathbf{T})$ with Fourier coefficients $\mathcal{F}_{\mathbf{T}}T(m) = 0$ for all $m<0$ (Schempp-Dreseler [6]) then $\mathfrak{D}'_0(\mathbf{T})$ is closed with respect to the weak dual topology $\sigma(\mathfrak{D}'(\mathbf{T}), \mathfrak{D}(\mathbf{T}))$ of $\mathfrak{D}'(\mathbf{T})$ and the injections

$$\mathcal{H}^q(\mathbf{T}) \hookrightarrow \mathcal{H}^p(\mathbf{T}) \to \mathfrak{D}'_0(\mathbf{T}) \qquad (1 \leq p < q \leq \infty)$$

are continuous linear mappings.

Suppose $p \in [1,\infty]$. If $f \in \mathcal{H}^p(D)$, $f \neq 0$, then $\log|\tilde{f}| \in L^1(\mathbf{T})$. It follows that for functions $f \in \mathcal{H}^p(D)$, $g \in \mathcal{H}^p(D)$ the equality $\tilde{f}(w) = \tilde{g}(w)$ on some subset Ω of $\mathbf{T}$ with measure $\lambda(\Omega) > 0$

implies f=g. This result can be rephrased in the following way: The functions $f \in \mathscr{H}^p(D)$ are uniquely determined by the values of $\tilde{f}$ on Ω. In a remarkable paper, D.J. Patil [3] has been concerned with the problem of recovering f by means of such "poor boundary informations". More precisely, he has shown how to construct explicitly approximations of the functions $f \in \mathscr{H}^p(D)$, $p \in]1,\infty[$, with respect to the norm $\|?\|_{\mathscr{H}^p(D)}$ by means of the restrictions $\tilde{f}|\Omega$; in this connection also see Patil [4] and Young [8]. Now it is here that by the Marcel Riesz and Newman theorems a very significant difference occurs between the cases $p \in]1,\infty[$ and $p = 1$. - It is the aim of the present article to deal with the case $p = 1$, i.e., with the study of Patil type approximations of functions $f \in \mathscr{H}^1(D)$ with respect to the Hardy norm $\|?\|_{\mathscr{H}^1(D)}$ which is not covered by Patil's paper [3] cited above.

2. TOEPLITZ OPERATORS WITH PERIODIC $\mathscr{L}^1$-SYMBOLS

Let $\chi_m: \mathbf{T} \ni w \rightsquigarrow e^{imw} \in \mathbf{C}^\times$ $(m \in \mathbf{Z})$ denote the characters of the torus group $\mathbf{T}$ and let

$$H = -i \sum_{m \in \mathbf{Z}} (\operatorname{sign} m)\chi_m$$

be the (periodic) Hilbert distribution. Observe that $H \in \mathscr{D}'(\mathbf{T})$ is a complex distribution of order exactly 1 such that $\sup_{m\in\mathbf{Z}} |\mathscr{F}_{\mathbf{T}} H(m)| = 1$. Clearly, for each distribution $T \in \mathscr{D}'(\mathbf{T})$ the conjugate distribution of T is given by the convolution $H * T$ on $\mathbf{T}$. Thus, $T \rightsquigarrow H * T$ is the Hilbert transformation on $\mathbf{T}$. If (cf. Schempp-Dreseler [6])

$$\varepsilon_o = \sum_{m\in\mathbf{Z}} \chi_m$$

denotes the Dirac measure located at the neutral element of $\mathbf{T}$, then the complex distribution

$$H_o = \frac{1}{2}(\chi_o+\varepsilon_o+iH)$$

of order exactly 1 gives rise to the weakly continuous linear projector

$$Q: \mathfrak{D}'(\mathbf{T}) \ni T \rightsquigarrow H_o * T \in \mathfrak{D}_o'(\mathbf{T})$$

of $\mathfrak{D}'(\mathbf{T})$ onto its closed vector subspace $\mathfrak{D}_o'(\mathbf{T})$. By the Marcel Riesz theorem Q restricts to a projector of $L^p(\mathbf{T})$ onto $\mathcal{H}^p(\mathbf{T})$, $p \in]1,\infty[$, which is continuous with respect to the norm topologies. The difficulty in the case p=1 is that the projector Q does not have a norm continuous restriction $L^1(\mathbf{T}) \to \mathcal{H}^1(\mathbf{T})$; indeed it is known by Newman's theorem [2] that there exists no continuous projector whatsoever of $L^1(\mathbf{T})$ onto $\mathcal{H}^1(\mathbf{T})$, i.e., that $\mathcal{H}^1(\mathbf{T})$ does not admit a closed complementary vector subspace of $L^1(\mathbf{T})$. Therefore the Toeplitz operators

$$P_\varphi: \mathcal{H}^1(\mathbf{T}) \ni f \rightsquigarrow Q(\varphi f) \in \mathcal{H}^1(\mathbf{T})$$

with symbols $\varphi \in L^\infty(\mathbf{T})$ do not need to be continuous with respect to the norm topology. However, P_φ acting on $\mathcal{H}^1(\mathbf{T})$ is the same as the linear projector Q acting on the vector space $\varphi.\mathcal{H}^1(\mathbf{T})$. Therefore, the main idea is to assure the existence and the norm continuity of the Toeplitz operator P_φ by means of some smoothness conditions on its symbol φ (cf. Theorem 3 infra).

Let Ω denote a non-empty open subset of $\mathbf{T}$ and let the functions $\rho \in \mathcal{C}^2(\mathbf{T})$ be real-valued with (compact) support contained in Ω. The $\mathcal{C}^2$-smoothness of ρ will be shown to be sufficient. - Suppose that ρ satisfies the following conditions:

(i) $\rho(w) \in [0,1]$ for each point $w \in \Omega$;
(ii) $M_\rho = \lambda(\{w \in \Omega\} \rho(w) = 1\})>0$.

For each value of the parameter $t \in]0,1[$ introduce the

function $\rho_t \in \mathcal{L}^2(\mathbf{T})$ ("inverse Schränkungstransformierte" of $t\rho$) according to the formula

$$\rho_t: \mathbf{T} \ni w \rightsquigarrow t\rho(w)/(1-t\rho(w)) \in \mathbb{R}_+.$$

Furthermore, for $t \in]0,1[$ define the $\mathbb{R}_+$-valued function

$$\varphi_t = 1+\rho_t = 1/(1-t\rho)$$

and, in addition, introduce the function

$$\theta_t = \log \varphi_t$$

which belongs to $\mathcal{L}^2(\mathbf{T})$. If $C_z: \mathbf{T} \ni w \rightsquigarrow 1/(1-w\bar{z}) \in \mathbb{C}$ denotes the Cauchy-Szegö kernel of the disc D located at the point $z \in D$, then the Cauchy transform

$$h_t: D \ni z \rightsquigarrow \langle\theta_t, C_z\rangle \in \mathbb{C} \qquad (t \in]0,1[)$$

of the function θ_t belongs to $\mathcal{H}^\infty(D)$. Some standard arguments based upon performing Abel-Poisson means and conjugate Abel-Poisson means of h_t, respectively, establish the following result:

THEOREM 1. <u>Retain the above notations - then, for</u> $t \in]0,1[$, <u>the boundary-value function</u> $\tilde{h}_t$ <u>of</u> $h_t \in \mathcal{H}^\infty(D)$ <u>satisfies</u>

$$\mathrm{Re}\, \tilde{h}_t \in \mathcal{L}^2(\mathbf{T}),\ \mathrm{Im}\, \tilde{h}_t \in \mathcal{L}^1(\mathbf{T}).$$

Given $f \in L^1(\mathbf{T})$ and any subarc $I \neq \emptyset$ of $\mathbf{T}$, by the average f_I of f on I is meant the complex number

$$f_I = \frac{1}{\lambda(I)} \int_I f(w)\, d\lambda(w).$$

Introduce the quantity

$$M(f) = \sup_{\lambda(I) \leq 1} \left(\frac{1}{\lambda(I)} \int_I |f(w) - f_I| d\lambda(w)\right) \in \mathbb{R}_+ \cup \{+\infty\}.$$

Then $f \in L^1(\mathbf{T})$ is said to have bounded mean oscillation on $\mathbf{T}$ whenever $M(f) < +\infty$. It is well known (Fefferman and Stein [1]) that the complex vector space

$$BMO(\mathbf{T}) = \{f \in L^1(\mathbf{T}) \mid M(f) < +\infty\}$$

of all functions with bounded mean oscillation on $\mathbf{T}$ may be identified with the topological dual of the Hardy space $\mathcal{H}^1(\mathbf{T})$ and that the norm $f \rightsquigarrow M(f) + |\mathfrak{F}_{\mathbf{T}} f(0)|$ on $BMO(\mathbf{T})$ induces the strong dual topology $\beta(BMO(\mathbf{T}), \mathcal{H}^1(\mathbf{T}))$.

Define the functions

$$\psi_t = e^{-\frac{1}{2}h_t} \qquad (t \in]0,1[)$$

of the space $\mathcal{H}^\infty(D)$. In view of Theorem 1 supra the associated boundary-value functions $(\tilde{\psi}_t)_{t \in]0,1[}$ are suffiently smooth on $\mathbf{T}$ in order to assure the following multiplier result which is at the basis of our investigations.

THEOREM 2. *The functions* $\mathrm{Re}\,\tilde{\psi}_t$ *and* $\mathrm{Im}\,\tilde{\psi}_t$ *are multipliers of the space* $BMO(\mathbf{T})$ *for all* $t \in]0,1[$, *i.e., we have*

$$(\mathrm{Re}\,\tilde{\psi}_t) \cdot BMO(\mathbf{T}) \subseteq BMO(\mathbf{T}), \qquad (t \in]0,1[)$$
$$(\mathrm{Im}\,\tilde{\psi}_t) \cdot BMO(\mathbf{T}) \subseteq BMO(\mathbf{T}).$$

Proof. See [5].

THEOREM 3. *The Toeplitz operators* $P_{\tilde{\psi}_t}$, $P_{\bar{\tilde{\psi}}_t}$ $(t \in]0,1[)$ *are continuous endomorphisms of the complex Banach space* $\mathcal{H}^1(\mathbf{T})$ *with respect to the norm topology such that*

$$\sup_{t\in]0,1[} ||P_{\tilde{\psi}_t}||<+\infty, \quad \sup_{t\in]0,1[} ||P_{\tilde{\tilde{\psi}}_t}||<+\infty$$

holds.

Proof. Suppose that the vector subspace

$$BMO_{\mathbb{R}}(\mathbf{T}) = \{f \in BMO(\mathbf{T}) \mid f = \text{Re } f\}$$

of the real vector space $L^2_{\mathbb{R}}(\mathbf{T})$ carries the norm $f \rightsquigarrow ||f||' = M(f) + |\mathfrak{F}_{\mathbf{T}} f(0)|$ inherited from the norm of the complex Banach $BMO(\mathbf{T})$. An application of the closed graph theorem shows that for each $t \in]0,1[$ the $\mathbb{R}$-linear multiplication operators

$$M_t: BMO_{\mathbb{R}}(\mathbf{T}) \ni f \rightsquigarrow (\text{Re } \tilde{\psi}_t)f \in BMO_{\mathbb{R}}(\mathbf{T}),$$
$$N_t: BMO_{\mathbb{R}}(\mathbf{T}) \ni f \rightsquigarrow (\text{Im } \tilde{\psi}_t)f \in BMO_{\mathbb{R}}(\mathbf{T})$$

are continuous endomorphisms of the real Banach space $BMO_{\mathbb{R}}(\mathbf{T})$. Let the real vector space Re $\mathcal{H}^1(\mathbf{T})$ be endowed with the norm $||?||_{Re}$ such that its complexification coincides with the complex Banach space $\mathcal{H}^1(\mathbf{T})$. Since the norm of $BMO(\mathbf{T})$ is equivalent to the norm of the dual $\mathcal{H}^1(\mathbf{T})'$ there exist real constants $k,k',k''\geq 0$ such that for all functions $g \in$ Re $\mathcal{H}^1(\mathbf{T})$ the following estimates hold:

$$\begin{aligned} ||(\text{Re}\psi_t)g||_{Re} &\leq k.\sup_{||f||'\leq 1} \left|\int_{\mathbf{T}} (\text{Re}\tilde{\psi}_t(w))g(w)f(w)d\lambda(w)\right| \\ &\leq k'.\sup_{||f||'\leq 1} ||g||_{Re}.||M_t(f)||' \qquad (t\in]0,1[) \\ &\leq k''.||M_t||.||g||_{Re}. \end{aligned}$$

It follows that for $t \in]0,1[$ the $\mathbb{R}$-linear operator M_t (and similarly the operator N_t) admits a continuous extension to the real Banach space Re $\mathcal{H}^1(\mathbf{T})$. A second application of the closed graph theorem shows the continuity of the Toeplitz operators $P_{\tilde{\psi}_t}$, $P_{\tilde{\tilde{\psi}}_t}$ $(t \in]0,1[)$. The estimate $\sup_{t \in]0,1[} ||\tilde{\psi}_t||_\infty \leq 1$ and the Fefferman-Stein duality prove the theorem. -

In view of the fact that $\tilde{\psi}_t \in \mathcal{H}^\infty(\mathbf{T})$, $\frac{1}{\tilde{\psi}_t} \in \mathcal{H}^\infty(\mathbf{T})$, and $\varphi_t = |\tilde{\psi}_t|^{-2}$ holds for $t \in]0,1[$, this result combined with the inverse mapping theorem yields

THEOREM 4. <u>The Toeplitz operators</u> $(P_{\varphi_t})_{t\in]0,1[}$ <u>are topological automorphism of</u> $\mathcal{H}^1(\mathbf{T})$ <u>with inverses</u> $P^{-1}_{\varphi_t} = P_{\tilde{\psi}_t} \circ P_{\bar{\tilde{\psi}}_t}$. <u>Moreover, the family</u> $(P^{-1}_{\varphi_t})_{t\in]0,1[}$ <u>is uniformly bounded on</u> $\mathcal{H}^1(\mathbf{T})$:

$$\sup_{t\in]0,1[} ||P^{-1}_{\varphi_t}||<+\infty.$$

After these functional analytic preparations concerning Toeplitz operators with sufficiently smooth symbols on $\mathbf{T}$ we are in a position to investigate norm approximations in the Hardy space $\mathcal{H}^1(D)$. The main results are collected in Theorem 5 infra.

3. PATIL TYPE APPROXIMATIONS WITH RESPECT TO $||?||_{\mathcal{H}^1(D)}$

An application of Harnack's inequality for the unit disc D yields the following estimate for each point $z \in D$:

$$\begin{aligned}
|\psi_t(z)| &= e^{-\frac{1}{2}\operatorname{Re} h_t(z)} \\
&= \exp\left(-\frac{1}{2}\int_{\mathbf{T}} \operatorname{Re}\left(\frac{w+z}{w-z}\right)\Theta_t(w)\,d\lambda(w)\right) \\
&\leq \exp\left(\frac{1}{2}\cdot\frac{1-|z|}{1+|z|}\int_{\Omega}\log(1-t\rho(w))\,d\lambda(w)\right) \qquad (t \in]0,1[) \\
&\leq (\sqrt{1-t})^{M_\rho(1-|z|)/(1+|z|)}
\end{aligned}$$

Since $M_\rho>0$, the family $(\psi_t)_{t \in]0,1[}$ converges pointwise in D to 0 as $t \to 1-$. Taking into account that the identity

$$P_{\varphi_t}^{-1}(C_z) = P_{\tilde{\psi}_t} \circ P_{\bar{\tilde{\psi}}_t}(C_z) = \overline{\psi_t(z)}\tilde{\psi}_t C_z$$

holds for $t \in]0,1[$ and all points $z \in D$, the estimate above combined with $\sup_{t\in]0,1[} ||\tilde{\psi}_t||_\infty \leq 1$ furnishes

$$\lim_{t\to 1-} ||P_{\varphi_t}^{-1}(C_z)||_1 = 0 \qquad (z \in D).$$

Thus, Theorem 4 supra and the well-known fact that $\{C_z \mid z \in D\}$ forms a total family in $\mathscr{H}^1(\mathbf{T})$ with respect to the norm topology imply

$$\lim_{t\to 1-} ||P_{\varphi_t}^{-1} f||_1 = 0$$

for all functions $f \in \mathscr{H}^1(\mathbf{T})$. The identities $P_1 = P_{\varphi_t}^{-1} \circ P_{\varphi_t} = P_{\varphi_t}^{-1} \circ (P_1 + P_{\rho_t})$ imply

$$P_{\varphi_t}^{-1} = P_1 - P_{\varphi_t}^{-1} \circ P_{\rho_t} \qquad (t \in]0,1[).$$

Therefore the following approximation theorem may be deduced.

THEOREM 5. *Let the function* $f \in \mathscr{H}^1(D)$ *be given. Starting with a function* $\rho \in \mathscr{C}^2(\mathbf{T})$ *that satisfies the conditions* (i), (ii) *supra, construct the families* $(\rho_t)_{t\in]0,1[}$ *and* $(\psi_t)_{t\in]0,1[}$ *as indicated above. Moreover, define for each* $t \in]0,1[$ *the function*

$$f_t: D \ni z \rightsquigarrow \psi_t(z) \int_\Omega \frac{\bar{\tilde{\psi}}_t(w)\rho_t(w)\tilde{f}(w)}{1-\bar{w}z}\, d\lambda(w) = \psi_t(z)\langle \tilde{f}, \rho_t \tilde{\psi}_t C_z\rangle.$$

Then the family $(f_t)_{t\in]0,1[}$ *belongs to the Hardy space* $\mathscr{H}^1(D)$ *and approximates* f *with respect to the norm topology, i.e.,*

$$\lim_{t\to 1-} ||f - f_t||_{\mathscr{H}^1(D)} = 0.$$

In particular, $\tilde{f}|\Omega = 0$ *implies* $f = 0$.

It follows from Theorem 5 that the family $(f)_t)_{t \in]0,1[}$ approximates $f \in \mathcal{H}^1(D)$ as $t \to 1-$ with respect to the topology of uniform convergence on the compact subsets of D. For this weaker result, see the paper [3] by Patil cited above which is based on families of Toeplitz operators with discontinuous symbols on $\mathbf{T}$.

For a different approach based on a theorem of Hardy and Littlewood, see the recent paper by Zarantonello [9]. Also see [5].

REFERENCES

1 Fefferman, C. and E.M. Stein, H^p spaces of several variables, Acta Math., 129 (1972), 137-193.

2 Newman, D.J., The non-existence of projections from L^1 to H^1. Proc. Amer. Math. Soc., 12 (1961), 98-99.

3 Patil, D.J., Representation of H^p-functions, Bull. Amer. Math. Soc., 78 (1972), 617-620.

4 Patil, D.J., Recapturing H^2-functions on a polydisc, Trans. Amer. Math. Soc., 188 (1974), 97-103.

5 Schempp, W., Approximations of holomorphic functions from boundary-values (to appear)

6 Schempp, W. and B. Dreseler, *Einführung in die harmonische Analyse*, Reihe "Mathematische Leitfäden", Verlag B.G. Teubner, Stuttgart 1980.

7 Schwartz, L., Sous-espaces hilbertiens d'espaces vectoriels topologiques et noyaux associés. (Noyaux reproduisants), J. Analyse Math., 13 (1964), 115-256.

8 Young, L.C., Some new stochastic integrals and Stieltjes integrals, Part II. Nigh-martingales, in: *Advances in probability and related topics*, Vol. 3 (P. Ney and S. Port, eds.), pp. 101-178, Marcel Dekker, New York 1974.

9 Zarantonello, S. E., A representation of H^p-functions with $0<p<\infty$. Pacific J. Math., 79 (1978), 271-282.

10 Zygmund, A., *Trigonometric series*, Volumes I and II, Cambridge University Press, Cambridge, London, New York, Melbourne 1977.

Walter Schempp
Lehrstuhl fuer Mathematik I
University of Siegen
Hoelderlinstrasse 3
D-5900 Siegen 21
Federal Republic of Germany

CONTINUOUS SELECTIONS FOR METRIC PROJECTIONS

M. Sommer

In this paper we study the problem of existence of continuous selections for metric projections. The problem of characterizing the spaces which admit continuous selections for metric projections from among the finite dimensional subspaces of C(X), X compact, has been posed by Lazar-Morris-Wulbert and also by Holmes. Here we are able to give a solution of this problem in the case that $X = [a,b]$, a real compact interval. Furthermore, we show some interesting properties of the weak Chebyshev subspaces of $C[a,b]$, since the theory of these subspaces is of fundamental importance to prove our characterization theorem.

1. INTRODUCTION

Let X be a compact metric space, and let C(X) be the space of all real-valued, continuous functions f on X under the uniform norm

$$||f||: = \sup\{|f(x)|: x \in X\}.$$

If G is an n-dimensional subspace of C(X), then for each $f \in C(X)$ we define

$$P_G(f): = \{g_o \in G: ||f-g_o|| = \inf\{||f-g||: g \in G\}\}$$

which is called the _set of best approximations_ of f from G. This defines a set-valued mapping P_G from C(X) onto 2^G which is called the _metric projection_ onto G. A continuous mapping s from C(X) onto G is called _continuous selection for the metric projection_ P_G (or, more briefly, continuous selection) if $s(f) \in P_G(f)$ for each $f \in C(X)$.

Lazar-Morris-Wulbert [7] have been the first to study the problem of existence of continuous selections. They have been able to characterize the spaces which admit continuous selections from among the one-dimensional subspaces of C(X). They have posed the problem of characterizing the higher dimensional subspaces of C(X). This question which is relevant to the convergence of algorithms for computing best approxi-

ISBN 0-12-213650-0

mations has also been raised in the book of Holmes [4].

With new methods and in the setting of weak Chebyshev subspaces, Nürnberger-Sommer [9] have been able to establish the existence of continuous selections for a class of finite dimensional weak Chebyshev subspaces of C[a,b] from which a result of Brown [3] for five-dimensional subspaces of C[-1,1] follows. Using the properties of the weak Chebyshev spaces, in further papers Nürnberger [8], Nürnberger-Sommer [10] and Sommer [11,12,14,15] have studied the problem of existence of continuous selections for P_G, in case G is a finite dimensional subspace of C[a,b].

Combining these results, in this paper we give the following characterization of the spaces G which admit continuous selections from among the finite dimensional subspaces of C[a,b] (Theorem 4.1):

> There exists a continuous selection for P_G if and only if the following conditions are satisfied:
> (i) G is weak Chebyshev
> (ii) No $g \in G, g \not\equiv 0$, vanishes on more than one interval
> (iii) The numbers of the boundary zeros of the elements $g \in G$ are bounded in a certain sense.

For proving this theorem we use the interesting result that many weak Chebyshev subspaces can be decomposed by a finite set of knots into Haar spaces or into very special weak Chebyshev spaces (Theorem 3.4). Such knots are very important to construct a continuous selection for P_G, in case the conditions (i) - (iii) are satisfied. The construction of such a selection is highly local and based on local alternation elements which are unique on certain subintervals of [a,b].

The nonexistence of a continuous selection, in case one of the conditions (i),(ii),(iii) is not satisfied, can be shown by using a fundamental lemma established by Lazar-Morris-Wulbert.

Recently, Blatt-Nürnberger-Sommer [2] have shown that each continuous selection constructed by Nürnberger and Sommer is even pointwise-Lipschitz-continuous and quasilinear.

2. PRELIMINARIES

In the following let G be always an n-dimensional subspace of C[a,b], n arbitrary.

DEFINITION 1. G is called a Haar space on a subset X of [a,b] if G satisfies the Haar condition on X, i.e. each $g \in G, g \not\equiv 0$, has at most n - 1 distinct zeros on X. G is called weak Chebyshev if each $g \in G$ has at most n - 1 sign changes on [a,b], i.e. there do not exist points $a \leq y_0 < y_1 < \dots < y_n \leq b$ such that $g(y_i)g(y_{i+1}) < 0$ for $i = 0,\dots,n-1$.

We denote the class of all n-dimensional weak Chebyshev subspaces of C[a,b] by W_n.

Jones-Karlovitz [5] have characterized the elements of W_n. For this characterization we need the following definition:

DEFINITION 2. If $f \in C[a,b]$, then $g_0 \in P_G(f)$ is called alternation element (= AE) of f if there exist n + 1 distinct points $a \leq t_0 < \dots < t_n \leq b$ such that

$$\varepsilon(-1)^i(f-g_0)(t_i) = ||f-g_0||, \quad i = 0,\dots,n, \quad \varepsilon = \pm 1 .$$

The points $t_0,\dots,t_n$ are called alternating extreme points of $f - g_0$. If [c,d] is any subinterval of [a,b], then $g_0 \in G$ is called local AE of f with respect to [c,d] if for each $g \in G$ $||f-g_0||_{[c,d]} \leq ||f-g||_{[c,d]}$ and if there exist m + 1 points $c \leq y_0 < \dots < y_m \leq d$ such that

$$\varepsilon(-1)^i(f-g_0)(y_i) = ||f-g_0||_{[c,d]}, \quad i = 0,\dots,m, \quad \varepsilon = \pm 1$$

where $m = \dim G|_{[c,d]}$.

Jones-Karlovitz have given the following characterization:

THEOREM 3. *The following conditions are equivalent:*

(2.1) $G \in W_n$

(2.2) *For each* $f \in C[a,b]$ *there exists at least one AE* $g_0 \in P_G(f)$.

3. DECOMPOSITION OF W_n INTO SUBCLASSES

In the next section we will give a characterization of the spaces which admit continuous selections from among the n-dimensional subspaces of $C[a,b]$. For proving this theorem we will use many properties of the elements of W_n. Therefore, in this section we deal with the class W_n.

Recently, Nürnberger [8] has obtained the following result:

THEOREM 1. *If* G *admits a continuous selection, then* $G \in W_n$.

Therefore, for studying the problem of characterizing such n-dimensional subspaces of $C[a,b]$ which admit continuous selections we have only yet to consider the case that $G \in W_n$. For considering this case it is very useful to divide W_n into certain subclasses.

At first we divide W_n into two subclasses:

$Y_n := \{G \in W_n$: no $g \in G, g \not\equiv 0$, vanishes on any interval$\}$

$Z_n := \{G \in W_n : G \notin Y_n\}$.

Arbitrarily many elements of Y_n can be constructed as follows: Let $G = \text{span}\,\{g_1,\ldots,g_n\}$ be a Haar space and $g_o \in C[a,b]$ such that g_o is nonnegative and does not vanish on any subinterval of $[a,b]$. Then $\tilde{G} := \text{span}\,\{g_o \cdot g_1, g_o \cdot g_2,\ldots,g_o \cdot g_n\} \in Y_n$.

We now want to divide the class Z_n into three further subclasses. For doing this we need the following notations: At first we give a condition concerning the length of zero intervals. We say that G satisfies *condition (I)* if there exists a constant $\delta > 0$ such that if $g \in G$ and $g \equiv 0$ on $[c,d] \subset [a,b]$ where $c,d \in \overline{\{x: g(x) \neq 0\}} \cup \{a,b\}$, then $d - c \geq \delta$.

Furthermore, a point $x_o \in [a,b]$ is called *vanishing* (resp. *non-vanishing*) with respect to G if for each $g \in G$ $g(x_o) = 0$ (resp. if there exists a function $g \in G$ with $g(x_o) \neq 0$).

In the following the term "with respect to G" will be omitted.

Then we define:

$Z_n^1 := \{G \in Z_n$: G satisfies condition (I) and all $x \in [a,b]$ are nonvanishing$\}$

Z_n^2: = {$G \in Z_n$: G satisfies condition (I) and there is at least one vanishing $x_o \in [a,b]$}

Z_n^3: = {$G \in Z_n$: G does not satisfy condition (I)}

EXAMPLES. (i) Let $m,k \in \mathbb{N}$ such that $m + k + 1 = n$ and $a = x_o < x_1 < \dots < x_{k+1} = b$ be a partition of $[a,b]$. Then the <u>space of spline functions</u> of degree m with the k fixed simple knots $x_1, x_2, \dots, x_k$ is defined by

$$S_{m,k}(x_1,\dots,x_k): = \text{span }\{1,x,\dots,x^m,(x-x_1)_+^m,\dots,(x-x_k)_+^m\}$$

where $(x-x_i)_+^m: = (x-x_i+|x-x_i|)^m/2^m$. Karlin-Studden [6,p. 18] have shown that $S_{m,k}(x_1,\dots,x_k) \in W_n$ and it immediately follows from the definition that $S_{m,k}(x_1,\dots,x_k) \in Z_n^1$.

(ii) Let $a = x_o < x_1 < \dots < x_s = b$ be a partition of $[a,b]$ and for $i = 0,\dots,s-1$ let $G^{i,i+1}$ be a Haar space with dimension $n_{i,i+1}$, $n_{i,i+1} \geq 1$, on $[x_i,x_{i+1}]$. Then the space G defined by

$$G: = \{g \in C[a,b]: g|_{[x_i,x_{i+1}]} \in G^{i,i+1} \quad , \ i = 0,\dots,s-1\}$$

is weak Chebyshev with dimension $n = \sum_{i=0}^{s-1} n_{i,i+1} - (s-1)$ as Bartelt has shown in [1]. Since each $x \in [a,b]$ is nonvanishing and condition (I) is satisfied, $G \in Z_n^1$. Such spaces G are denoted by <u>continuously composed Haar spaces</u>.

(iii) Let $m \in \mathbb{N}$ and $a= x_o < x_1 < \dots < x_{n+1} = b$ be a partition of $[a,b]$. Then the space G defined by

$$G: = \text{span }\{(x-x_1)_+^m,(x-x_2)_+^m,\dots,(x-x_n)_+^m\}$$

is an element of W_n and it is easy to show that $G \in Z_n^2$.

For examples of elements of Z_n^3 see Bartelt [1]. This subclass of W_n is in my opinion without any importance to the Approximation Theory.

The subclass Z_n^1, however, is very important, since the spline spaces are elements of Z_n^1 . These spaces are the prototypes of weak Chebyshev spaces.

The spline spaces and also the continuously composed Haar spaces have the interesting property that they can be decomposed by certain points into Haar spaces. We will now show that even each element of Z_n^1 has this property and, furthermore, that each element of Z_n^2 can be decomposed by cer-

tain points into such weak Chebyshev spaces whose nonzero elements do not vanish on intervals.
In the following we will denote such points by _knots_.

For proving the next theorem we need two lemmas which might be of independent interest.

LEMMA 2. _If_ $G \in W_n$, _then for any subinterval_ $[c,d]$ _of_ $[a,b]$ _the space_ $\tilde{G}: = G|_{[c,d]}$ _is weak Chebyshev with dimension_ m.

This lemma has been shown by Sommer [13, Theorem 1.4].

LEMMA 3. _If_ $G \in Y_n$ _and each_ $x \in [a,b]$ _is nonvanishing, then_ G _is a Haar space on_ $[a,b)$ _and on_ $(a,b]$.

A proof of this lemma can be found in Stockenberg [16, Theorem 3.1].

We are now in position to prove the following result:

THEOREM 4. _If_ $G \in Z_n^1$ _(resp._ $G \in Z_n^2$_), then there exists a minimal set of knots_ $a = x_0 < x_1 < \ldots < x_s = b$ _such that for_ $i = 0,\ldots,s-1$ _the spaces_ $G|_{[x_i,x_{i+1}]}$ _are Haar spaces with dimension_ $n_{i,i+1}$ _(resp. for_ $i = 0,\ldots,s-1$ _the spaces_ $G|_{[x_i,x_{i+1}]}$ _are weak Chebyshev spaces with dimension_ $n_{i,i+1}$ _and no_ $g \in G|_{[x_i,x_{i+1}]}$, $g \not\equiv 0$, _vanishes on a subinterval of_ $[x_i,x_{i+1}]$).

Proof. If $G \in Z_n^1 \cup Z_n^2$, then a minimal set of knots for G is very important to construct continuous selections. Therefore, we want to give here a detailed description how to get a minimal set of knots.

First. We will define a minimal set of points $\{y_i\}_{i=0}^r$ such that for $i = 0,\ldots,r-1$ no $g \in G|_{[y_i,y_{i+1}]}$, $g \not\equiv 0$, vanishes on a subinterval of $[y_i,y_{i+1}]$. Here we will follow the lines of Bartelt [1, Theorem 3].

Let $\delta > 0$ be given by condition (I) for any $G \in Z_n^1 \cup Z_n^2$. For all $\tilde{x},\tilde{\tilde{x}} \in [a,b]$ we define the subset $G(\tilde{x},\tilde{\tilde{x}})$ of G by

$$G(\tilde{x},\tilde{\tilde{x}}): = \{g \in G: g|_{[\tilde{x},\tilde{\tilde{x}}]} \not\equiv 0,\ g \equiv 0 \text{ on some subinterval of } [\tilde{x},\tilde{\tilde{x}}]\}.$$

We set $y_0: = a$. Since there is at least one $g \in G$, $g \not\equiv 0$, vanishing on some subinterval of $[a,b]$, there exists an integer k such that $G(a,a+k\delta) \neq \emptyset$, while for each $l = 1,\ldots,k-1$ $G(a,a+l\delta) = \emptyset$.

Let $g \in G(a,a+k\delta)$ be given. Then there exists a subinterval $[c,d] \subset [a,a+k\delta]$ on which g vanishes identically. Let $[c_g,d_g]$

denote the largest subinterval of $[a,b]$ containing $[c,d]$ such that $g \equiv 0$ on $[c_g,d_g]$. Then, since for each $l = 1,\dots,k-1$ $G(a,a+l\delta) = \emptyset$,

either $c_g = a$ and $a + (k-1)\delta \le d_g < a + k\delta$

or $a + (k-1)\delta \le c_g < a + k\delta \le d_g$.

Let $C_1 := \inf \{c_g : g \in G(a,a+k\delta),\ c_g > a\}$

$D_1 := \inf \{d_g : g \in G(a,a+k\delta),\ c_g = a\}$

and $y_1 := \min \{C_1, D_1\}$.

Then by Lemma 3 in [1], $C_1 > a$ and by condition (I), $D_1 > a$. Hence $a < y_1 < a+k\delta$.

For defining a further point y_2, we must distinguish the following cases:

(i) $y_1 + \delta \le b$.

Then, if there exists a function $g \in G(y_1,b)$, there exists an integer k such that $G(y_1,y_1+k\delta) \neq \emptyset$, while for each $l = 1,\dots,k-1$ $G(y_1,y_1+l\delta) = \emptyset$.

Let $g \in G(y_1,y_1+k\delta)$ be given. Then $g \equiv 0$ on some subinterval $[c,d]$ of $[y_1,y_1+k\delta]$. As defined above let $[c_g,d_g]$ denote the largest subinterval of $[a,b]$ containing $[c,d]$ such that $g \equiv 0$ on $[c_g,d_g]$. Then

either $c_g \le y_1 \le y_1 + (k-1)\delta \le d_g < y_1 + k\delta$

or $y_1 \le y_1 + (k-1)\delta \le c_g < y_1 + k\delta \le d_g$.

We set $C_2 := \inf \{c_g : g \in G(y_1,y_1+k\delta),\ c_g > y_1\}$

$D_2 := \inf \{d_g : g \in G(y_1,y_1+k\delta),\ c_g \le y_1\}$

and $y_2 := \min \{C_2, D_2\}$.

Then by Lemma 3 in [1], $C_2 > y_1$ and by Lemma 4 in [1], $D_2 > y_1$. Hence $y_1 < y_2 < y_1+k\delta$.

If $G(y_1,b) = \emptyset$, then we set $y_2 := b$ and we are ready, since neither $G|_{[a,y_1]}$ nor $G|_{[y_1,b]}$ contain any function g vanishing on intervals.

(ii) $y_1 + \delta > b$.

Then we only consider the subset $G(y_1,b)$ of G. If $G(y_1,b) \neq \emptyset$, then we can conclude as in (i) and obtain a further knot y_2 satisfying $y_1 < y_2 < b$. If $G(y_1,b) = \emptyset$, then we set $y_2 := b$ and we are ready.

Using the preceding arguments we finally get a set of points $a = y_0 < y_1 < \dots$ such that for each i no $g \in G|_{[y_i,y_{i+1}]}$, $g \not\equiv 0$, vanishes on a subinterval of $[y_i,y_{i+1}]$.

We will now show that this set is finite.

Assume that there are infinitely many distinct such y_i's. Then we may assume that there exists an $j \in \mathbb{N}$ such that infinitely many y_i's are contained in $[y_j, y_j+k\delta]$ where $k \in \mathbb{N}$ is defined as above. Then, since $y_i = \min\{C_i, D_i\}$ where for $i > 2$ C_i and D_i are defined analogously as C_2 and D_2, it follows that for infinitely many integers i, $y_i = C_i$ or for infinitely many i, $y_i = D_i$. We may assume that the former case occurs. This implies that $y_i = C_i$ for $i \in J$ where J is a set of infinitely many integers. Let $\bar{y} := \sup\{y_i : i \in J\}$. Then $y_i < \bar{y}$ for all $i \in J$. By definition of y_i for each $i \in J$ there exists a function g_i vanishing identically on $[c_{g_i}, d_{g_i}]$ such that $\bar{y} > c_{g_i} > y_i$. Then it follows from Lemma 4 in [1] that $\sup\{c_{g_i} : i \in J\} < \bar{y}$, since $y_j < y_i < c_{g_i} < \bar{y} \leq y_j+k\delta \leq d_{g_i}$ for each $i \in J$.

But this contradicts the definition of $\bar{y}$, since $C_i = y_i < c_{g_i}$ for each $i \in J$.

Analogously we can study the case that $y_i = D_i$ for infinitely many integers.

Thus we have got a set of points $a = y_o < y_1 < \ldots < y_r = b$ such that for each $i = 0,\ldots,r-1$ $G(y_i, y_{i+1}) = \emptyset$. This implies that $G|_{[y_i,y_{i+1}]}$ contains no nonzero function g which vanishes on a subinterval of $[y_i, y_{i+1}]$. Furthermore, it immediately follows from the construction of these points that there is no smaller set of points with that property. Hence, $\{y_i\}_{i=0}^{r}$ is a minimal set.

<u>Second</u>. We will now define a minimal set of knots such that the statements of this theorem will be satisfied.

First of all, by Lemma 3.2 for each $i = 0,\ldots,r-1$ the space $G|_{[y_i,y_{i+1}]}$ is weak Chebyshev.

Now we distinguish:

(i) $G \in Z_n^2$. Then for each $i = 0,\ldots,r$ we set $x_i := y_i$ and by the first part of this proof the set $\{x_i\}_{i=0}^{r}$ guarantees the statement of this theorem.

(ii) $G \in Z_n^1$. By definition of Z_n^1 for each $x \in [a,b]$ there exists a $g \in G$ such that $g(x) \neq 0$.

Let for each $i = 0,\ldots,r-1$ $m_i := \dim G|_{[y_i,y_{i+1}]}$. Since the subspaces $G|_{[y_i,y_{i+1}]}$ are weak Chebyshev with dimension m_i and no $g \in G|_{[y_i,y_{i+1}]}$, $g \not\equiv 0$, vanishes on a subinterval of $[y_i, y_{i+1}]$ and each $x \in [y_i, y_{i+1}]$ is nonvanishing, it follows

from Lemma 3.3 that for each $i = 0,\ldots,r-1$ the space $G|_{[y_i,y_{i+1}]}$ is a Haar space on $[y_i,y_{i+1})$ and on $(y_i,y_{i+1}]$.
If $G|_{[y_i,y_{i+1}]}$ is even a Haar space on the closed interval $[y_i,y_{i+1}]$, then we are ready with this interval. But if for some $i \in \{0,\ldots,r-1\}$ $G|_{[y_i,y_{i+1}]}$ does not satisfy the Haar condition on $[y_i,y_{i+1}]$, then we divide the interval $[y_i,y_{i+1}]$ into the two intervals $[y_i,\tilde{y}_i]$ and $[\tilde{y}_i,y_{i+1}]$ where $\tilde{y}_i: = (y_i+y_{i+1})/2$. Then the spaces $G|_{[y_i,\tilde{y}_i]}$ and $G|_{[\tilde{y}_i,y_{i+1}]}$ are Haar spaces with dimension m_i.
Using these arguments in each interval $[y_i,y_{i+1}]$ on which $G|_{[y_i,y_{i+1}]}$ does not satisfy the Haar condition we get a further point $\tilde{y}_i$. If we denote all the points y_i and $\tilde{y}_i$ by $\{x_i\}_{i=0}^{s}$ arranged ascendingly, then we obtain a minimal set of knots such that for each $i = 0,\ldots,s-1$ $G|_{[x_i,x_{i+1}]}$ is a Haar space.
This completes the proof.

Theorem 3.4 shows that each $G \in Z_n^1$ can be decomposed by a finite set of knots into Haar spaces. Therefore, we may denote the class Z_n^1 by the class of the _generalized spline spaces_.

4. THE CHARACTERIZATION THEOREM

In Section 3 we have defined for each $G \in Z_n^1 \cup Z_n^2$ a set of knots $\{x_i\}_{i=0}^{s}$ such that for each $i = 0,\ldots,s-1$ no $g \in G|_{[x_i,x_{i+1}]}$, $g \not\equiv 0$, vanishes on a subinterval of $[x_i,x_{i+1}]$. Then it is an immediate consequence of this definition that for each $G \in Y_n$ we may define the points $x_0: = a$, $x_1: = b$ as knots, since by definition of Y_n no $g \in G|_{[x_0,x_1]} = G$, $g \not\equiv 0$, vanishes on a subinterval of $[a,b]$.
Hence, in the following for each $G \in Y_n$ we always define the set $\{x_0,x_1\}$ as a set of knots.

For formulating and also for proving the characterization theorem we also need certain knots for each $G \notin W_n$ resp. for each $G \in Z_n^3$. In both cases we will now define certain sets of knots which in general do _not_ guarantee the same properties as the knots of G, in case $G \in Y_n \cup Z_n^1 \cup Z_n^2$. But one will see that they are only important to prove the next theorem. We set:

If $G \notin W_n$, then we set $x_o := a$, $x_1 := b$ and denote these points by the knots of G.

If $G \in Z_n^3$, then we distinguish two cases:
(i) If there exists a function $g \in G$ such that g vanishes on two separated intervals, i.e. there exist intervals $[c_1, d_1]$ and $[c_2, d_2]$ where $d_1 < c_2$ such that $g \equiv 0$ on $[c_1, d_1] \cup [c_2, d_2]$ while $g \not\equiv 0$ on $[d_1, c_2]$, then we set $x_o := a$, $x_1 := b$ and denote these points by the knots of G.
(ii) If there does not exist any $g \in G$ vanishing on two separated intervals, then by the proof of Theorem 2.5 in Sommer [15] there exists a $g_o \in G$, $g_o \not\equiv 0$, such that $g_o \equiv 0$ on a certain interval $[\bar{x}, \bar{\bar{x}}]$ and $|\text{bd } Z(g_o)| \geq m + 1$. Here we denote by $Z(g) := \{x: g(x) = 0\}$, by bd A the boundary of A, by $|A|$ the number of the elements of A and by
$m := \dim \{g \in G: g \equiv 0 \text{ on } [\bar{x}, \bar{\bar{x}}]\}$. In this case we set $x_o := a$, $x_1 := \bar{x}$, $x_2 := \bar{\bar{x}}$, $x_3 := b$ and denote these points by the knots of G. One can see that in this case the choice of the knots depends on the choice of a certain function $g_o \in G$.

Now we are in position to give a complete characterization of the spaces which admit continuous selections from among the n-dimensional subspaces of $C[a,b]$.

THEOREM 1. _Let G be an n-dimensional subspace of_ $C[a,b]$ _and_ $\{x_i\}_{i=0}^{s}$ _be a minimal set of knots. Then the following conditions_ (4.1) _and_ (4.2) _are equivalent:_

(4.1) _There exists a continuous selection for_ P_G

(4.2) (i) _G is weak Chebyshev_
(ii) _No_ $g \in G$ _vanishes on two separated intervals_
(iii) _For all_ $i, j \in \{0, \ldots, s\}, i < j$, _and each_ $g \in \{g \in G: g \equiv 0 \text{ on } [x_i, x_j]\}$:
$|\text{bd } Z(g)| \leq \dim \{g \in G: g \equiv 0 \text{ on } [x_i, x_j]\}$
(iv) _For each_ $i \in \{0, \ldots, s-1\}$ _and each_ $g \in G|_{[x_i, x_{i+1}]}$, $g \not\equiv 0$: $|Z(g)| \leq \dim G|_{[x_i, x_{i+1}]}$

The proof of this theorem is very complicated. It uses many properties of weak Chebyshev spaces. Here we will only give a short sketch of the proof. We will do this by giving corresponding characterization theorems for the subclasses of W_n as have been defined in Section 3. These results have been shown by Nürnberger [8], Nürnberger-Sommer [9,10] and Sommer [11,12,14,15]. Combining these results one can obtain the proof of Theorem 4.1.

At first we want to remark that the nonexistence of a continuous selection, in case one of the conditions (4.2)(i) - (4.2)(iv) is violated, has been shown in those papers by using a fundamental lemma of Lazar-Morris-Wulbert [7].

Now we choose the following notations:

Let $\{x_i\}_{i=0}^{s}$ be a minimal set of knots for G according to Theorem 3.4 resp. to the definitions at the beginning of this section.Then for all $i,j \in \{0,\ldots,s\}, i < j$, we set:

$$\bar{G}_{ij} := \{g \in G:\ g \equiv 0 \text{ on } [x_i,x_j]\},\quad m_{ij} := \dim \bar{G}_{ij}$$

$$G^{ij} := G|_{[x_i,x_j]},\quad n_{ij} := \dim G^{ij}$$

It immediately follows from Theorem 3.1 that condition (4.1) implies condition (4.2)(i). Therefore, we have only to consider the case that $G \in W_n$. We will now study this case by using the subclasses of W_n as have been defined in Section 3.

LEMMA 2. <u>Let</u> $G \in Y_n$. <u>Then the following conditions are equivalent</u>:

(4.3) <u>There exists a continuous selection for</u> P_G

(4.4) $|Z(g)| \leq n$ <u>for each</u> $g \in G,\ g \not\equiv 0$

<u>Proof</u>. By Theorem 2.3 for each $f \in C[a,b]$ there exists at least one AE $g_0 \in P_G(f)$. In addition to this Nürnberger-Sommer [9,Theorem 8] have proved that for each $f \in C[a,b]$ there exists <u>exactly</u> one AE $g_f \in P_G(f)$ if and only if $G \in Y_n$ and $|Z(g)| \leq n$ for each $g \in G,\ g \not\equiv 0$. Using this result they have been able to show [9,Corollary 9] that if condition (4.4) is satisfied, then the selection s defined by $s(f) := g_f$ is continuous. Hence, condition (4.4) implies condition (4.3). Furthermore, this selection is the unique continuous selection as also have been shown in [9]. Conversely, Sommer [11] has shown that condition (4.3) implies condition (4.4).

One can see that if $G \in Y_n$, then condition (4.4) is as

same as condition (4.2)(iv), since in this case there are only the two knots $x_o = a$ and $x_1 = b$.

LEMMA 3. Let $G \in Z_n^3$. Then there does not exist any continuous selection for P_G.

Proof. If there exists a function $g \in G$ such that g vanishes on two separated intervals, then the statement has been shown in [15, Lemma 2.3].
If there does not exist any function in G having two separated zero intervals, then by Theorem 2.5 in [15] there exist a function $g_o \in G$ and two points $\bar{x}, \bar{\bar{x}} \in [a,b]$, $\bar{x} < \bar{\bar{x}}$, such that $g_o \equiv 0$ on $[\bar{x}, \bar{\bar{x}}]$ and $|\mathrm{bd}\ Z(g_o)| \geq m + 1$ where $m = \dim \{g \in G: g \equiv 0 \text{ on } [\bar{x}, \bar{\bar{x}}]\}$. In this case the statement of this lemma has been shown in [15, Lemma 2.4].

Therefore, if $G \in Z_n^3$, then condition (4.2)(ii) or condition (4.2)(iii) is violated. Remember that in the second case we have to use the knots $x_o = a$, $x_1 = \bar{x}$, $x_2 = \bar{\bar{x}}$, $x_3 = b$.

LEMMA 4. Let $G \in Z_n^1 \cup Z_n^2$ and $\{x_i\}_{i=0}^s$ be a minimal set of knots. Then the following conditions (4.5) and (4.6) are equivalent:

(4.5) There exists a continuous selection for P_G

(4.6) (i) No $g \in G$ vanishes on two separated intervals
(ii) For all $i,j \in \{0,\ldots,s\}, i < j$, and each $g \in \bar{G}_{ij}$: $|\mathrm{bd}\ Z(g)| \leq m_{ij}$
(iii) For each $i \in \{0,\ldots,s-1\}$ and each $g \in G^{i,i+1}, g \not\equiv 0$: $|Z(g)| \leq n_{i,i+1}$

Proof. We distinguish:
First: $G \in Z_n^1$. Then by Theorem 3.4 for each $i \in \{0,\ldots,s-1\}$ the spaces $G^{i,i+1}$ are Haar spaces and, therefore, condition (4.6)(iii) is always satisfied.
The nonexistence of a continuous selection, in case one of the conditions (4.6)(i), (4.6)(ii) is violated, has been shown in [12, Lemma 3.2 and Lemma 3.3].
Conversely, let condition (4.6) be satisfied. In this case Sommer [12, Lemma 3.7] has been able to construct a continuous selection for P_G. Here we will only give a sketch of this construction:
Let $f \in C[a,b]$ and $g_o \in P_G(f)$ arbitrary. Then by Lemma 2.5 in

[12] there exists a subinterval $[x_p,x_{p+1}]$ of $[a,b]$ such that $g = g_o$ on $[x_p,x_{p+1}]$ for all $g \in P_G(f)$. Starting with this interval we now construct a selection step by step:

(i) Local approximation. We approximate $f - g_o$ in $[x_{p+1},b]$ by $\bar{G}_{o,p+1}$. Since condition (4.6)(ii) implies that each space $\bar{G}_{ij}$ is weak Chebyshev with dimension m_{ij}, Theorem 2.3 guarantees the existence of a local AE (see Definition 2.2) $g_1 \in P_{\bar{G}_{o,p+1}}(f-g_o)$. Then it is easy to prove that $g_o+g_1 \in P_G(f)$. Furthermore, it has been shown in [12]:

(ii) Uniqueness of local AE's on $[x_{p+1},x_{p+2}]$. If $g_1,\bar{g}_1 \in P_{\bar{G}_{o,p+1}}(f-g_o)$ are two local AE's for approximation in $[x_{p+1},b|$, then

$$g_1 = \bar{g}_1 \text{ on } [x_{p+1},x_{p+2}].$$

(iii) Independence of the choice of best approximations. If $\tilde{g}_o \in P_G(f), \tilde{g}_o \neq g_o$, and $\tilde{g}_1 \in P_{\bar{G}_{o,p+1}}(f-\tilde{g}_o)$ is a local AE for approximation in $[x_{p+1},b]$, then

$$\tilde{g}_o + \tilde{g}_1 = g_o + g_1 \text{ on } [x_p,x_{p+2}].$$

The method of the local approximation will be continued in $[x_{p+2},b]$ in the following way:
We approximate $f - g_o - g_1$ in $[x_{p+2},b]$ by $\bar{G}_{o,p+2}$. Theorem 2.3 guarantees the existence of a local AE $g_2 \in P_{\bar{G}_{o,p+2}}(f-g_o-g_1)$. As has been shown in (ii) all these AE's coincide on $[x_{p+2},x_{p+3}]$ and as has been shown in (iii) $\tilde{g}_o+\tilde{g}_1+\tilde{g}_2 = g_o+g_1+g_2$ on $[x_p,x_{p+3}]$ for any choice of $g_o,\tilde{g}_o$, $g_o+g_1,\tilde{g}_o+\tilde{g}_1$. Furthermore, $g_o+g_1+g_2 \in P_G(f)$.

Using the same kind of arguments as above for the intervals $[x_{p+3},b],\ldots,[x_{s-1},b],[a,x_p],[a,x_{p-1}],\ldots,[a,x_1]$ we finally obtain a function $\bar{g}$ defined by

$$\bar{g}: = g_{-p}+g_{-p+1}+\ldots+g_{-1}+g_o+g_1+\ldots+g_{s-1-p} \in P_G(f)$$

where for each $i \in \{1,\ldots,p\}$ g_{-i} is a local AE in $P_{\bar{G}_{p+1-i,s}}(f-g_o-g_{-1}-\ldots-g_{-i+1})$ for approximation in $[a,x_{p+1-i}]$ and for each $i \in \{1,\ldots,s-1-p\}$ g_i is a local AE in $P_{\bar{G}_{o,p+i}}(f-g_o-g_1-\ldots-g_{i-1})$ for approximation in $[x_{p+i},b]$.

We set: $s(f): = \bar{g}$

Following the preceding arguments one can see that we want to

select a function $\bar{g} \in P_G(f)$ which is in a certain sense "the best" under the best approximations of f.
By (iii) the function s(f) is independent of the particular choice of $g_o, g_o+g_1, \ldots, g_o+g_1+\ldots+g_{s-2-p}, g_{-1}+g_o, \ldots, g_{-p+1}+\ldots+g_o$.
This property and the local uniqueness of the local AE's ensure that s is continuous.

<u>Second</u>: $G \in Z_n^2$. The nonexistence of a continuous selection, in case one of the conditions (4.6)(i),(ii),(iii) is violated, has been shown in [15].
Conversely, let condition (4.6) be satisfied. In this case the construction of a continuous selection for P_G is much more complicated than in the case that $G \in Z_n^1$, since in general there does not exist any subinterval of [a,b] on which all best approximations of a function $f \in C[a,b]$ coincide as has been shown in [15]. Therefore, this shows an essential difference between the behavior of the elements of Z_n^1 and of the elements of Z_n^2. But we have shown in [15] that for each $f \in C[a,b]$ there is a subinterval $[x_i, x_j]$ of [a,b] on which all AE's of f coincide. Starting with such an interval we have been able to construct a continuous selection also in this case. The construction is slightly different from the construction established in the case that $G \in Z_n^1$. For details see [15].

Now it is easy to show that Theorem 4.1 immediately follows from Theorem 3.1 and the Lemmas 4.2, 4.3 and 4.4.

<u>REMARK</u>. (i) As has been shown in [15] by some examples, <u>none</u> of the conditions (4.2)(i) - (4.2)(iv) can be omitted.
(ii) For constructing a continuous selection for P_G, in case $G \in Z_n^2$ and condition (4.2) is satisfied, it is essential that the set of knots $\{x_i\}_{i=0}^s$ needed for the construction is a <u>minimal</u> set. This has also been shown in [15].

Theorem 4.1 is in our opinion very appropriate to look for continuous selections for P_G, in case G is chosen as an element of some interesting subclasses of W_n. Such subclasses are the class of the spline spaces $S_{m,k}(x_1,\ldots,x_k)$ and also the class of the continuously composed Haar spaces (for definition see the examples in Section 3).

Applying Theorem 4.1 to these classes we obtain the following characterizations as we have shown in [12, Theorem 4.1 and Theorem 4.3]:

THEOREM 5. *Let* $S_{m,k}(x_1,\ldots,x_k)$ *be a space of spline functions of degree* m *with the* k *fixed simple knots* $x_1,\ldots,x_k$. *Then there exists a continuous selection for* $P_{S_{m,k}}$ *if and only if* $k \leq m+1$.

This result has directly been proved by Nürnberger-Sommer [10].

The next result concerns the class of the continuously composed Haar spaces G. Remember that G is defined by

$$G = \{g \in C[a,b] : g|_{[x_i,x_{i+1}]} \in G^{i,i+1}\ ,\ i = 0,\ldots,s-1\}$$

where $a = x_0 < x_1 < \ldots < x_s = b$ is a partition of $[a,b]$ and for each $i = 0,\ldots,s-1$ $G^{i,i+1}$ is a Haar space with dimension $n_{i,i+1}$, $n_{i,i+1} \geq 1$, on $[x_i,x_{i+1}]$. Then $\dim G = n = \sum_{i=0}^{s-1} n_{i,i+1} - (s-1)$.

THEOREM 6. *Let* $G \in W_n$ *be a continuously composed Haar space. Then there exists a continuous selection for* P_G *if and only if* $n_{ij} \leq 2$ *for all* $i,j \in \{1,\ldots,s-1\}, i < j$.

REMARK. The dimensions n_{01} and $n_{s-1,s}$ of the Haar spaces G^{01} and $G^{s-1,s}$ are not needed in the preceding characterization. Therefore, the existence resp. the nonexistence of a continuous selection depends only on the "inner" dimensions n_{ij} where $i,j \in \{1,\ldots,s-1\}$.

Finally we give a result recently proved by Blatt-Nürnberger-Sommer [2]:

THEOREM 7. *Let* s *be such a continuous selection for* P_G *as has been constructed in Lemma 4.2 resp. in Lemma 4.4. Then* s *is pointwise-Lipschitz-continuous and quasilinear.*

Here s is called *pointwise-Lipschitz-continuous* (resp. *quasilinear*) if for each $f \in C[a,b]$ there exists a constant $K_f > 0$ such that for each $\tilde{f} \in C[a,b]$ $||s(f)-s(\tilde{f})|| \leq K_f||f-\tilde{f}||$ (resp. if for each $f \in C[a,b]$, $g \in G$ and constants c,d $s(cf+dg) = cs(f)+dg$).

This result shows that each continuous selection

constructed in Lemma 4.2 resp. in Lemma 4.4 is even pointwise-Lipschitz-continuous which is a strong property of the metric projection.

REFERENCES

1 Bartelt, M. W., Weak Chebyshev sets and splines, J. Approximation Theory, 14 (1975), 30-37.

2 Blatt, H. P., G. Nürnberger and M. Sommer, Pointwise-Lipschitz-continuous selections for the metric projection, preprint.

3 Brown, A. L., On continuous selections for metric projections in spaces of continuous functions, J. Functional Analysis, 8 (1971), 431-449.

4 Holmes, R. B., A Course On Optimization And Best Approximation, Lecture Notes 257, Springer Verlag, Berlin-Heidelberg-New York 1972.

5 Jones, R. C. and L. A. Karlovitz, Equioscillation under nonuniqueness in the approximation of continuous functions, J. Approximation Theory, 3 (1970), 138-145.

6 Karlin, S. and W. J. Studden, Tchebycheff Systems, Interscience Publishers, New York 1966.

7 Lazar, A. J., P. D. Morris and D. E. Wulbert, Continuous selections for metric projections, J. Functional Analysis, 3 (1969), 193-216.

8 Nürnberger, G., Nonexistence of continuous selections for the metric projection, SIAM J. Math. Anal., to appear.

9 Nürnberger, G. and M. Sommer, Weak Chebyshev subspaces and continuous selections for the metric projection, Trans. Amer. Math. Soc., 238 (1978), 129-138.

10 Nürnberger, G. and M. Sommer, Characterization of continuous selections of the metric projection for spline functions, J. Approximation Theory, 22 (1978), 320-330.

11 Sommer, M., Nonexistence of continuous selections of the metric projection for a class of weak Chebyshev spaces, Trans. Amer. Math. Soc., to appear.

12 Sommer, M., Characterization of continuous selections for the metric projection for generalized splines, SIAM J. Math. Anal., to appear.

13 Sommer, M., Weak Chebyshev spaces and best L_1-approximation, J. Approximation Theory, to appear.

14 Sommer, M., Continuous selections of the metric projection for 1-Chebyshev spaces, J. Approximation Theory, 26 (1979), 46-53.

15 Sommer, M., Continuous selections for the metric projection, preprint.

16 Stockenberg, B., On the number of zeros of functions in a weak Tchebyshev-space, Math. Z., 156 (1977), 49-57.

M. Sommer
Institut für Angewandte Mathematik
Universität Erlangen-Nürnberg
Martensstraße 3
8520 Erlangen
West Germany

PROBLEMS

The following problems were compiled during and after the symposium by Klaus Höllig.

C. de Boor.

1. Prove that the inverse of a bounded and boundedly invertible totally positive band matrix (infinite or biinfinite) is checkerboard (see the contribution of C. de Boor to this proceedings).

2. Prove that for fixed k, least squares approximation from $S_{k,\underline{t}}$:= splines of order k with knot sequence $\underline{t}$ can be bounded as a map on C independent of $\underline{t}$ (see C. de Boor, The quasi-interpolant as a tool in elementary polynomial spline theory, in "Approximation Theory", G.G. Lorentz ed., Academic Press, (1973), 269-273, and C. de Boor, On a max-norm bound for the least squares spline approximant, in "Approximation and Function Spaces", Z. Ciesielski ed., (Proc. of Gdansk Conf., Poland, 1979)).

3. Prove that for fixed k, spline interpolation at knot averages, explicitly, at the points $\tau_i := (t_{i+1}+\ldots+t_{i+k-1})/(k-1)$, all i, can be bounded as a map on C independent of the knot sequence $\underline{t}$ (see C. de Boor, On bounding spline interpolation, J. Approximation Theory 14(1975), 191-203).

4. Prove that a biinfinite banded totally positive matrix A is boundedly invertible if and only if the linear system $Ax = ((-1)^i)$ has a unique bounded solution (see C.A. Micchelli, Infinite spline interpolation, in Proc. Conf. Approximation Theory at Siegen, Germany, 1979, G. Meinardus ed.).

ISBN 0-12-213650-0

Yu. Brudnyi.

1. Find a constructive proof of the estimate

$$r_n(f,p) = O(n^{-\alpha}), \qquad f \in \Lambda^{\alpha}_{q^+},$$

where $q = q(\alpha,p) := (\alpha+1/p)^{-1}$ (see the contribution of Yu. Brudnyi in this proceedings).

2. Characterize the class of function $f \in L_p$ for which

$$\overline{\lim_{n\to\infty}} (r_n(f,p))^{1/n} < \infty .$$

R. DeVore.

1. Find the best inequality of the form

$$E_n(f) \leq C_1\omega_2(f,C_2n^{-1}),$$

where $E_n(f)$ is the error in approximating f in $C[-\pi,\pi]$ by trigonometric polynomials of degree $\leq n$. The inequality $E_{n-1}(f) \leq \frac{1}{2}\bar{\omega}(f,\pi/n) \leq \omega(f,\pi/n)$ is the famous result of Korneicuk.

2. If r is a non-negative integer, let C_r^* be the smallest constant for which the inequality

$$K(f,t^r,L_p[0,1], W_p^r[0,1]) \leq C_r^* \, \omega_r(f,t)_{L_p}$$

holds with K the Peetre K-functional (see R. DeVore, Degree of approximation, in Inequalities II, eds. G. G. Lorentz, C.K. Chui, L. L. Schumaker, Academic Press, 1976). Find the asymptotic behavior of the sequence (C_r^*) as $r \to \infty$.

K. Höllig.

Find a constructive proof for B. S. Kashin's surprising result about the n-width of the embedding $W_2^r[0,1] \to L_\infty[0,1]$, $r > \frac{1}{2}$, i.e.

$$d_n(W_2^r[0,1], L_\infty[0,1]) \sim n^{-r}.$$

B. S. Kashin.

Determine the exact order of $d_n(B_2^{n^2}, \ell_\infty^{n^2})$, $n \to \infty$ (see the contribution of B. S. Kashin to this proceedings).

A. Melkman.

For fixed $k \in N$ compute $\overline{\lim_{n\to\infty}}\, n^{\frac{1}{2}}\, d_n(\ell_1^{kn}, \ell_\infty^{kn})$ (see also the contribution of B.S. Kashin to this proceedings).

C.A. Micchelli.

1. Consider Kergin's map

$$H_n f(x) := \sum_{j=0}^{n} \int_{[x^0,\ldots,x^j]} D_{x-x^0} \cdots D_{x-x^{j-1}} f,$$

$$x^i \in I^k := [-1,1]\times\cdots\times[-1,1].$$

What is the smallest domain Ω_k containing I^k such that whenever f is analytic in a neighborhood of $\overline{\Omega}_k$ then $H_n f$ converges to f uniformly for any sequence of vectors $P^n = \{x^{0,n},\ldots,x^{n,n}\} \subseteq I^k$? In the paper "Multivariate B-splines and Lagrange interpolation" (to appear in the Rocky Mountain J.), we prove the convergence of $H_n f$ for $\Omega_k = \{(z_1,\ldots,z_k): |z_i| \le 3, i=1,\ldots,k\}$. In one dimension ($H_n$ is Lagrange interpolation), the minimal set is $\{z: |z-1| \le 2$ or $|z+1| \le 2\}$ (see Approximate Calculation of Integrals, by V.I. Krylov).

2. Let $x^o,\ldots,x^n \in I^k$ be in general position, i.e. every subset of k+1 points forms a k-dimensional simplex, then H_n (see Problem 1) is defined for $f \in C^{k-1}(I^k)$. Does there exist a sequence of points P^n such that $H_n f$ converges uniformly whenver $f \in C^k(I^k)$? When k=1, the Chebyshev nodes are known to have this property.

3. If $\lambda, x \in R^k$, $g \in C(R^1)$ and $\Lambda \subseteq R^k$, define

$$M_g(\Lambda) := \text{linear span } \{g(\lambda\cdot x): \lambda \in \Lambda\}.$$

i) Is it true that $\dim M_g(R^k) < \infty$ iff g is a polynomial?

ii) For a fixed compact subset K of R^k, determine the closure of $M_g(\Lambda)$ in $C(K)$.

One result of this type is the following. If g is entire and $N_g := \{j: g^{(j)}(0) \neq 0\}$, then for any set Λ with positive measure

$$\overline{M_g(\Lambda)} = S := \text{closed linear span of } \{Q: Q \text{ is a homogeneous polynomial of degree } j, \text{ with } j \in N_g\}.$$

This is easy to prove. From

$$g(x) = \lim_{m\to\infty} \sum_{\substack{j\in N_g \\ j\le m}} \frac{g^{(j)}(0)}{j!} x^j,$$

it follows that $\overline{M_g(\Lambda)} \subseteq S$. Conversely, if μ is any signed Borel measure on K then $f(\lambda) := \int_K g(\lambda\cdot x)\,d\mu(x)$ is an entire function. If $f(\lambda) = 0$ on , then since measure $\Lambda > 0$, $f \equiv 0$. However, $D^\alpha f(0) = g^{(|\alpha|)}(0) \int_K x^\alpha d\mu(x)$ and thus $\int_K Q\,d\mu(x) = 0$ for all $Q \in S$. Thus $S = \overline{M_g(\Lambda)}$ as asserted.

A. Pinkus.

1. Let $S_{n-1,r}$ denote the space of smooth splines of degree n-1 with r variable knots. Determine the best constant C_p depending on p such that

$$\inf_{s\in S_{n-1,r}} \|f-s\|_p \le C_p \|f^{(n)}\|_\infty, \quad 1 \le p \le \infty.$$

For $p=\infty$ it is known that C_∞ coincides with the best constant in approximation by splines with r fixed knots $\xi_1^*,\ldots,\xi_r^*$ and the corresponding perfect spline of degree n is the extremal function for the problem.

2. Let $\Delta := \{z: |z| \le 1\} \subset C$. For each choice of $\underline{z} = (z_0,\ldots,z_n)$, $z_i \ne z_j$, $i \ne j$, let $\ell_k(z;\underline{z})$ denote the fundamental polynomial of degree n such that $\ell_k(z_i;\underline{z}) = \delta_{ik}$, $i,k = 0,\ldots,n$. The problem of minimum norm polynomial interpolation on the unit disc is one of minimizing

$$\max_{|z|\le 1} \sum_{k=0}^{n} |\ell_k(z;\underline{z})|.$$

It is known that if the z_i are restricted to lie on $\partial\Delta$, i.e. $|z_i| = 1$, $i = 0,\ldots,n$, then an optimal choice are the roots of unity. Prove that this result remains valid if the z_i are permitted to lie in Δ.

K. Scherer.

1. Find conditions within the class of entire functions (or analytic in a region around an interval [a,b]) which are necessary and sufficient for pure polynomial approximation $E_n(f;[a,b])$ in the Chebyshev norm to be optimal for best approximation $\operatorname{dist}(f,P^N)$ by piecewise polynomials with variable degrees and knots and total number N=n+1 of parameters. In Dahmen-Scherer, Journal of Approximation Theory, 26(1979), 1-14, it is conjectured that by

$$V := \{f \in C[a,b]: \ E_n(f,I_1) \sim E_n(f,I_2)\cdot(|I_1|/|I_2|)^n \quad \text{for all } I_1 \subseteq I_2 \subseteq [a,b]\}$$

such conditions are given.

2. Find classes of functions (as large as possible) for which best approximation by rational functions p_n/q_n in the Chebyshev norm, p_n,q_n of degree n, and $\operatorname{dist}(f,P^n)$ have asymptotically the same rate of decrease. Begin with a fixed rate, e.g. $a^{\sqrt{n}}$ for some $0<a<1$ (see the contribution of DeVore and Scherer in this proceedings which combine with results of Gonchar to show that x^α, $\alpha > 0$ belong to such a class).

3. Calculate the K-functional for all combinations of two Besov spaces B^α_{pq}, $1\le p,q\le\infty$, $\alpha>0$.

H.J. Schmid.

Let I be a centrally symmetric Radon measure with compact support and

$$\sum_{j=1}^{N} c_j f(x^{(j)}), \quad c_j>0, \quad x^{(j)} \in \mathbb{R}^n$$

be a cubature formula for I which is exact for all polynomials of degree $\le m = 2k-1$, $k\in\mathbb{N}$ It was shown by H.M. Möller (see I.P. Mysovskikh's contribution to this proceedings) that $\frac{1}{2}k(k+1) + [\frac{k}{2}]$ is a lower bound for the number of knots N. When I is the product integral

$$I(f) = \int_{-1}^{1} \int_{-1}^{1} f(x_1,x_2)\sqrt{1-x_1^2}\ \sqrt{1-x_2^2}\ dx_1 dx_2$$

then for each $k \in \mathbb{N}$ there is a formula for which the lower bound is attained. However for the integral

$$I(f) = \int_{-1}^{1} \int_{-1}^{1} f(x_1,x_2)\ (1-x_1^2)(1-x_2^2)\ dx_1 dx_2$$

there exists no such formula attaining the lower bound for $k=5$. Improve Möller's estimate for the lower bound.

A. Sharma.

Does there exist a 3-row incidence matrix $E:=(e_{ij})$ with $\sum_{i,j} e_{ij} = n+1$ satisfying the strong Polya condition which is not regular for the cube roots of unity?

R. Sharpley.

Calculate the K-functional for the couple (H^1, BMO) where H^1 is the Hardy space and BMO is the space of functions of bounded mean oscillation.

P.W. Smith.

Given $f \in C^\infty[0,1]$ with $f^{(k)} \geq 0$, does there exist for each $n>0$ a polynomial of degree n interpolating f at n+1 points such that $P_n^{(k)} \geq 0$? In the meantime, A. Pinkus has shown that this is the case when $k = 0,1$.

D. Wulbert.

Let Λ be a subset of the integers Z such that $n \in \Lambda$ implies that $-n \in \Lambda$ and F_Λ denote the L_2 projection onto $T_\Lambda :=$ closed span $\{e^{i\lambda x} : \lambda \in \Lambda\}$. Give a concrete characterization of those sets Λ for which F_Λ has the unique minimum norm among all projections from $C_{2\pi}$ onto T_Λ. This is the case for $\Lambda_n := \{\lambda : |\lambda| \leq n\}$ or $Z-\Lambda_n$ or $Z-\{0,\pm r,\ldots,\pm nr\}$ for example. On the other hand F_Λ is not the unique minimum norm projection when $\Lambda = \{\pm r,\ldots,\pm nr\}$ or $\Lambda = Z-\{0,\pm 2,\pm 3,\pm 4\}$ (see Unique minimality of Fourier projections, by S.D. Fisher, P.D. Morris and D.E. Wulbert).